CHEMISTRY AND BIOCHEMISTRY
OF LEGUMES

CHEMISTRY
AND BIOCHEMISTRY
OF LEGUMES

Edited by

S.K. ARORA

EDWARD ARNOLD

First published in India
by Oxford & IBH Publishing Co.,
66 Janpath, New Delhi 110 001

First published in the United Kingdom in 1983
by Edward Arnold (Publishers) Limited,
41 Bedford Square, London WC1B 3DQ

ISBN 0 7131 2854 2

Printed in India by Oxonian Press Pvt. Ltd.,
Faridabad

Preface

Legumes, the protein-rich crops, hold a great promise in meeting the protein needs of vulnerable groups. They have found an extremely wide area of utilisation in human foods, industrial application and in animal feeds. The extent of the importance of legumes is evident by the fact that recently the National Academy of Sciences, Washington brought out a book on tropical legumes describing legumes as resources for the future. It dealt mostly with the underexploited legumes. The United Nations has a separate cell, "Protein Advisory Group", which coordinates and disseminates scientific knowledge in this area.

This book is intended for postgraduate students and professional workers interested in the chemistry and biochemistry of legumes. Biochemistry can and does contribute to the understanding and solution of problems involved in many of the more specialised aspects like plant biology, genetics, phytopathology, toxicology, food science and nutrition.

This book will help students and research workers in these diverse fields by providing them with a ready source of chemical information directly applicable to their specific field. The book also contains up-to-date chemical information on guar (*Cyamopsis tetragonoloba*) which will be valuable to both researchers and guar industrialists. Researches leading to better understanding of the chemistry of legumes viz. protein, lipid, carbohydrates, toxic constituents and flavour problems will open up new avenues for a major expansion of the food uses of these legumes. The practical solutions to nutritive problems in food legumes will ultimately come from the field plots of plant breeders through an interdisciplinary team approach.

Each topic has been comprehensively presented. To assist research workers references pertinent to the original literature have been included.

The vast increase in knowledge in the field of modern biochemistry and molecular biology has made it impossible for one person to gather all the information required in this era of specialisation. I had therefore asked my colleagues to assist with contributions in their specialised fields.

I wish to make a special acknowledgement not only to the enthusiasm with which the contributors to this treatise agreed to accept their tasks, but also to the speed and energy with which they completed them. I am deeply indebted to all of them. I also wish to thank Prof. Dr. W. Grosch, Technical University, Munich, West Germany for his encouragement and guidance and to the Alexander von Humboldt Foundation, Bonn for the award of "Senior Research Fellowship" which enabled me to undertake the task more efficiently. I am equally grateful to the Haryana Agricultural University for their encouragement.

S.K. ARORA

Contents

Contributors

S.K. ARORA, Haryana Agricultural University, Hissar 125004, India

M.L. DIXIT, Haryana Agricultural University, Hissar 125004, India

Y.P. GUPTA, Department of Biochemistry, Indian Agricultural Research Institute, New Delhi 110012, India

E. JOLIVET, Laboratoire d'Étude du Metabolisme intermédiaire et de Nutrition minérale, I.N.R.A., route de Saint-Cyr, 78000 Versailles, France

IRVIN E. LIENER, Department of Biochemistry, University of Minnesota, St. Paul, Minnesota 55108, USA

J. MOSSÉ, Laboratoire d'Étude des Protéines, I.N.R.A., route de Saint-Cyr, 78000 Versailles, France

J.C. PERNOLLET, Laboratoire d'Étude des Protéines, I.N.R.A., route de Saint-Cyr, 78000 Versailles, France

N.R. REDDY, Utah State University, Logan, USA

D.K. SALUNKHE, Vice-Chancellor, Mahatma Phule Agricultural University, Rahuri, 413722, Dist. Ahmednagar, Maharashtra, India

S.K. SATHE, Utah State University, Logan, USA

U.C. SHUKLA, Faculty of Agriculture, University of Maiduguri, Maiduguri, Nigeria

J. SMARTT, Department of Biology, Building 44, The University, Southampton, SO9 5NH, England

1
Legume Carbohydrates

S.K. ARORA

1.1. Introduction

Carbohydrates have played a leading role in the advancement of civilization. Agriculture was essential in the evolution of mankind and the cultivation of crops rich in carbohydrates accelerated the standard of living. Besides, plant carbohydrates are of great economic importance as the source of food, paper and gums. The latest developments in organic chemistry, enzymology, biophysics and physiology have influenced the knowledge of carbohydrates. Since early times legumes are grown for carbohydrates and proteins. The carbohydrates of legumes, especially the monosaccharide, glucose and fructose, are sources of energy. The legume seeds are also rich sources of galactosides of sucrose and galactomannose which are present in the endosperm of the seeds.

1.2. Monosaccharides

Only a relatively few of the monosaccharides occur free in plants as most of them are found as the components of glycosides or as units in various oligosaccharides or polysaccharides. Of the monosaccharides, D-glucose, D-mannose, D-galactose and D-fructose are the most widely distributed. Glucose is by far the most common; it is found free and in combination with many important plant substances such as the oligosaccharides, sucrose and raffinose, and the polysaccharides such as starch and cellulose and in many important glycosides. Fructose is the only abundant ketose. It is found free in a pyranose structure and in a combined state as in sucrose and inulin, in the furanose form. D-mannose and D-galactose occur principally in the form of polymer galactomannan in the leguminosae family. Only L-galactose is found as a constituent of agar. The pentoses D-xylose, D-ribose and L-arabinose do not occur free. D-xylose is found in woody material such as xylans, arabinose as arabans and in gums, ribose as a constituent of nucleotides.

In normal sugars, the replacement of the hydroxyl group by hydrogen

1

forms deoxysugars, the most important of them being 2-deoxy, D-ribose and the sugar component of deoxy-ribonucleic acid (DNA). Sugars with a methyl group instead of a primary alcohol group in the terminal position are glycosides and gums. Amongst the simplest monosaccharides are the trioses glyceraldehyde and dihydroxyacetone, both of which have an important function in the metabolism of the living cell. These are formed from the more simple member glycerol on oxidation.

The most important tetroses found in plants are D-erythrose and D-erythrulose:

These are the pentoses:

1.3. Oligosaccharides

Polymeric sugars with less than ten monosaccharides in the molecule are known as oligosaccharides. The monosaccharides are joined through glycosidic bonds with the elimination of $n-1$ molecules of water. By far the most important disaccharide occurring in plants is sucrose, others are maltose, cellobiose, gentiobiose, lactose, melibiose. The trisaccharides are raffinose, planteose, etc. The oligosaccharides are reducing and non-reducing.

1.3.1. REDUCING SUGARS

(1) MALTOSE

Maltose (4–0–α-D-glucopyranosyl glucoside) is found in many plants in small amounts. It is suspected that maltose to some extent might be the hydrolysis product of starch during its extraction. Quiller and Bourdon (1956) reported the occurrence of exceedingly high quantities of maltose in soybean tissues. Petioles, stems, and roots contained 45, 38 and 14 per cent, respectively. Glucose, fructose and sucrose were also found, but galactose was lacking either in free or bound form.

Maltose

(2) LACTOSE

Lactose (4–0–β-D-galactopyranosyl-D-glucopyranose) is a relatively rare and unexpected sugar in higher plants. Its presence in the chicle plant, *Achros sapota*, in which it was first found in 1871 was later confirmed by Venkatarman and Reithel (1958). Wallenfels and Lehmann (1957) identified from the

Lactose

grains of *Ceratonia*, lactose and through the column chromatography found other disaccharides ceratose which is fructosyl-glucose and primeverose 6–0–β-D-Xylopyranosyl-D-glucose.

Primeverose

1.3.2. NON-REDUCING SUGARS

The main emphasis will be on the saccharides of raffinose family which are present in the leguminosae family.

(1) RAFFINOSE

Raffinose is a non-reducing sugar without any food value, unless it has been hydrolyzed by strong acids into its components. This occupies the second position next to sucrose in the abundance of occurrence in the plant kingdom. Johnston (1843) was the first to obtain crystalline raffinose from an extract of the tissue of *Eucalyptus manna*. It invariably occurs in association with its higher homologues, for example, stachyose, verbascose and ajugose. Raffinose is the galactoside of sucrose. Methylation analysis by Haworth et al. (1923) and through enzymic and chemical methods by French (1954) has confirmed its structure.

Raffinose occurs only at low concentration in the leaves of leguminous plants, but accumulates in the storage organs, seeds and roots, during the process of development. Its content increases as the moisture content decreases during maturation of seeds. Most of the leguminous crops are drought resistant and contain these oligosaccharides in higher amount. Sautarius and Milde (1977) suggested that the affinity of the hydroxyl group of a sugar for water may influence the micro-environment of the labile, membrane bound proteins, thus stabilizing them against different stress conditions. Levitt (1972) reported that sugars may also bind directly to the protein molecule and prevent their denaturation. The sugars of this family are responsible for the production of flatus in man. Hardinge et al. (1965) and Cristofaro et al. (1974) reported that the important food legumes contain appreciable amount of oligosaccharides (Table 1.1).

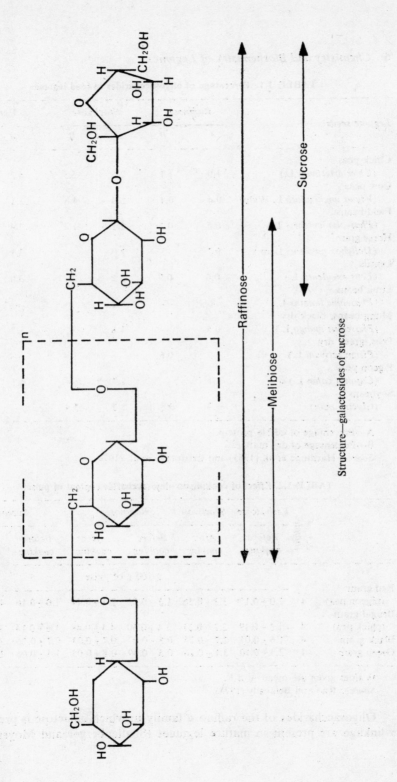

Sucrose

Raffinose

Melibiose

Structure—galactosides of sucrose

TABLE 1.1. Percentage of oligosaccharides in food legumes

Legume seeds	Raffinose		Stachyose		Verbascose	
	A	B	A	B	A	B
Chick peas						
(*Cicer arietinum* L.)	1.0	1.1	2.5	2.5	4.2	—
Cow peas						
(*Vigna unguiculata* L. Walp)	0.4	0.4	2.0	4.8	3.1	0.5
Field beans						
(*Phaseolus vulgaris* L.)	0.5	0.2	2.1	1.2	3.4	4.0
Horse gram						
(*Dolichos uniflorus* Lam.)	0.7	—	2.0	—	3.1	—
Lentils						
(*Lens exculenta* L.)	0.6	0.9	2.2	2.7	3.0	1.4
Lima beans						
(*Phaseolus lunatus* L.)	—		0.2	—	—	—
Mung beans, black dry						
(*Phaseolus mungo* L.)	0.5	—	1.8	—	3.7	—
Peas, green dry						
(*Pisum sativum* L.)	—	0.6	—	1.9	—	2.2
Pigeon peas						
(*Cajanus cajan* L.)	1.1	—	2.7	—	4.1	—
Soybeans						
(*Glycine max*)	1.9	0.8	5.2	5.4	—	—

A—Percentage of edible protein.
B—Percentage of dry matter.
Source: Hardinge et al. (1965) and Cristofaro et al. (1974).

TABLE 1.2. Effect of cooking on oligosaccharide content of pulses

	No. of samples	Verbascose + Stachyose		Raffinose		Sucrose	
		Before cooking	After cooking	Before cooking	After cooking	Before cooking	After cooking
		g/100 g of pulse					
Red gram							
(pigeon pea)	4	2.0±0.15[a]	3.3±0.56	1.3±0.22	2.4±0.44	1.0±0.16	4.4±0.37
Bengal gram							
(chick pea)	4	1.3±0.19	2.7±0.27	2.4±0.30	4.4±0.48	1.6±0.14	3.1±0.92
Black gram	4	1.6±0.07	2.7±0.25	0.5±0.07	0.7±0.03	0.7±0.26	2.6±0.63
Green gram	4	2.4±0.10	3.1±0.26	0.5±0.09	0.8±0.05	1.3±0.05	2.7±0.59

[a]Values given are mean ± S.E.
Source: Rao and Belavady (1978).

Oligosaccharides of the raffinose family in which galactose is present in α-linkage are present in mature legumes (Shallenberger and Moyer 1961)

TABLE 1.3. Effect of germination on oligosaccharide content of pulses

	Germination period			
	0 h	24 h	48 h	72 h
	g/100 g of pulse			
Red gram (4)[a]	2.0[b]	1.0	0.5	0.2
Verbascose + Stachyose	(1.6–2.3)	(0.6–1.6)	(0.3–0.7)	
Raffinose	1.3	0.8	0.4	0.2
	(0.7–1.7)	(0.6–0.9)	(0.3–0.7)	(0–0.4)
Bengal gram (4)	1.1	0.6	0.3	0.1
Verbascose + Stachyose	(0.8–1.4)	(0.4–0.7)	(0.1–0.5)	(0–0.2)
Raffinose	2.4	1.1	0.6	0.3
	(1.9–3.0)	(0.8–1.3)	(0.2–0.9)	(0.1–0.5)
Black gram (3)	1.7	0.9	0.3	0.1
Verbascose + Stachyose	(1.6–1.8)	(0.7–1.0)	(0.1–0.5)	(0–0.3)
Raffinose	0.5	0.3	0.1	0.1
	(0.4–0.5)	(0.2–0.3)	(0.1–0.2)	(0–0.2)
Green gram (3)	2.4	0.6	0.3	0.3
Verbascose + Stachyose	(2.1–2.6)	(0.4–0.7)	(0.2–0.4)	(0.2–0.4)
Raffinose	0.6	0.4	0.3	0.2
	(0.5–0.7)	(0.2–0.5)	(0.1–0.4)	(0.1–0.3)

[a]Numbers in parentheses indicate number of samples analyzed.
[b]Values given are mean with range in parentheses.
Source: Rao and Belavady (1978).

have been shown to be responsible for flatulence following the consumption of these beans (Steggerda, 1968). Murphy (1973) made extensive studies on the oligosaccharide content and the flatulence producing capacity of soybeans, Rao and Belavady (1978) reported the oligosaccharide content of red gram (*Cajanus cajan*), Bengal gram (*Cicer arietinum*), black gram (*Phaseolus mungo*) and green gram (*Phaseolus aureus*). Red gram and green gram had significantly higher amounts of verbascose and stachyose than did Bengal gram and black gram. Bengal gram had higher amounts of raffinose. They further observed a significant increase in the oligosaccharide content of all pulses on cooking, while germination decreased the oligosaccharide contents. Varietal differences in different pulses is given in the Tables 1.2 and 1.3.

Depending upon the amount of oligosaccharide flatulence degree is varied. Legumes being rich in protein, is a source of dietary protein for a large percentage of the world population, has attracted the attention of plant breeders, nutritionists and technologists to lessen the flatulence factors of food legumes. The possible approaches are:

(1) Genetic manipulation to breed varieties low in these oligosaccharides;
(2) Use of enzymes to hydrolyze the oligosaccharides (Reynolds 1974);

(3) Soaking and germinating the seeds before cooking (Calloway et al. 1971; Kim et al. 1973); and

(4) Use of solvents for the removal of oligosaccharide before processing. Raffinose had also been related to the viability of seeds (Oveharov and Koshelev 1974).

(2) STACHYOSE

Next to raffinose in the homologous series of galactosides of sucrose is stachyose. It is widely present in most of the leguminous seeds and is economically important. The stachyose and its homologs occur in the leaves and accumulate during winter, at a concentration parallel to that of sucrose. At this concentration they provide frost resistance to the plant. Its structure was determined through enzymic and chemical methods (Onuki 1932, 1933; Herissey et al. 1951; Laidlaw and Wylam 1953). French (1954) has reviewed the chemistry of stachyose. He investigated the hydrolytic products of stachyose as a means of determining the type linkages between the D-galactose and D-glucose units in the molecule. They found that the partial hydrolysis of stachyose by an almond emulsion preparation gave D-galactose and raffinose, sucrose and galactobiose. Hydrolysis and periodate oxidation confirm the presence of a 1:6 linkage between D-galactose and D-glucose units in stachyose. Based on extensive studies they proposed stachyose is 0-α-D-galactopyranosyl-(1→6)-0-α-D-galactopyranosyl (1→6)-0-α-D-glucopyranosyl-β-D-fructofuranoside.

Kasai and Kawamura (1966) reported the sugars from soybean by dextran gel filtration. The soybeans used were from both defatted and autoclaved flour and determined their physical constants like melting point and specific rotation (Table 1.4). Kawamura's results on a dry basis for 45 whole soybean

TABLE 1.4. Melting points and specific rotations of sugars and acetates of sugar from raw and autoclaved soybean flakes

	Melting point, °C		Specific rotation (α_D)	
	Found	*Literature*	*Found*	*Literature*
Sugars				
Sucrose	176–180	184–185	+66.7/24° (H_2O)	+66.5° (H_2O)
Raffinose	118–120	118–120	+123.4/20° (H_2O)	+123.1° (H_2O)
Stachyose	170–172	170	+146.2/20° (H_2O)	+146.3° (H_2O)
Acetates of Sugars				
Sucrose octaacetate	87	87	+59.0/20° ($CHCl_3$)	+59.6° ($CHCl_3$)
Raffinose hexaacetate	98–101	99–101	+97.4/16° (EtOH)	+92.0 (EtOH)
Stachyose tetraacetate	94–95	95–96	+120.1/16° (EtOH)	+120.2° (EtOH)

Source: Kasai and Kawamura (1966).

are sucrose, 4.5 per cent, raffinose, 1.1 per cent, stachyose, 3.7 per cent, and traces of arabinose and glucose for a total of 9.3 per cent sugars. For the Japanese beans the values are sucrose, 5.7 per cent, raffinose, 1.1 per cent, stachyose, 4.1 per cent, and traces of arabinose and glucose for a total of 10.9 per cent sugars.

1.4. Cyclitols

1.4.1. MYO-INOSITOL

Cyclitols are aliphatic alcohols in which the hydroxyl group is directly attached to the atom of carbon in cycle. The principal alcohols are the derivatives of cyclohexane. The important cyclitol is inositols, while others have taxonomic interest. The structure of different cyclitols present in legumes is given below.

myo-inositol

1.4.2. (+) QUERCITOL

In 1955 Plouvier isolated (+) quercitol from *Pterocarpus lutens* for the first time. It is desoxy inositol.

(+) Quercitol

1.4.3. (+) PINITOL

It is the methylated derivative of (+) inositol. The first evidence of its presence in plants was reported by Charaux (1922) in *Ceratonia siliqua*. Schweizer et al. (1978) found in soybean, chickpeas, lentils and other beans the

(+) Pinitol

presence of pinitol along with galacto-pinitol. Their amount ranged from 0.2 to 0.9 per cent and 0.03 to 0.8 per cent of pinitol and galacto-pinitol on dry weight basis respectively. Soybean especially was found to contain appreciable amounts of pinitol.

1.5. Galactosides of Inositol

Like the galactosides of sucrose, the galactosides of myo-inositol are also reported in *Leguminosae* family. Brown and Serro (1953) reported first the presence of galactinol-0-α-D-galactopyranosyl-myo-inositol.

Galactinol

It is accompanied by digalactosido-inositol. The work of Kandler and his school has shown that in number of families, the galactinol is accompanied along with other oligosaccharide or with galactosides or sucrose.

Galactinol occurs in the leaves in an amount similar to that of raffinose and stachyose. In the roots and rhizomes the amount of galactinol decreases compared to the amount of other oligosaccharide. Galactinol was found in all those plants which also contain raffinose and stachyose (Senser and Kandler 1967). Sioufi et al. (1970) identified the UDP-glucose, UDP-galactose and GDP-mannose in *Trigonella foenum graecum* seedlings. A study of the variation of galactomannan, verbascose, stachyose, galactinol, α-galactosidase and β-mannanase during the germination has been made. Sugar nucleotides are shown to be essential factors in the metabolic pathways leading to the synthesis of the cell-wall polysaccharides. Hassid (1967) showed that galactinol is formed earlier than stachyose and raffinose and proposed the reactions:

(1) UDP-Gal + Myo-inositol $\xrightarrow{\text{transferase}}$ Galactinol + UDP

(2) UDP-Gal + Sucrose $\xrightarrow{\hspace{1cm}}$ Raffinose + UDP

(3) Galactinol + Raffinose $\xrightarrow{\text{transferase}}$ Stachyose + Myo-inositol

These oligosaccharides are synthesized at the end of the ripening period in *Phaseolus vulgaris* (Tanner 1969). The enzyme transferase requires Mn^{2+} for its activity and this could not be replaced by Mg^{2+}.

1.6. Galactosides of Pinitol

Schweizer et al. (1978) found a new disaccharide which they named galacto-pinitol in leguminous seeds specially in soybean. Pinitol and galactopinitol were also identified in chickpea, lentils and beans. The structure assigned is given below.

Galactopinitol

They proved to be the isomers of α-D-galactopyranosyl pinitol. The identi-ty of the monosaccharide units of the galactopinitols was proved by acid and enzymatic hydrolysis (α-galactosidase) yielding equal amounts of galactose and pinitol shown to be identical and further confirmed by gas chromatogra-phy, mass spectrometry of their TMS derivatives. Permethylation analysis of the disaccharides furnished additional proof since, after methylation, hydro-lysis, reduction and acetylation, 2,3,4,6, tetra-O-methyl-galactinol-1, 5-diace-tate and pentamethyl acetyl-cyclitol were identified by their MS fragmentation pattern.

1.6.1. The Possible Role of Galactopinitol

Galactinol was believed to occur in all higher plants containing the sugars of the raffinose family (Senser and Kandler 1967) and its role in the biosynthesis of raffinose, stachyose and verbascose has been established insofar as it acts as a transfer-intermediate of galactose (Senser and Kandler 1967; Lehle and Tanner 1973). Schweizer et al. (1978) could not find galactinol in soybean, though in chickpeas it occurs along with galactopinitol whereas in other seeds only galactinol was found. They concluded that galactopinitol might also act as a transfer-intermediate. Galactopinitol, being α-galactosides, might also be responsible for flatulence problems.

1.7. Polysaccharides

Polysaccharides are defined as polymers of a large number of monosacchari-des, usually more than 20 monosaccharides joined through glycosidic links. Each glycosidic link is formed by condensation with the removal of one water molecule between the hydroxyl on the α or β anomer of the hemiacetal (re-ducing) carbon (generally C-1) of a monosaccharide and the hydroxyl on any other carbon of a second monosaccharide. The sugars may be joined in linear or branched chains.

Legume polysaccharides are generally classified in term of (1) structure; (2) function in the plant; (3) distribution in the plant; and (4) method of isolation from plant tissues, though these overlap sometime. In this chapter, they will be discussed under the following heads:

1.7.1. STRUCTURAL POLYSACCHARIDES

These include both the fibre components—cellulose of cell walls and non-fibrous polymers such as the hemicelluloses and pectic substances which are believed to form an amorphous matrix around these fibres or to act as a cement between the cells.

(1) PECTIC SUBSTANCES

In legumes three types of pectic substances are present: (a) pectin, a polymer based on the chains of D-galacturonic acid; (b) galactan, a galactose polysaccharide; (c) araban, an arabinose polysaccharide. Pectic substances are extracted from plant tissues with hot dilute oxalate or chelating agent such as ethylene diamine tetra acetic acid (EDTA).

(a) *Pectins*

Most pectins contain other sugars present either as side chains or actually within the uronic acid chains. They are simple homologous polymers of D-galacturonic acid in α-1–4 linked linear chains (Aspinall et al. 1968). Herbaceous legumes often contain large amounts of pectins in their leaves and stem. Hirst et al. (1959) reported the presence of 10–12 and 6–9 per cent of pectin in lucerne (*Medicago sativa*) leaves and stems, respectively. The content varies with species, the pasture grasses contain generally less than 1 per cent of pectic polysaccharides in their leaves (Bailey 1964). The seed pectic polysaccharides are structurally more complex than the leaf and stem pectins. Aspinall et al. (1966 and 1967 a, b, c) reported the polygalacturonide pectic fractions from soybean.

(b) *Galactan*

The purification of galactan is usually hampered due to the presence of other pectic substances. Tadros and Kamel (1952) extracted pectic substances from *L. termis* seeds which contained D-galactose, L-arabinose and D-galacturonic acid but could not be separated into araban and galactan, *Lupinus alba* seeds had galactan having a more complex structure rich in 1→3 and 1→6 linked D-galactosyl units.

A galactan isolated from *Strychnos nux vomica* seeds (Andrews et al. 1954) had similar basic structure of β-1-4 linked D-galactosyl units but was also branched through 1→3 and 1→6 links. Aspinall et al. (1967 a, b) extracted a complex acidic pectin polysaccharides from soybean seed meal which afforded 4-0-β-D galactopyranosyl-D-galactose, the polymer homologous tri, tetra, penta and hexasaccharides. The enzymic hydrolysis of partially degraded polysaccharide furnished 3-0-β-D-xylopyranosyl-D-galacturonic acid and higher acidic oligosaccharides containing chains of 1-L linked β-D-galacto-pyranose residues.

(c) *Araban*

Hirst and Jones (1939, 1947) isolated pure araban from the seeds of peanut, and Narasaki and Fujimoto (1965) from soybean. Araban are always associated with other pectic substances. Structural studies by Whistler and Smart (1953) indicate araban to be a highly branched polysaccharide based on a chain of 1→5 linked L-arabinofuranosyl units with branches on the C-3 atoms.

(2) NON-CELLULOSIC POLYSACCHARIDES

The occurrence of these polysaccharides in leguminosae is considered as a part of woody and non-woody tissues separately. Stewart (1966) gave a good account of its chemistry. Not much work on the woods of edible legumes has been done except *Prosopis Juliflora* (Sands and Nutter 1935) and *Robinia pseudacacia* (Anderson et al. 1940). 4-0-methyl glucuronoxylans and glucomannans are widely distributed. In the stem, leaves, seed pods, seed hulls and seeds the interest of structural carbohydrates is particularly from ruminant digestion point of view. In storage legumes the cell-wall monosaccharide levels during growth have been reported in lucerne, clovers, guar. On a dry weight basis, lucerne stem contained D-xylose (8–10 per cent), L-arabinose (2–3 per cent), D-galactose (1–2 per cent) and D-mannose (1–2 per cent), whilst the leaves contained less percentage of these saccharides (Hirst et al. 1959). Nevins et al. (1967) reported the changes in levels of L-rhamnose, L-fucose, L-arabinose, D-xylose, D-mannose, D-galactose and D-glucose in cell wall of various parts of growing plants of *Phaseolus vulgaris, P. aureus* and *Glycine max*. Aspinall et al. (1967 a, b) conducted extensive work on the polysaccharides of soybean and identified the monosaccharides composition and determined its structure. Monosaccharide compositions of hemicellulose fractions from several legumes are listed in Table 1.5 and those from legume seeds, peanut, pea, seed meal, *Vicia faba* and *Glycine max* are given in Table 1.6.

In legume leaves and stems the hemicellulose-A xylans are probably conventional glucurono-xylans of the type generally found in angiosperms, L-arabinose, mannose and some galactose are also present. The pentosan-β fractions all contain a possible associated glucan which can be enzymically

TABLE 1.5. Composition of hemicellulose fractions

Species	Fraction	Monosaccharide composition (%)				
		D-Xyl	L-Ara	D-Gal	Uronic	Other
Glycine max (L.)	Hemicellulose A	93	—	—	7	—
Merr. stalks	Pentosan-B[a]	89	8	—	4	
	Heteroglycan-B	3	24	34	25	D-Glc 7. L-Rha 7 D-Man traces
Trifolium pratense	Hemicellulose-A	95	—	—	5	—
L. stalks	Pentosan-B	88	10	—	1	—
	Heteroglycan-B	17	28	35	21	—
	Galactoglucoman-nan	16	—	12	—	D Glc 48, D-Man 24
Medicago sativa	Hemicellulose-A	93	—	—	7	—
L. stalks	Pentosan-B	90	8	—	2	—
	Heteroglycan-B	3	34	31	22	D Glc 9
Lolium perenne	Hemicellulose-A	85	13	—	2	—
L. leaves	Pentosan-B	83	16	—	0.5	—
	Heteroglycan-B	63	24	8	5	—

[a]All pentosan-B fractions had 5–10% D-glucose, believed to be present as a separate β glucan, associated with them. Results listed were calculated on a glucose free basis.
Source: Whistler and Gaillard (1961), Gaillard (1965), Gaillard and Bailey (1968).

hydrolysed to β 1→4 and β 1→3 linked disaccharides (Bailey and Gaillard 1965). Seed polysaccharides contain glucuronoxylans, uronic acid free pentosans and possible galacto glucomannan.

(3) CELLULOSE

Cellulose is a linear polymer of β 1→4 linked D-glucose units, repeating units 4–0-β-D-glucopyranosyl-D-glucopyranose, cellobiose. It is of high molecular weight with degrees of polymerization of up to 9,000–10,000 glucosyl units reported for wood and plant fibres (Goring and Timell 1962). Leaf cellulose particularly from dicotyledons plants may be lower in the degree of polymerization.

Cellulose concentrations in pasture legumes are often measured because of interest in ruminant digestion of cellulose. On dry weight basis 8–30 per cent of cellulose, depending on maturity in stems and 5–10 per cent in leaves is present in such feed legumes as *Medicago sativa* (Hirst et al. 1959), *Trifolium repens* (Bailey 1964), cowpeas (Das et al. 1975 a), *Cyamopsis tetragonoloba* (Das et al. 1975 b), Moth (*Phaseolus aconitifolius*) (Arora et al. 1975) and *Trigonella foenum graecum* (Das et al. 1979) (Table 1.7).

Cellulose is present in all plants and is probably the structural polysac-

TABLE 1.6. Non-cellulosic structural polysaccharides

Species and tissues	Polymer	Structure	Monosaccharide	Reference
Soybean seed hulls	Hemicellulose-A xylan	β-1-4-D-Xyl-chains single Glur 1-2 side units	D-Xyl, D-Glur 95: 3–4%	Aspinall et al. (1966)
	Galactoglu-comannan	β-1-4-D-Man chain	D-Man, D-Gal, D-Glc 23: 2 : 1	—ibid—
	Hemicellulose-B	Highly branched	D-Xyl, L-Ara, D-Gal, D-Glc, 4-0-Me-D-Glur	Sannella and Whistler (1962)
Peanut, shells	Hemicellulose-A	β-1-4-D-Xyl chain 1-2 uronic side units	D-Xyl, D-Glur	Radhakrishna-murty and Srinivasan (1959)
	Hemicellulose-B	Glur-ara units linked to β-1-4-D-Xyl chain	D-Xyl, L-Ara, D-Glur	—ibid—
	Hemicellulose-B_2	Highly branched	D-Xyl, L-Ara, L-Rha, D-Gal, D-Glur	—ibid— (1960)
Seed meals				
Soybean	5% alkali soluble	—	D-Xyl, L-Ara, D-Glc, D-Gal, L-Fuc, L-Rha, D-Glur	Kawamura and Narasaki (1961)
Vicia faba	Hemicellulose-B	—	D-Xyl, L-Ara, D-Glc, L-Rha, L-Fuc, D-Gal, D-Glur	—ibid— (1958)
Lupinus luteus L.	—	—	L-Rha, D-Gal, L-Ara, D-Glur D-Xyl	Tomoda and Kitamura (1967)

charide which contributes most to the rigidity and strength of plant structures.

Cellulose is present in plant tissues as fibres composed of microfibrils which consist of cellulose chains aligned along the microfibril axis and held together by intermolecular and intra-molecular hydrogen bonds. The regular arrangement gives an x-ray pattern indicative of crystallinity in a large part of the microfibril.

1.7.2. STORAGE POLYSACCHARIDES

(1) STARCH
Starch is widely distributed in plant leaves and in seeds, roots and tubers.

TABLE 1.7. Percentage range of structural carbohydrates of leguminous forages

Species	Neutral detergent fibre	Acid detergent fibre	Hemicellulose	Cellulose	Reference
Guar (*Cyamopsis tetragonoolba* L.)	31.2–44.4	26.1–37.4	3.6–9.2	16.6–28.1	Das et al. (1975b)
Cowpea (*Vigna unguiculata*)	46.76–58.21	32.32–41.26	10.0–25.87	20.77–29.46	Das et al. (1975a)
Moth (*Phaseolus aconitifolius*)	38.24–43.95	30.10–34.40	6.94–9.84	20.04–24.32	Arora et al. (1975)
Metha (*Trigonella foenum graecum*)	40.9–43.8	34.8–37.8	4.82–8.86	28.6–31.4	Das et al. (1979)
Berseem (*Trifolium alexandrinium*)	53.00	39.95	13.05	26.85	
Lucerne (*Medicago sativa*)	46.00	32.00	14.00	21.50	
Senji (*Melilotus alba*)	54.70	42.70	12.00	28.57	

Starch is composed of two closely related polysaccharides: amylose and amylopectin. Amylose is a linear polymer of α 1→4 linked D-glycosyl units while amylopectin is a modification of amylose highly branched through 1→6 links. Amylose stains blue with iodine and amylopectin violet.

The starches of peas and beans have been extensively tested. Seed starches are classified either as cereal starches or tuber starches depending upon attaining powers and physical properties. Generally, it is mentioned that legume seeds are starch free and should be regarded with some suspicion. The species of *Phaseolus, Vigna, Lathyrus, Lens, Pisum* and *Vicia* and *Arachis* do contain starch. Kawamura et al. (1955) noted that the starch present in minute amounts in soybean seeds only gave a positive iodine test. As early as 1913, Reichert examined starch grains mainly by microscopy and staining reaction, from the seeds of seventeen legume species along with other species. The starch grains from legume species were smooth, ovoid, elliptical or kidney shaped except of wrinkled pea seeds, which has rosettes type. Recently, Arora and Das (1976) reported the presence of 50.66 to 67 per cent starch in cowpea. Cowpea contained both types of starches with low and high amylose contents. Its physical properties have not been worked out. The amylose content ranged from 20.85 to 48.72 per cent and as per classification by Kawamura they are largely cereal type starches. Pea starches are divided into two main classes depending upon whether the seeds are smooth or wrinkled. McCready et al. (1950) reported that the wrinkled seeds had 34 per cent of total starch, while smooth peas had 42 per cent.

Smooth seed peas had 29–35 per cent amylose and wrinkled seed pea contain 66–90 per cent amylose (Peat et al. 1948; McCready et al. 1950; Potter et al. 1953). The molecular weights of amylose from smooth as well as wrinkled pea seeds are similar, but amylopectin portion had higher molecular weight (2,000,000) in smooth peas, while amylopectin from wrinkled seeds had 140,000 (Potter et al. 1953). Greenwood and Thomson (1962) compared the various physical properties of smooth pea, wrinkled pea and broad bean (*Vicia faba*) starches and some of the results are tabulated below:

TABLE 1.8. Properties of whole starches

	Average granule size (μ)	Protein[a] (%)	Iodine[b] affinity	Amylose[c] (%)	Gelation temperature range
Broad bean	30	0.16	4.5	24	64–67
Pea (Smooth seeded)	30	0.19	6.6	35	>98
Pea (Wrinkled seeded)	40	0.23	12.5	66	>98

[a]Calculated from (% N × 6.25).
[b]Expressed as mg of iodine bound per 100 mg of starch.
[c]Calculated from [(Iodine affinity of starch/iodine affinity of amylose) × 100].
Source: Greenwood and Thomson (1962).

(2) SEED GALACTOMANNAN

Almost all important food plants produce seeds containing starch as the carbohydrate reserve. This starch serves as the principal food stored for use by the embryonic plant in its initial growth stages. Many plant seeds, however, contain polysaccharides food reserves that are not starch but are the polymers of other sugar molecules such as galactose and mannose. These polymers also have excellent hydrocolloids properties. The important leguminous seeds which are of industrial value for their gum are guar (*Cyamopsis tetragonoloba*) (Das and Arora 1978); Locust bean gum (Rol 1973); and tamarind (Rao and Srivastava 1973).

Plant galactomannans are reserve polysaccharides composed of linear chains of (1→4) linked β-D-mannopyranosyl residues having single stubs of α-D-galactopyranosyl groups joined by 1→6 linkages along the chain. Most of these leguminous crops are cultivable and the seeds are harvested to produce gum in the industry. Galactomannans are located in the endospermic part of the seeds and have taxonomic importance.

In addition to the seed galactomannans, the leaf and stem tissues of red clover (*Trifolium pratense*) have been shown to contain a galactoglucomannan (Gaillard and Bailey 1968; Buchala and Meier 1973).

(a) *Guar Gum*

The guar plant was not used for its seed gum until the World War II, but now the seeds are produced by conventional agricultural practices. Das and Arora (1978) have recently reviewed the chemistry of guar seed. An average composition of the various seed components of guar is given in Table 1.9.

TABLE 1.9. Composition of the components of guar seed

Seed part	Protein (%) (N × 6.25)	Ether extract (%)	Ash (%)	Mois- ture (%)	Crude fibre (%)	Type of carbo- hydrate
Hull (14–17 %)	5	0.3	4	10	35.0	D-glucose
Endosperm (35–42%)	5	0.6	0.6	10	1.5	Galacto- mannan
Germ (43–47 %)	55.3	5.2	4.6	10	18.0	D-glucose

Source: Goldstein et al. 1973.

Composition and Structure: Guar gum is a galactomannan, known as guaran. The crude gum is a greyish-white powder and may contain small amounts of proteinaceous matter. An analysis of a commercial guar gum is as follows (Glicksman 1969):

Galactomannan	78–82	per cent
Water	10–13	per cent
Protein	4–5	per cent
Crude fibre	0.5–2.0	per cent
Ash	0.5–0.9	per cent
Fat	0.5–0.75	per cent

Artaud et al. (1975) have reported the neutral sugar, fatty acid and mineral substance composition of guar gum. In a recent study, Artaud et al. (1977) observed that palmitic, oleic and linoleic acids predominated in the gum.

The results of methylation analysis (Rafique and Smith 1950; Ahmed and Whistler 1950) and partial hydrolysis experiments (Whistler and Stein 1951; Whistler and Durso 1951) have shown that guar gum contains a β-D $(1\rightarrow4)$ linked D-mannan backbone to which single α-D-galactosyl stubs are attached at 0–6 of certain of the D-mannosyl residues. Methylation and subsequent hydrolysis gave mixtures of 2, 3, 4, 6, tetra-O-methyl-galactose, 2, 3, 6, tri-O-methyl-D-mannose, and 2-3-di-O-methyl-D-mannose, where as partial hydrolysis yielded the disaccharides 6-0-α-D galactopyranosyl and 4-0-β-D-manno-pyranosyl-D-mannose. The periodate oxidation (Rafique and Smith 1950; Moe et al. 1947) have supported the above results. The possible repeating unit for guar gum has been suggested as shown on p. 19.

For an extensive discussion of the chemistry of the gum, the readers are referred to the monographs by Whistler and Smart (1953) and Smith and Montgomery (1959). Further information about the structure is forthcoming

Guaran

from the enzymatic studies. The biodegradative hydrolysis of a galactoman-nan requires the presence of at least three enzymes in the germinating seeds (Reese and Shibata 1965). α-D-galactosidase for removal of the $(1\rightarrow6)$ α-D-galactose side chains, β-D-mannanase for fission of the $(1\rightarrow4)$-β-D-mannan backbone into oligosaccharides, and β-D-mannonsidase for the complete hydrolysis of the D-manno-oligosaccharides to D-mannose. These enzymes are discussed in the subsequent part of this chapter. These enzymes have been reported (Courtis and Dizet 1963; Whistler and Smith 1952; Sehgal et al. 1973) in guar seed. These enzymes exist in multiple forms having different pH optima. Two forms of α-galactosidase (McCleary and Matheson 1974) and three β-mannanase (Lee 1967) have been reported in guar and these enzymes have been used in the study to find structure. It has been shown that there is preliminary removal of D-galactosyl groups followed by scission of the mannan chain and concomitant release of D-manno-oligosaccharides. Reese and Shibata (1965) have found that the residue remaining after hydro-lysis of guaran had a galactose-mannose ratio of 1 : 2 by using fungal β-man-nanase. McCleary and Matheson (1975) have suggested that for effective hydrolysis at least two contiguous unsubstituted anhydromannose units are required. The three possible extremes of structures for a galactomann having mannose : galactose ratio of 2 : 1 could be (a) regular arrangement of side chains, (b) random distribution of side chains and (c) a structure in which the side chains occur in blocks. It has been indicated by Hoffman et al. (1976) that random distribution of galactose branches occurs in guaran.

The fibre structure of ordered films of guar gum was examined by Palmer and Ballantyne (1950). These authors suggested that the material had an orthorhombic unit cell with $a = 15.5$ Å, $b = 10.3$ Å (fibre axis), and $c = 8.65$ Å. The presence of meridional reflections on alternate layer-lines suggested a two-fold screw axis parallel to the galactomannan chains and lying between adjacent sheets. They proposed a structure in which the chains were arrang-ed in sheets, with the side chains and the planes of the pyranose ring lying in the planes of the sheets, and the side chains of adjacent sheets pointing in the opposite directions. The x-ray fibre diffraction patterns indicated that the D-galactose side chains may not necessarily affect the conformation of the main chain, and supported extended ribbon-like conformation as the ordered structure. Koleske and Kurath (1964) also concluded that the D-galactose side chains have little, if any, effect on the conformation of the main chain while studying the molecular weight of acetylated guar gum.

Audsley and Fursey (1965) studied the gum structure by electron micro-scopy and found that the smallest fibre diameter was 4 nm of the evaporated film.

Molecular weight of guar galactomannan has been reported (Hui and Neukon 1964) from 2 to 1.5×10^5 by chemical method and 1.9×10^6 by sedi-mentation analysis. Deb and Mukherjee (1963) assumed spherical shape of the molecules of the guar gum and obtained molecular weight of 1.7×10^6 by using light scattering method. This value seems to be large and may be due to the aggregation. Hoyt (1966) reported molecular weight of $2-2.2 \times 10^6$. Using molecular weight as obtained by the chemical method of Hui and Neukom and 2/3 of the molecule as a mannan backbone, it becomes evident that the mannan chain is of about 1000 anhydro mannose units, since retardation required a pore size of 5000 Å and β-$(1\rightarrow4)$ linked anhydromannose unit has a length of 5.2 Å (Sundararajan and Rao 1970). This indicates that the polymer in aqueous solution has the mannan backbone in a fully extended rod conformation. Sundararajan and Rao (1970) have also predicted this from the theoretical calculations. McCleary and Matheson (1975) while investigating structural changes using viscometry, which is a sensitive method of detecting differences in rodlike molecules, also indicated extended shape in aqueous solution.

(b) *Locust Bean Gum*

Locust bean gum contains about 88 per cent of D-galacto-D-mannoglycan. The ratios of D-galactose to D-mannose slightly differ as reported by various workers as 27 : 33; 20 : 80; 18 : 82; 16 : 84 and 14 : 86. These differences might be due to different analytical methods employed by them or due to different varieties of the seeds as effected by environment, location, age, etc. Locust bean gum may be classified chemically as a galactomannan and is a high molecular weight carbohydrate polymer made up of a large number of mannose and galactose units linked together.

The locust bean gum is a linear chain of β-D-manno-pyranosyl units linked $(1\rightarrow4)$ with every fourth and fifth D-mannopyranosyl units substituted on C-6 with an α-D-galactopyranosyl unit. It differs from guar gum only in the smaller number of D-galactosyl units as side chains.

Both guar gum and locust bean gum are neutral polysaccharides. The pH has little effect on the viscosity. Various derivatives have been made from both these gums to suit a particular end-use. The long straight chain nature of the molecule, combined with its regular side branching sets it apart from other natural polymers. Because of these unique properties, these gums have proven to be versatile natural colloids and base polymers from which many derivatives are produced.

(3) Seed "Amyloid"

Vogel and Schleiden (1839) extracted polysaccharides with hot water from the seeds of *Tamarindus indica* and named amyloid, which stained blue with iodine. The polysaccharide has assumed an industrial importance and is present to the extent of 50 per cent in the decorticated seeds of tamarind cotyledons. Kooiman (1961) did extensive work on its structure and reported to contain D-glucose, D-xylose and D-galactose in the proportion of 4 : 2.99 : 1.31.

The presence of L-arabinose in the tamarind kernel product is controversial. Savur and Sreenivasan (1946, 1948), Rao et al. (1945–46), Kahn and Mukherjee (1959), and Kooiman (1961) had not been able to detect the presence of L-arabinose while Srivastava and Singh (1967), Chakravarti et al. (1961) and Damodaran and Rangachari (1945) detected the presence of L-arabinose when tamarind seeds polysaccharide is hydrolysed under mild condition (0.01 N HCl), the first sugar detected is L-arabinose.

On the basis of research, Kooiman (1961) postulated a highly branched structure based on a linear chain of β-1→4 linked D-glucosyl units (i.e. a cellulose chain); α-D-xylosyl units are attached to the C–6 carbons of three out of every four D-glucosyl units and β-D-galactosyl units in turn linked to the C-2 carbons of some of the D-xylosyl units.

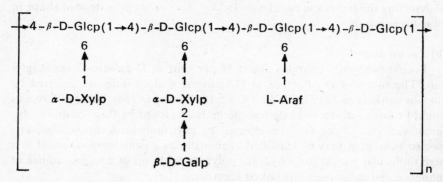

Structure proposed for tamarind seed polysaccharide (Srivastava and Singh 1967).

Sequential distribution of branches along the main chain is uncertain due to the controversy about the presence of L-arabinosyl unit in tamarind seed polysaccharide.

Kooiman (1960) examined also non-leguminous species and found a positive test in seeds from 134 species from 15 dicotyledonous families, which shows that a positive amyloid reaction is not only confined to legumes. Rao and Srivastava (1973) have written an extensive review on tamarind and enumerated all the possible applications.

Uses

The tamarind seed polysaccharide have numerous applications in food, pharmaceutical and textile industries. The polysaccharide forms jellies with sugar concentrates over a wide pH range and is an excellent substitute for fruit pectins in jams and jellies and marmalades (Anon 1941). It might find use in the treatment of colitis, diarrhoea, dysentery and other intestinal disorders. In combination with gum tragacanth it finds use in textile printing pastes.

Tawakley and Bhatnagar (1953) and Savur (1955, 1956a, b) have enumerated the other uses of these polysaccharides. It is used in insecticidal preparations for emulsifying the active principles and mineral oils. It is used in sizing yarns as it gives them a protective coating.

1.8. Biosynthesis

1.8.1. BIOSYNTHESIS OF OLIGOSACCHARIDES

(1) BIOSYNTHESIS OF RAFFINOSE

Bourne, Pridham and Walter (1962) showed that raffinose could be synthesized by an enzyme preparation from dormant broad bean (*Vicia faba*) seeds with a mixture of sucrose, α-D-galactose 1-phosphate and UTP as substrates using UDP-D-galactose labelled with ^{14}C in the D-galactose moiety and sucrose in the presence of an enzyme preparation from mature broad beans, a direct transfer of D-galactose-^{14}C to sucrose could be effected resulting in the production of raffinose (Pridham and Hassid 1965). The amount of ^{14}C-labelled D-galactose incorporated into the raffinose was 33 per cent and 39 per cent after 1 and 2.5 hours, respectively. The percentage incorporation of labelled carbon was quite high, indicating that this mechanism functions in maturing seeds, in spite of the crude enzyme preparation.

The other possibility that raffinose is also formed by the transfer of D-galactose from low energy donors to sucrose should not be overlooked. Courtois et al. (1961) demonstrated the synthesis of raffinose and planteose, using D-galactose as the donor in the presence of α-galactosidase preparation.

(2) BIOSYNTHESIS OF STACHYOSE

Tanner and Kandler (1966, 1968) obtained a soluble enzyme from ripening seeds of dwarf beans (*Phaseolus vulgaris*) which transfers D-galactose with high yield from galactino (0-α-D-galactopyranosyl-(1$\rightarrow$1)-myo-inositol to raffinose giving rise to stachyose and myo-inositol. It was also found that galactinol is a major galactoside constituent in the bean during a certain maturation period and that its formation precedes stachyose.

The amount of stachyose formed was found to be proportional to the amount of enzyme added, and the reaction was linear with time until 35 per cent of D-galactose-^{14}C of the labelled galactinol had been transferred to raffinose, after which the rate decreased.

$$\text{Galactinol} + \text{Raffinose} \rightleftarrows \text{Stachyose} + \text{Myo-inositol.}$$

The enzyme appears to be specific with regard to the acceptor molecule and is not stimulated by Mg^{++}, ADP or UDP. Its pH optimum is 7.0 and is inhibited by *p*-chloromercuribenzoate. The crude extract from the ripening bean seeds also contains an enzyme which transfers the labelled D-galactose from UDP-D-galactose-^{14}C but not from ADP-D-galactose-^{14}C to myo-inositol.

On the basis of these results, the pathway for stachyose synthesis is postulated to be as follows:

(a) UDP-D-galactose + Myo-inositol ⇌ Galactinol + UDP
(b) UDP-D-galactose + Sucrose → Raffinose + UDP
(c) Galactinol + Raffinose → Stachyose + Myo-inositol.

1.8.2. BIOSYNTHESIS OF POLYSACCHARIDES

The biosynthesis of polysaccharides is well documented in number of reviews/books, the emphasis will be only for galactomannans in this chapter being an industrially important carbohydrate.

BIOSYNTHESIS OF GALACTOMANNANS

Apart from the detailed research-work by Reid and Meier (1970) in fenugreek seeds very little attention has been paid to the biosynthesis of galactomannans in legumes. It has been generally concluded that D-galactose is stored in the seeds in the form of galactomannan, by random attachment of D-galactosyl groups to a main chain of D-mannan (Andrews et al. 1952). Henderson et al. (1958) pointed out that a plausible way whereby a galactomannan may be synthesized is by trans-D-galactosylation as shown:

$$
\begin{array}{c}
\text{G} \\
| \\
\text{Raffinose} + \dots \text{M-M-M} \dots \rightleftharpoons \dots \text{M-M-M} \dots + \text{Sucrose}
\end{array}
$$

D Mannan polymer   galactomannan

Reid and Meier (1970) observed that the polysaccharide is present at all stages except in the very young seeds, the maximum yield was obtained from 9–10 weeks old seeds. The mannose/galactose ratio remains constant throughout the stages of maturation which was 1.2. In addition both free D-mannose and D-manno-oligosaccharides, which might act as primers for the synthesis, are absent during development. The results also suggest that D-galactose and D-mannose units are deposited simultaneously to form the galactomannan.

Reid and Meier (1970) further observed the presence of sucrose, myo-inositol, D-glucose, galactinol and stachyose in immature seed, stachyose is present in abundance even after maturation in fenugreek seeds. Sioufi et al. (1970) proposed the following pathway for the formation of stachyose:

UDP-D-galactose + Myo-inositol → Galactinol + UDP

UDP-D-galactose + Sucrose → Raffinose + UDP

Galactinol + Raffinose → Stachyose + Myo-inositol.

Sioufi et al. (1970) also detected the presence of UDP-D-galactose and GDP-D-mannose in fenugreek seeds. Reid and Meier (1970) and Courtois (1974) suggested a similar mechanism of D-galactosyl transfer may be involved in the biosynthesis of galactomannan, with the participation of GDP-D-mannose. Many workers (Strominger et al. 1972; Richards et al. 1972; Kauss 1972; Tanner et al. 1972; Hassid 1972) have shown the participation of D-mannosyl-lipid intermediates in the biosynthesis of D-mannose containing oligo and polysaccharides.

1.9. Hydrolytic Enzymes

1.9.1. α-D-GALACTOSIDASES (MELIBIASE, α-D-GALACTOSIDE GALACTOHYDROLASE EC 3.2.1.22)

α-Galactosidases were first reported in 1895. The glycosidases have specific hydrolytic role in formation of glycosidic bonds. Recent interest has been on their mechanism of reaction and its existence in multiple forms. Dey and Pridham (1972) have reviewed this topic in detail. Galactose containing oligo- and polysaccharides are common carbohydrate reserves in leguminous seeds particularly guar, locustbean, alfalfa, etc. Besides its hydrolytic role the enzyme α-galactosidase is important in the metabolism of galactolipids (Sastry and Kates 1964) and in the function of chloroplast membranes (Bamberger and Park 1966; Gatt and Baker 1970).

STRUCTURE AND PROPERTIES OF α-GALACTOSIDASES

α-galactosidases catalyse the following reversible reaction.

CH₂OH ... O ... HO ... H ... H ... OH ... H ... O–R ... H ... OH   + H–OH ⇌   CH₂OH ... O ... HO ... H ... H ... OH ... H ... OH ... H ... OH   + ROH

α-D-galactoside   D-galactose

R may be alkyl, aryl, monoglycosyl or polyglycosyl groups. The equilibrium of this catalysed reaction normally favours hydrolysis. Tanner and Kandler (1968) isolated the enzyme from *Phaseolus-vulgaris*, which transgalactosylated galactinol to raffinose more readily than the hydrolysis of these galactosides. Dey and Pridham (1972) reported the *de novo* synthesis of oligosaccharides when various high concentrations of galactose were incubated with the enzyme. Only few studies on the chemical and physical nature of the enzymic protein have been reported. Dey and Pridham (1969) and Dey et al. (1971) prepared a homogeneous preparation of α-galactosidase from *Vicia faba* seeds and *Vicia sativa* seeds (Petek et al. 1969).

α-galactosidase occurs in multimolecular form. The first report appeared in 1961 (Petek and Dong). Dey and Pridham (1968, 1969) showed that the extracts of dormant *V. faba* seeds contained two α-galactosidase.

Dey and Pridham (1972) obtained the two molecular forms of α-galacto-sidase (I and II) from the *Vicia faba* mature seeds. They were purified 3660 and 337 fold, respectively. They were homogeneous in nature and had mole-cular weights of 209,000 and 38,000 respectively. Enzyme I could be disso-ciated with six inactive protein fractions of varying molecular weights when treated with 6 M urea (Dey and Pridham 1969). They reported their kine-tics and nature of active site of enzyme I. Enzyme I but not II was also observed to hydrolyse galactomannans. Pridham and Dey (1974) have men-tioned this in detail. Interestingly, the α-D-galactosidases isolated from guar seeds removed all of the D-galactose residues from guaran leaving a residue of water insoluble D-mannan (Lee 1967).

Various workers have studied the activity of α-D-galactosidases which has been shown to increase during germination with concomitant depletion of α-D-galactosidic reserve carbohydrate in different leguminous crops, guar (Lee 1967), soybean, lucerne and carob (McCleary and Matheson 1974). Mc-Cleary and Matheson (1974) showed the presence of a multimolecular form of α-D-galactosidase in the germinated seeds of carob, guar, lucerne and soybean, and designated A, B, C. They achieved the separation of the isoen-zymes by DEAE-cellulose chromatography. The enzymes A and B, confined to the cotylendon-embryo part of the seeds, are mainly responsible for the hydrolysis of α-D-galactosidic oligosaccharides. Enzyme C increases during germination and is highly specific for the hydrolysis of galactomannan.

1.9.2. β-D-MANNANASE (ENDO β-D-MANNANASE, (1–4)-β-D-MANNAN MANNOHYDROLASE EC 3.2.1.78)

This enzyme catalyzes the hydrolysis of:
  (i) β-D (1–4) mannopyranosyl linkages of mannans.
  (ii) galactomannans, glucomannans and galacto glucomannans and D-manno oligosaccharide.
  (iii) β-D-Man p-(1–4)-β-D-Glc p and β-D-Glc p-(1–4)-D-Man p link-ages.

The presence of β-D mannanase in leguminous crops has been reported by Dekker and Richards (1976); McCleary and Matheson (1975); Whistler et al. (1950); Villarroya and Petek (1976). Dekker and Richards (1976) characteriz-ed various β-D-mannanase, but its properties had been hampered due to its purification problem. Beaugiraud and Percheron (1964) purified partially β-D-mannanase from germinating fenugreek seeds. Lee (1967) resolved multiple forms of β-D-mannanase from germinated guar seeds. Villarroya and Petek (1976) obtained the purified β-D-mannanase from lucerne and reported its properties.

Most β-D-mannanases are endoacting enzymes, as evidenced by a rapid fall in viscosity of polymeric substrates producing a series of D-manno-oligosaccharides of different degree of polymerisation. Dey (1978) has dealt the β-D-mannanase in detail.

1.9.3. β-D-Mannosidase

This enzyme catalyzes the following reactions:

β-D-mannoside   D-mannose

where R may be alkyl, aryl or glycosyl group.

In legumes the work of McCleary and Matheson (1975) showed the presence of two forms of β-D-mannosidase in the seeds of guar, lucerne, carrot and honey-locust; β-D-mannosidases are exohydrolases which remove β-D (1–4)-linked D-mannosyl groups from the non-reducing end of their substrate. The activity in the galactomannan seeds is low, but it would be sufficient to degrade the D-manno-oligosaccharide produced by the action of β-D-mannanase on endogenous galactomannans.

Hylin and Sawai (1964) obtained crystalline depolymerase from *Leucaena leucocephala*, which converts its galactomannan to D-galactose and D-mannose. Reid and Meier (1973) showed the presence of β-D-mannanase in fenugreek at all stages of germinations.

1.10. Industrial Applications

The patent literature pertaining to the industrial exploitation of gum from 1948 to 1962 was collected by Saxena (1965) and a selection from the literature after that period is compiled in Table 1.10.

It may be seen from Table 1.10 that guar gum is, indeed, versatile product and enjoys a wide usage in different industries because of its ability, at relatively low concentrations, to form very viscous solutions that are only slightly affected by other factors. This property of viscosity production of the gum when dispersed in water serves the gum as thickeners, suspending agents,

TABLE 1.10. Industrial applications of guar gum

Product	*Form of gum*	*Function*	*Reference*
1	2	3	4
1. FOOD			
Ice cream	Carboxymethylated guar gum	Thickening and gelling agents	Yueh and Schilling (1971)
Ice cream	Guar gum with carrageenan and 0-carboxymethyl cellulose	Stable thixotropic stabilizer-emulsifier system	Leo and Bielskis (1968)
Liquid egg	Guar gum with carboxymethyl cellulose and tamarind seed gum	Thickening and improver	Bai et al. (1978)
Acidified dairy products	Guar gum with or without xanthan gum	Improves consistency	Yamatani (1978)
Fat containing whips	Guar gum/locust bean gum	—	Little (1968)
Instant corn grits	Guar gum with whipable albumin and edible fat	To create pudding and cream dressing of cakes	Lenderink (1969)
Water ices	Guar gum	—	Collins and Hyldon (1970)
Bakery filling	Guar gum	Freeze thaw and heat stabilizer	Rubenstein (1967)
Mayonnaise and sauces	Guar gum	Thickening agent	Messina (1967)
Protein low bread	Guar gum	Binder	Reinders and Gotlieb (1969)
Icing stabilizers	Guar gum	Stabilizer	Erba (1970)
Cheese	Guar gum	Stabilizer	Svolos (1971)
	Guar gum with xanthan and carob gum		Kovacs and Igoe (1976)
2. BUILDING			
Plaster	Guar gum	Thickening agent	Benz (1965)
Concrete	Guar gum	Water proofing	Compernass et al. (1970)
Foamed cement material	Guar gum	Foam stabilizer	Plunguian and Cornwell (1976)

Application	Material	Function	Reference
3. FIRE-FIGHTING			
Air drop, forest-fire control	Guar gum with decyl sulphate and ammonium phosphate solution	Provides viscosity stability	Morgenthaler (1972)
Forest-fire fighting	Guar gum with glycerol and ethylene glycol	Dispersions	Anonymous (1967)
4. EXPLOSIVES			
Explosive gel	Guar gum	Improved resistance to water and aging	Yancik et al. (1972)
Blasting agent	Guar gum with transition metal ions	Increases viscosity	Goffart (1968) Chrisp (1967)
Blasting slurry	Guar gum with major portions of inorganic oxidizer salts	Thickening agent	Cook (1968)
Explosive mixture	Cyanoethyl ether of guar gum	Thickening agent	Jordan (1972)
Extrudable primer	Guar gum	Increases viscosity	Andrew (1977)
Gel explosive	Guar gum	Gelling agent	Machacek (1978)
Gel explosive	Guar gum with cross linking agents	Gelling agent	Machacek (1979)
Slurry explosives	Guar gum or hydroxypropyl	Thickening agent	Matts and Seto (1979)
Blasting agent	Guar gum with cross-linking agents	Thickening agent	Wheeler (1978)
Nitroalkane explosives	Nitrate esters of guar gum	Thickening agent	Carroll and Griffith (1978)
5. PETROLEUM WELL DRILLING			
Oil bores	Borates cross-linked guar gum	Stable superelasitc liquid with lessened temp. sensitivity for control of lost circulation in oil field drilling operation	Walker (1965)
Oil, gas water bores	—	For plugging leaks	Black and Melton (1966) Horner and Walker (1965)

(Contd.)

1	2	3	4
Oil, gas water bores	Guar gum-Xanthomonas campestris mixtures	Stabilisation of cross-linkage	Browning et al. (1972)
Rotary drilling	Guar gum with methylene blue	Indicator of fluid life	Mogg (1970)
Fracturing solution	Hydroxyalkyl ether derivatives of guar gum	Reduces friction and increases permeability	Jordan (1969)
Oil bore	Guar gum with laminar silicate, starch and swelling clay	For reduction of fluid loss	Kuhn and Brown (1968)
Drilling	Guar gum with friction reducers	Thickener	Lummus and Randell (1969)
Petroleum wells	Guar gum with fused potassium pyroantimonate	—	Holtmyer et al. (1972)
Fracturing fluid	Guar gum with boric acid and magnesium oxide	Provides viscosity stability	Free (1976)
Well stimulator	Derivatized guar gum	Gelling agent	Githens and Burnham (1977)
Drilling mud	Gum ether	Thickening agent	DeMartino (1977)
Fracturing well	Potassium permanganate cross-linked guar gum	—	Holtmyer et al. (1977)
Drilling fluid	Hydroxyalkyl guar gum	Good fluid loss properties	Jackson (1979)
Fracturing fluid	Guar gum and hydroxyethyl cellulose	Gelling agent	Chatterji (1979)
Petroleum well fracturing	Guar gum/copolymer	Thickening agent	DeMartino (1979)
Petroleum well blow out	Hydroxypropyl guar gum	Thickening agent	Arwine and Ely (1978)
6. PAPER			
Additive Sizing	Oxidised guar gum	Improves wet strength of paper	Opie and Keen (1966)
	Guar gum	Crosslinked with transition metals	Chrisp (1967)
Paper/paperboard	Aminoethyl gum	Increases dry strength of paper	Nordgren (1967)
	Modified guar gum	Retention aid prior to the sheet forming operation	Anonymous (1969)

Application	Form	Function	Reference
Paper making	Quaternary ammonium guar gum	Imparts dry strength	General Mills Inc. (1968)
Paper and Textile	Modified guar gum	Floatation aid	Hahn et al. (1969)
	Guar gum	Improves strength	Kent (1969)
	Guar gum formate	Flocculent and sizing agent	Trapasso (1977)
7. PHARMACEUTICALS			
Vitamin B12 prep. Preparation for treatment of gastrointestinal ulcers and diarrhoea	Guar gum	Stability	Nuernberg (1969)
	Guar gum	Synergistic activity with bismuth salt	Laboratories Dausse (1970)
Sustained release drugs	Guar gum		Nuernberg et al. (1972)
Microencapsulation of drugs	Guar gum	High resorptivity	Nuernberg et al. (1971)
For gastrointestinal disorders	Guar gum		Synthelabo (1971)
Tablet prep.	Guar gum	Dry binder	Merck (1965)
Dietetic comp.	Guar gum	Malnutrition treatment	Boissier and Mendy (1966)
Lozenges	Guar gum		Christenson and Huber (1971)
Anti-inflammatory drugs	Guar gum		Negrevergne (1969)
Tablet Formation	Guar gum/starch combination		Nath and Gaitonde (1976)
Anti-inflammatory drugs	Guar gum	Suspending agent	Trivedi and Parikh (1972)
Tablet additive	Guar gum	Disintegrant or binder	Sakr and Elasbbagh (1976)
			—ibid— (1977)
8. PHOTOGRAPHY			
Processing solution	Hydrolysed guar gum	Binder	Kragh and Greenway (1968)
	Guar gum	Thickener	Friedel (1973)
9. COSMETICS			
Hair setting	Guar splits treated with steam	Thickener	Keen (1969)
	Guar gum/urea and sulphite	Thickener	Gillette Co. (1965)
10. COAL MINING			
	Guar gum with boric acid/borax	For shock impregnation of coal seams	Knop (1970)

(Contd.)

1	2	3	4
11. ORE REFINING			
Iron scrap	Aminoethyl gum; Gum with sulphuric acid	In settling fine particles	Nordgren (1967); Société de Prayon (1972)
Copper electro refining	Guar gum	Colloidal flocculent	Vereecken and Winand (1976)
12. TEXTILES			
Textile print	Guar gum with acrylamide	Stabilizer	Nordgren (1968)
Fire and shrinkproof textiles	Guar gum with chloroisocya-nuric acid	—	Bouvet (1968)
Polyester printing	Derivatives guar gum	Thickener	Alsberg and Dawson (1973)
Printing fluid	Guar gum and xanthomonas gum	Printing paste thickener	Jordan and Carter (1973)
Textile printing	Modified guar gum	—	Mudki and Warty (1976a, b)
13. AGRICULTURE			
Insecticidal comp.	Guar gum	—	Hattangadi et al. (1970)
Coating for fertilizer	Guar gum with fatty acid deriv. and kaolin	Prevention of the granules	Sarrade-Locheur (1971)
Poultry	Guar gum	For decreasing cholesterol levels in chicks	Creger et al. (1970)
—do—	Guar gum	Decreased nitrogen retention, fat absorption and metabolizable energy	Kratzer et al. (1967)
14. SHOE INDUSTRY	Guar gum	Improves the properties of the substrate through better fibri-lation and reaggregation of the collagen fibres	Tu and Appleton (1969)

	Material	Application	Reference
15. CERAMICS			
Ceramics articles	Guar gum with ceramic powders	Paster for moulded articles	Norwalk (1968)
16. ELECTRICALS			
Cable cores	Gum with isobutylene and carboxymethyl cellulose	Electrical insulator	Arendt et al. (1970)
17. TOBACCO			
Tobacco products	Guar gum	Adhesive	Halter and Fiore (1963)
Tobacco sheet	Guar gum	Reduces irritation properties	Mold and Killianos (1968)
Tobacco sheet	Guar gum	Strengthening agent	Rothmans of Pall Mall (1969)
18. MEDICINAL			
Dietary fibre	Guar gum along with other fibrous material	Protective effect	Huang et al. (1978)
Mixed in the diet	Guar gum or pectin	Reduces urinary glucose excretion, cholesterol	Iwasaki et al. (1978)
Dietary fibre	Guar gum	Hypocholesterolmic agent	Jenkins et al. (1979)
Mixed diet	Guar gum	Antidiabetic agent	Jenkins et al. (1977)
Mixed diet	Guar gum	Effects bacterial counts in faeces	Muenzner and Harmuth-Hoene, (1978)
19. ANALYTICAL			
Selective resin	Cross-linked guar gum	Analytical resin for separation of boron	Bhatnagar and Mathur (1977)
Beads	Guar gum	Purification of lectins	Bywater (1978)
Gel filtration	Cross-linked guar gum	Chromatographic separation	Gupta et al. (1979)
20. MISCELLANEOUS			
	Guar gum with sodium sulphate	Flocculating agent	Jordan (1969)

(Contd.)

1	2	3	4
Production of sodium carbonate	Substituted guar gum	Clarifier	Lobunez et al. (1976)
Aluminium reflectors	Guar gum	In increasing reflecting power	Mosier (1970)
Turbulent flow between pipe walls	Guar gum	Reduction of resistance	Kilian (1970)
Galactose binding lectins	Cross-linked guar gum	Immunosorbent	Lonngren et al. (1976)
Civil disorder control	Guar gum	Lubricant for concrete surface	Whitsitt et al. (1970)

emulsion stabilizers, film formers, binders or just water retention agents (commonly known as hydrocolloids) in the food industry for soups, desserts, pie-fillings, sauces, mayonnaise, etc. (Glicksman 1969). On hydration, the guar gum forms a thick mucilaginous paste which is impervious to further water and thus is used as a waterproofing agent in explosives and plugging composition for leaking wells in the mining industry. It is also used as a flocculent or settling agent in the concentration and purification of ores, and as well as in the treatment of industrial waters. The gum is used in oil well fracturing because of the electrolyte compatibility property of the gum.

The major use in the paper industry is to increase the wet strength of paper. It is also used in textiles as a thickening and sizing agent. It is used as a thickening agent in various pharmaceutical preparations. It is a suitable base for delayed-release drugs because of slow hydration. The gum is used as a binder or disintegrator for tablets and as such is also used as a laxative. It is also used in various cosmetics (Chudzikowski 1971). In the manufacture of both paste and shaving creams it imparts slip for extruding the paste from the tube without applying excessive pressure. It is used as a protective colloid in creams and lotions. Owing to its compatibility it mixes well with most detergent systems and hence used in the manufacture of shampoos and cleansers. Dry face mask mixes have also been prepared using guar gum.

From the plant breeder's point of view, it is the variation within the genotypes which is important for the manipulation in a breeding programme. Incorporation of any desired characters into an improved variety will depend on the type of genetic control, the amount of genetic variation available and its heritability including stability over a range of environments and seasons.

Guar gum has diversified uses in various industries and hence breeding efforts should be concentrated on developing new high-yielding varieties with a high gum content. A concerted effort is needed to find the variability in the physical characters of gum in the genetic material. A positive correlation between seed yield and gum percentage and negative correlation between seed weight and gum percentage has been observed (Menon et al. 1970; Mital et al. 1971). Studies have also been made (Lal and Gupta 1977) to establish relationships between different characters and the physical and chemical characters for evolving some simple criteria for the use of plant breeders to make selections in the field for a high gum content. But these studies are neither conclusive nor enough to draw definite conclusions and hence more work is required in this direction.

1.11. Conclusions

This chapter has aimed at defining the chemical nature of carbohydrates. The legumes have a good amount of non-reducing oligosaccharides, which are responsible for the flatulence. This has already been discussed. Stachyose provides a frost resistance to the plants. The reserve polysaccharides, specially galactomannans of the structure, found in these seeds are almost confined to the *Leguminosae*. Galactomannan has assumed wide industrial application specially in explosives and oil drilling, besides its use in food, pharmaceuticals, textiles and paper. Recently it has found enormous uses in medicines in lowering the blood cholesterol. The chemical nature of galactomannans, the enzymes responsible for its degradation and biosynthesis has been dealt with in detail. The variations in these polysaccharides may become of chemotaxonomic value in the future.

The structural polysaccharides are more or less common to all higher plants. The leguminosae family in general contains less hemicellulose as compared to cereals/gramineae family. No attempt has been made to discuss the structural polysaccharides from the point of view of their role in the digestibility in the ruminants. In all leguminous forages, with the delay in harvesting, the *in vitro* dry matter digestibility decreases, while in the moth (*Phaseolus aconitifolius*) it increases with the delay in harvesting. The hemicellulose content decreases while cellulose content remains constant. The variability in well-defined polysaccharides fractions has been discussed instead of the ill-defined fractions as crude fibre and nitrogen free extract.

Further investigations on the structure of seed amyloids from tamarind seeds needs to be confirmed with regard to the controversy about the presence of L-arabinose. The subfamily *Mimsoideae*, also rich in soluble polysaccharides, has been largely neglected, needs further investigations. *Leucaena leucocephala* contains 25–30 per cent galactomannan type gum, but no systematic studies have been conducted to confirm its structure.

Lastly, the industrial applications of polysaccharides specially galactomannans have been discussed and compiled for the benefit of the industries and researchers. The new uses of this gum warrant further research.

References

Ahmed, Z.F. and Whistler, R.L. 1950. The structure of guaran. *J. Amer. Chem. Soc.* **72**: 2524–2525.

Alsberg, F.R. and Dawson, P.R. 1973. Polyester printing thickeners. *Textilveredlung.* **8**: 365–370 (*Chem. Abstr.* **79**: 80168e, 1973).

Anderson, E., Seeley, M., Stewart, W.T., Redd, J.C. and Westerbeke, D. 1940. The origin and composition of the hemicelluloses obtained from hardwoods. *J. Biol. Chem.* **135**: 189–98.

Andrews, P., Hough, L. and Jones, J.K.N. 1952. Mannose-containing polysaccharide. I. The Galactomannans of lucerne and clover seeds. *J. Amer. Chem. Soc.* **74**: 4029–4032.

Andrews, P., Hough, L. and Jones, J.K.N. 1954. The galactan of *Strychnos nux-vomica* seeds. *J. Chem. Soc*. 806–812.

Andrew, E.A. 1977. Priming mix with minimum viscosity change. *U.S. Patent* 4,133,707 (*Chem. Abstr*. **90**: 154325e, 1979).

Anonymous. 1941. Daurala Sugar Works. *Indian Patent* 28,409.

Anonymous. 1967. Dissolution of hydrophilic organic colloids. *Fr. Patent* 1,485,729 (*Chem. Abstr*. **69**: 46333p, 1968).

Anonymous. 1969. Food additives: Components of paper and paper board in contact with aqueous and fatty foods. *Fed. Regist*. **34** (1977), 14429 (*Chem. Abstr*. **71**: 110544q. 1969).

Arendt, I.P. Wappler, Goetwe, Z and Schmidt, P. 1970. Telecommunications cable comprising a cable core of wires insulated with synthetic resinous material surrounded by a sheath. *U.S. Patent* 3,538,235 (*Chem. Abstr*. **74**: 77141g. 1971).

Arora, S.K. and Das, B. 1976. Cowpea as potential crop for starch. *Die Stärke* **28**: 158–160.

Arora, S.K., Das, B., Paroda, R.S. and Luthra, Y.P. 1975. Structural carbohydrates, influencing the *in vitro* digestibility of moth (*Phaseolus aconitifolius* Jacq.). *Forage Res*. **1**: 107–113.

Artaud, J., Derbesy, M. and Extienne, J. 1977. Fatty materials in grain colloids. III. Difference among tamarind. *Ann. Falsif. Expert. Chim*. **70**: 39–44 (*Chem. Abstr*. **86**: 185961F, 1977).

Artaud, J., Extienne, J. and Cas, M. 1975. Recent progress in the analysis of gums and hydrosoluble natural vegetable colloids. I. Guar gum. *Ann. Falsif. Expert. Chim*. **68**: 9–27 (*Chem. Abstr*. **86**: 2370v, 1977).

Arwine, L.C., and Ely, J.W. 1978. Polymer use in blow out control. *J. Pet. Technol*. **30**: 705–711.

Aspinall, G.O., Hunt, K. and Morrison, I.M. 1966. Polysaccharides of soybeans. II. Fractionation of a hull cell-wall polysaccharides and the structure of a xylan. *J. Chem. Soc*. 1945–1949.

Aspinall, G.O., Begbie, R., Hamilton, A. and Whyte, J.N.C. 1967a. Polysaccharide of soybean. III. Extraction and fractionation of polysaccharide from cotyledon meal. *J. Chem. Soc*. 1065–1070.

Aspinall, G.O., Cottrell, I.W., Egan, S.V., Morrison, I.M. and Whyte, J.N.C. 1967b. Polysaccharides of soybeans. IV. Partial hydrolysis of the acidic polysaccharides complex from cotyledon meal. *J. Chem. Soc*. 1071–1080.

Aspinall, G.O., Craig, J.W.T. and Whyte, J.N.C. 1968. Lemon peel pectin. I. Fractionation and partial hydrolysis of water soluble pectin. *Carbohydr. Res*. **7**: 442–452.

Aspinall, G.O., Hunt, K. and Morrison, I.M. 1967c. Polysaccharides of soybean. V. Acidic polysaccharides from the hulls. *J. Chem. Soc*. 1080–1086.

Audsley, A. and Fursey, A. 1965. Examination of polysaccharide flocculent and flocculated kaolinite by electron microscopy. *Nature* **208**: 753–754.

Bai, H.M., Ahn., J.K., Yoow, Y.H. and Kim, H.U. 1978. A study on the development of the mixed stabilizer for ice-cream manufacture. *Hanguk Chuksan Hakhoe Chi.* **20**: 436–445.

Bailey, R.W. 1964. Pasture quality and ruminant nutrition. I. Carbohydrate composition of ryegrass varieties grown as sheep pasture. *N.Z.J. Agric. Res.* **7**: 496–507.

Bailey, R.W. and Gaillard, B.D.E. 1965. Carbohydrases of the rumen *Ciliate epidin acaudatum. Biochem. J.* **95**: 758–766.

Bamberger, E.S. and Park, R.B. 1966. Effect of hydrolytic enzymes on the photosynthetic efficiency and morphology of chloroplasts. *Plant Physiology* **41**: 1591–1600.

Beaugiraud, S. and Percheron, F. 1964. *Compt. Rend.* **259**: 3879–3881.

Benz, G. 1965. Thickening agent for dispersions of powdered plaster and other hydraulic materials. *Ger. Patent* 1,206,777 (*Chem. Abstr.* **64**: 6290c, 1966).

Bhatnagar, R. and Mathur, N.K. 1977. A new boron-selective resin derived from guar gum. *Talanta* **24**: 466–467.

Black, H.N. and Melton, L.L. 1966. Temporary well plugging agent with a predetermined life. *U.S. Patent* 3, 227, 212 (*Chem. Abstr.* **66**: 7941b, 1966).

Boissier, J.R. and Mendy, F. 1966. Dietetic compositions, *Fr. M.* 3789 (*Chem. Abstr.* **66**: 84901t, 1967).

Bourne, E.J., Pridham, J.B. and Walter, M.W. 1962. The biosynthesis of galactosylsucrose derivatives. *Biochem. J.* **82**: 44.

Bouvet, J. 1968. Fire and shrinkproofing of keratinic textiles. *Fr. Patent* 1,512,574 (*Chem. Abstr.* **70**: 58875a, 1969).

Brown, R.J. and Serro, R.F. 1953. Isolation and identification of O-α-D-galactopyranosyl-myo-inositol and of myo-inositol from juice of the sugar beet (*Beta vulgaris*). *J. Amer. Chem. Soc.* **75**: 1040.

Browning, W.C., Perricone, A.C. and Elting, K.A.C. 1972. Galactomannan composition and the prevention of destruction of its thixotropic structure by ligands. *U.S. Patent* 3,677,961 (*Chem. Abstr.* **77**: 128458n, 1972).

Buchala, A.J. and Meier, H. 1973. Galactoglucomannan from the leaf and stem tissues of red clover (*Trifolium pratense*). *Carbohydr. Res.* **31**: 87–92.

Bywater, R. 1978. Purification of lectins on beaded polysaccharide materials. *Chromatogr. Synth. Biol. Poly.* **2**: 325–329.

Calloway, D.H., Hickey, C.A. and Murphy, E.L. 1971. Reduction of intestinal gas-forming properties of legumes by traditional and experimental food processing methods. *J. Food Sci.* **36**: 251–255.

Carroll, W.K. and Griffith, G.L. 1978. Nitrate esters of galactomannan gums. *U.S. Patent* 4,112,220 (*Chem. Abstr.* **90**: 139329r, 1979).

Chakraverti, I.B., Nag, S. and MacMillian, W.G. 1961. Isolation, purification and fractionation of tamarind kernel polysaccharide. *J. Sci. Ind. Res. Sect.* D. **20**: 380.

Charaux, C. 1922. Presence of aucubin in seeds of *Veronica hederaefolia* L.

Bull. Soc. Chim. Biol. **4**: 568–70.

Chatterji, J. 1979. Composition of treating low temperature subterranean well formations. *U.S. Patent* 4,144,179 (*Chem. Abstr.* **91**: 23716f, 1979).

Chudzikowski, R.J. 1971. Guar gum and its applications. *J. Soc. Cosmet. Chem.* **22**: 43–60.

Chrisp, J.D. 1967. Galactomannan gum gels for explosive compositions and for textile or paper sizing. *U.S. Patent* 3,301,723 (*Chem. Abstr.* **67**: 92485 w, 1967).

Christenson, G.L. and Huber, H.E. 1971. Long lasting troche containing guar gum. *U.S. Patent* 3,590,117 (*Chem. Abstr.* **75**: 91314 t, 1971).

Cook, M.A. 1968. Thickened aqueous blasting agents containing entrapped gas. *U.S. Patent* 3,382,117 (*Chem. Abstr.* **69**: 11881o. 1968).

Courtois, J.E. and LeDizet, P. 1963. Action of some enzyme preparations on galactomannans of clover and gleditsia. *Bull. Soc. Chim. Biol.* **45**: 731. (*Chem. Abstr.* **59**: 14295a, 1963).

Courtois, J.E. Petek, F. and Dong, T. 1961. Synthesis of planteose by the transferring action of the α-galactosidase of plantago seeds. *Bull. Soc. Chim. Biol.* **43**: 1189–96.

Courtois, J.E. 1974. In: *Plant Carbohydrate Biochemistry.* Pridham, ed. p. 1. Academic Press, London.

Collins, J.T. and Hyldon, R.G. 1970. Instant corn grits. *U.S. Patent* 3,526,512 (*Chem. Abstr.* **73**: 119427s, 1970).

Compernass, J., Gruenberger, E. and Schmidt, F.E. 1970. Dressing sand, gravel, and other minerals for use as concrete additives. *Ger. Patent* 1,953,159 (*Chem. Abstr.* **74**: 67306a, 1971).

Creger, C.R., DeGuzman, V.A. and Cough, J.R. 1970. Effect of guar gum on cholesterol levels of chicks. *Nutr. Rep. Int.* **2**: 243–247. (*Chem. Abstr.* **74**: 62104y, 1971.)

Cristofaro, E., Mottu, F. and Wubrmann, J.J. 1974. In: *Sugars in Nutrition.* Sipple and McNutt, eds. pp. 313–336. Academic Press, New York.

Damodaran, M. and Rangachari, P.N. 1945. Tamarind seed "pectin". *Curr. Sci.* **14**: 203–204.

Das, B., Arora, S.K. and Luthra, Y.P. 1975. Variability in structural carbohydrates and *in vitro* digestibility of forages. 2. Cowpeas (*Vigna sinensis*). *Cuban J. Agric. Sci.* **9**: 61–66.

Das, B., Arora, S.K. and Luthra, Y.P. 1975b. Variability in structural carbohydrates and *in vitro* digestibility of forages. 3. Guar (*Cyamopsis tetragonoloba*). *J. Dairy Sci.* **58**: 1347–1351.

Das, B. and Arora, S.K. 1978. Guar seed—Its chemistry and industrial utilization of gum. In: *Guar—Its Improvement and Management.* Paroda and Arora, eds. pp. 80–102. Indian Society of Forage Research, Hissar.

Das, B., Arora, S.K. and Luthra, Y.P. 1979. Variability in structural carbohydrates and *in vitro* digestibility of forages. 4. Fenugreek (*Trigonella foenum graecum*). *Anim. Feed Sci. & Technol.* **4**: 17–22.

Deb, S.K. and Mukerjee, S.N. 1963. Molecular weight and dimensions of guar gum from light scattering in solution. *Indian J. Chem.* **1**: 413–414.

Dekker, R.F.H. and Richards, G.N. 1976. Hemicellulases: Their occurrence, purification, properties and mode of action. *Adv. Carbohydrate Chem. Biochem.* **32**: 277–352.

De Martino, R.N. 1977. Polygalactomannan ethers. *Ger. Offen.* 2,652,395 (*Chem. Abstr.* **87**: 41053J, 1977).

De Martino, R.N. 1979. Thickening agent containing a polygalactomannan gum and a copolymer of an olefinically unsaturated dicarboxylic acid anhydride useful in hydraulic well treating. *U.S. Patent* 4,143,00 (*Chem. Abstr.* **90**: 206999z, 1979).

Dey, P.M. and Pridham, J.B. 1968. Multiple forms of α-galactosidase in *Vicia faba* seeds. *Phytochemistry* **7**: 1737–1739.

Dey, P.M. and Pridham, J.B. 1969. Substrate specificity and kinetic properties of α-galactosidases from *Vicia faba*. *Biochem. J.* **115**: 47–54.

Dey, P.M., Khalaque, A. and Pridham, J.B. 1971. Further observations on the α-galactosidase activity of *Vicia faba* seeds. *Biochem. J.* **124**: 27.

Dey, P.M. 1978. Biochemistry of plant galactomannans. *Adv. Carbohydrate Chem. Biochem.* **35**: 341–376.

Dey, P.M. and Pridham, J.B. 1972. Biochemistry of α galactosidase. *Advances Enzymol.* **36**: 91–130.

Erba Carlo, S.P.A. 1970. Low protein bread. *Brit. Patent* 1,181,949 (*Chem. Abstr.* **72**: 120276c, 1970).

Free, D.L. 1976: Fracturing subterranean formation. *U.S. Patent* 3,974,007 (*Chem. Abstr.* **86**: 31745b, 1977).

French, D. 1954. The raffinose family of oligosaccharides. *Adv. Carbohydrate Chem.* **9**: 149–184.

Friedel, M. 1973. Viscous photographic processing solutions. *Ger. Offen.* 2,204,870 (*Chem. Abstr.* **79**: 120473, 1973).

Gaillard, B.D.E. 1965. Comparison of the hemicellulose from plants belonging to two different plant families. *Phytochemistry* **4**: 631–634.

Gaillard, B.D.E. and Bailey, R.W. 1968. The distribution of galactose and mannose in the cell-wall polysaccharides of red clover (*Trifolium pratense*) leaves and stems. *Phytochemistry* **7**: 2037–2044.

Gatt, S. and Baker, E.A. 1970. Purification and separation of α- and β-galactosidases from spinach leaves. *Biochem. Biophys. Acta* **206**: 125–135.

General Mills Inc. 1968. Quaternary ammonium galactomannan gum ethers. *Brit. Patent* 1,136,842 (*Chem. Abstr.* **70**: 69446n, 1969).

Gillette Co. 1965. Hair setting preparations. *Neth. Appl.* 6,410,355 (*Chem. Abstr.* **63**: 6782b, 1965).

Githens, C.J. and Burnham, J.W. 1977. Chemically modified natural gum for use in well stimulation. *Soc. Pet. Eng. J.* **17**: 5–10. (*Chem. Abstr.* **87**: 25591 C, 1977.)

Glicksman, M. 1969. Gum Technology in the Food Industry. p. 130. Academic Press, New York.

Goldstein, A.M., Alter, E.N. and Seaman, J.K. 1973. Guar gum. In: *Industrial Gums*. Whistler and Bemiller, eds. pp. 303–321. Academic Press, New York, London.

Goring, D.A.I. and Timell, T.E. 1962. Molecular weight of native celluloses. *Tappi* **45**: 454–460.

Goffart, P.R. 1968. Safety explosives. *Fr. Patent* 1,533,471 (*Chem. Abstr.* **71**: 23404q, 1969).

Greenwood, C.T. and Thomson, J. 1962. Physicochemical studies on starches fractionation and characterization of various plant origins. *J. Chem. Soc.* 222–229.

Gupta, K.C., Sahni, M.K., Rathaur, B.S., Narang, C.K. and Mathur, N.K. 1979. Gelfiltration medium derived from guar gum. *J. Chromatogr.* **169**: 183–190.

Hahn, D.J., Jende, J.J., Rich, T.F. and Tremel, F.J. 1969. New floatation aid in a paper mill white water system. *Eng. Bull. Purdue Univ. Eng. Ext. Sr. No.* 135 (Pt. 2) 1348–1356 (*Chem. Abstr.* **74**: 79306p, 1971).

Halter, H.M. and Fiore, J.V. 1963. Water resistant reconstituted tobacco product. *U.S. Patent* 3,106,212 (*Chem. Abstr.* **60**: 2059c, 1964).

Hardinge, M.G., Swarner, J.B. and Crooks, H. 1965. Carbohydrates in foods. *J. Amer. Diet. Assoc.* **46** (3): 197–204.

Hassid, W.Z. 1967. Transformation of sugars in plants. *Ann. Rev. Pl. Physiol.* **18**: 253–280.

Hassid, W.Z. 1972. In: *Biochemistry of the Glycosidic Linkage*. Piras and Pontis, eds. p. 315. Academic Press, London.

Hattangadi, U.R., Shah, V.R. and Hattangadi, D.S. 1970. Insecticidal compositions. *Indian Patent* 116,080 (*Chem. Abstr.* **75**: 4617C, 1971).

Haworth, W.N., Hirst, E.L. and Ruell, D.A. 1923. Constitution of raffinose. *J. Chem. Soc.* **123**: 3125–3131.

Henderson, M.E., Hough, L. and Painter, T.J. 1958. Mannose containing polysaccharides. V. The isolation of oligosaccharides from Lucerne and Fenugreek galactomannans. *J. Chem. Soc.* 3519–3522.

Herissey, H., Wickstrom, A. and Courtois, J.E. 1951. Action of periodic acid on non-reducing tri-saccharides and tetrasaccharides. II. Gentianose. *Bull. Soc. Chim. Biol.* **33**: 1768–1776.

Hirst, E.L. and Jones, J.K.N. 1939. Pectic substances. II. Isolation of an araban from the carbohydrate constituents of the peanut. *J. Chem. Soc.* 452–453.

Hirst, E.L. and Jones, J.K.N. 1947. VI. Structure of the araban from *Arachis hypogaea*. *J. Chem. Soc.* 1221–1225.

Hirst, E.L., Mackenzie, D.J. and Wylam, C.B. 1959. Analytical studies on the carbohydrates of grasses and clovers. X. Changes in carbohydrate composition during the growth of lucerne. *J. Sci. Food. Agric.* **10**: 19–26.

Hoffmann, J., Linderberg, G. and Painter, T. 1976. The distribution of the D-galactose residue in guaran and locust bean gum. *Acta Chem. Scand.* Ser. B. **30**: 365–366 (*Chem. Abstr.* **85**: 43686a, 1976).

Holtmyer, M.D., Githens, C.J. and Tinsley, J.M. 1972. *Ger. Patent* 2,058,629 (*Chem. Abstr.* **77**: 142077i, 1972).

Holtmyer, M.D., Githens, C.J. and Tinsley, J.M. 1977. Compositions for fracturing well formations. *U.S. Patent* 4,021,355 (*Chem. Abstr.* **87**: 41640y, 1977).

Horner, V.V. and Walker, R.E. 1965. Gels for plugging oil and gas bore-holes. *U.S. Patent* 3,208,524 (*Chem. Abstr.* **64**: 5036, 1966).

Hoyt, J.W. 1966. Friction reduction as an estimator of molecular weight. *J. Polym. Sci. Part. B.* **4**: 713–716.

Huang, C.T.L., Gopalkrishna, G.S. and Nichols, B.L. 1978. Fibre, intestinal sterols and colon cancer. *Am. J. Clin. Nutr.* **31**: 516–526.

Hui, P.A. and Neukom, H. 1964. Properties of galactomannans. *Tappi* **47**: 39–52.

Hylin, J.W. and Sawai, K. 1964. The enzymatic hydrolysis of *Leucaena glauca* galactomannan. Isolation of crystalline galactomannan depolymerase. *J. Biol. Chem.* **239**: 990–992.

Iwasaki, Y., Mukano, S., Higuchi, T., Malda, T., Kono, N., Handa, R., Hiraoka, R., Sasaki, I. and Inoue, R. 1978. Pectin and guar gum in the treatment of diabetes mellitus. *Igaku No. Ayumi* **106**: 468–470.

Jackson, J.M. 1979. Clay-free well bore fluid. *U.S. Patent* 4,140,639 (*Chem. Abstr.* **90**: 189574n, 1979).

Jenkins, D.J.A., Wolever, T.M.S., Hokaday, T.D.R., Leeds, A.R., Howarth, R., Bacow, S., Apling, E.C. and Dilawari, J. 1977. Treatment of diabetes with guar gum: reduction of urinary glucose loss in diabetics. *Lancet* **2**: 779–780.

Jenkins, D.J.A., Leeds, A.R., Slavin, B., Manu, J. and Japson, E.M. 1979. Dietary fibre and blood lipids reduction of serum cholesterol in type II hyperlipidemia by guar gum. *Am. J. Clin. Nutr.* **32**: 16–18.

Johnston, J.F.W. 1843. *Phil. Mag.* **23**: 14–18.

Jordon, W. 1969. Sedimentation agent. *Ger. Patent* 1,297,048 (*Chem. Abstr.* **71**: 105113h, 1969).

Jordon, W.A. 1972. Thickened nitroparaffins. *U.S. Patent* 3,666,577 (*Chem. Abstr.* **77**: 50941t, 1972).

Jordon, W.A. and Carter, W.H. 1973. Blends of Xanthomonas and guar gum. *U.S. Patent* 3,765,915 (*Chem. Abstr.* **80**: 61374y, 1974).

Kasai, T. and Kawamura, S. 1966. Soybean oligosaccharides. Isolation by gel filtration and identification by acetylation. *Kagawa Univ. Fac. Tech. Bull.* 18, No. 1.

Kauss, H. 1972. In: *Biochemistry of the Glycosidic Linkage.* Piras and Pontis, eds. p. 221. Academic Press, London.

Kawamura, S., Kobayashi, T., Oshima, M. and Mino, M. 1955. Carbohydrates of soybeans. V. A hemicellulose fraction soluble in hot water: Isolation and component sugar determinations. *Bull. Agri. Chem. Soc. Japan* **19**: 69–76.

Kawamura, S. and Narasaki, T. 1958. Fucose as a constituent of hemicellulose B, from broad bean (*Vicia faba*) seeds. *Bull. Agri. Chem. Soc. Japan* **22**: 436–437.

Kawamura, S. and Narasaki, T. 1961. Studies on the carbohydrates of soybean. VI. Component sugars of fractionated polysaccharides, especially identification of fucose in some hemicelluloses. *Agri. Biol. Chem.* **25**: 527–531.

Keen, J.L. 1969. Low odor low taste galactomannan gums. *U.S. Patent* 3,455,899 (*Chem. Abstr.* **71**: 122595j, 1969).

Kent, S.E. 1969. Starch and gum additive compositions and use thereof in paper making. *U.S. Patent* 3,471,362 (*Chem. Abstr.* **72**: 4487y, 1970).

Khan, N.A. and Mukherjee, B.D. 1959. The polysaccharide in tamarind seed kernel. *Chem. Ind.* (London) 1413.

Killan, F.P. 1970. Reduction of resistance of the linear macromolecules in boundary layers. *Umschau* **70**: 345–346 (*Chem. Abstr.* **73**: 36022e, 1970).

Kim, W.J., Smit, C.J.B. and Nakayama, T.O.M. 1973. Removal of oligosaccharides from soybeans. *Lebensm. Wiss. Technol.* **6**: 201–204.

Knop, F.B. 1970. Production of coal and/or minerals by the shock impregnation process. *S. African Patent* 6,906,946 (*Chem. Abstr.* **73**: 111681z, 1970).

Koleske, J.V. and Kurath, S.F. 1964. Configuration and hydrodynamic properties of fully acetylated guaran. *J. Polym. Sci. Part A.* **2**: 4123–4149.

Kooiman, P. 1960. On the occurrence of amyloids in plant seeds. *Acta Bot. Nedr.* **9**: 208–219.

Kooiman, P. 1961. A method for the determination of amyloid in plant seeds. *Rec. trav. Chim.* **79**: 675–678.

Kragh, A.M. and Greenway, E.W. 1968. Guar gum binder for photographic emulsions. *Brit. Patent* 1,126,798 (*Chem. Abstr.* **69**: 101767k, 1968).

Kratzer, F.H., Rajaguru, R.W.A.S.B. and Vohra, P. 1967. The effect of polysaccharides on energy utilization, nitrogen retention and fat absorption in chickens. *Poultry Sci.* **46**: 1489–1493.

Kuhn, D.A. and Brown, J.L. 1968. Loss control additive for subterranean fracturing fluids. *U.S. Patent* 3,408,296 (*Chem. Abstr.* 13224w, 1969).

Kovacs, P. and Igoe, R.S. 1976. Xanthan gum galactomannan system improves functionality of cheese spreads. *Food Prod. Dev.* **10**: 32–58 (*Chem. Abstr.* **86**: 28557m, 1977).

Laboratories Dausse, S.A. 1970. Guar gum bismuth salt medicament for treating gastro intestinal diseases. *Fr. M.* 7794 (*Chem. Abstr.* **76**: 131509q, 1972).

Laidlaw, R.A. and Wylam, C.B. 1953. The structure of stachyose. *J. Chem. Soc.* 567–571.

Lal, B.M. and Gupta, O.P. 1977. Studies on galactomannans in guar and some correlations for selecting genotype rich in gum content. *Proc. First Guar Res. Workshop*, CAZRI, Jodhpur, pp. 124–130.

Lee, S.R. 1967. Purification and properties of enzymes which attack guar gum. *Diss. Abstr.* B. **27**: 2626.

Lehle, L. and Tanner, W. 1973. The function of myo-inositol in biosynthesis of raffinose: Purification and characterisation of galactinol; sucrose-6 galactosyl transferase from *Vicia faba* seeds. *Europ. J. Biochem.* **38**: 103–110.

Lenderink en Co. N.V. 1969. Edible fat containing whips. *Neth. Patent* 6,907,938 (*Chem. Abstr.* **72**: 99390y, 1970).

Leo, A.J. and Bielskis, E. 1968. Pumpable food stabiliser-emulsifier incorporating readily dispersible hydrophilic colloids. *U.S. Patent* 3,396,039 (*Chem. Abstr.* **69**: 66286d, 1968).

Levitt, J. 1972. *Responses of Plants to Environmental Stress*. Academic Press, New York.

Little, L.L. 1968. Acidified dairy products. *U.S. Patent* 3,370,955 (*Chem. Abstr.* **68**: 94742x, 1968).

Lobunez, W., Kim, N.K. and Rau, E. 1976. Clarifier process for producing sodium carbonate. *U.S. Patent* 3,981,686 (*Chem. Abstr.* **85**: 19489t, 1976).

Lonngren, J., Goldstein, I.J. and Bywater, R. 1976. Cross-linked guaran: a versatile immunoabsorbent for D-galactopyranosyl binding lectins. *FEBS Lett.* **68**: 31–34.

Lummus, J.L. and Randall, B.V. 1969. Low friction drilling fluids. *U.S. Patent* 3,472,769 (*Chem. Abstr.* **71**: 126926r, 1969).

Machacek, O. 1978. Water gel explosive composition. *Brit. Patent* 1,518,409 (*Chem. Abstr.* **90**: 106629k, 1979).

Machacek, O. 1979. Water gel explosives. *Brit. Patent* 1,518,563 (*Chem. Abstr.* **90**: 106630d, 1979).

Matts, T.C. and Seto, P.F.L. 1979. Slurry explosives. *Ger. Offen.* 2,829,559 (*Chem. Abstr.* **90**: 139801p, 1979).

McCleary, B.V. and Matheson, N.K. 1974. α-D-Galactosidase activity and galactomannan and galactosyl sucrose oilgosaccharide depletion in germinating legume seeds. *Phytochemistry* **13**: 1747–1757.

McCleary, B.V. and Matheson, N.K. 1975. Galactomannan structure and β-mannanase and β-mannosidase activity in germinating legume seeds. *Phytochemistry* **14**: 1187–1194.

McCready, R.M., Guggolz, K., Silveira, V. and Owens, H.S. 1950. Determination of starch and amylose in vegetables. *Anal. Chem.* **22**: 1156–1158.

Menon, U., Dube, M.M., and Bhargava, P.D. 1970. Gum content variations in guar (*Cyamopsis tetragonoloba* L.). *Indian J. Heredity* **2**: 55–58.

Merck, E. 1965. Omission of granulation in tablet production. *Neth. Patent*

6,504,974 (*Chem. Abstr.* **64**: 14042g, 1966).

Messina, B.T. 1967. Heat and freeze resistant bakery filling. *U.S. Patent* 3,352,688 (*Chem. Abstr.* **68**: 11824b, 1968).

Mital, S.P., Thomas, T.A., Dabas, B.S. and Lal, B.M. 1971. Gum content as related to seed yield and other characters in guar. *Indian J. Gen. Pl. Breed.* **31**: 228–232.

Moe, O.E., Miller, S.E. and Iwan, M.H. 1947. Investigation of the reserve carbohydrates of leguminous seeds. 1. Periodate oxidation. *J. Amer. Chem. Soc.* **69**: 2621–2625.

Mogg, J.L. 1970. Guar gum drilling fluids. *U.S. Patent* 3,515,667 (*Chem. Abstr.* **73**: 37157q, 1970).

Mold, J.D. and Killianos, A.G. 1968. Use of humectants and guar gum in the production of tobacco sheet. *U.S. Patent* 3,379,198 (*Chem. Abstr.* **69**: 16913k, 1968).

Morgenthaler, W.W. 1972. Stabilized ammonium phosphate solutions comprising a galactomannan gum and a metal salt. *U.S. Patent* 3,634,234 (*Chem. Abstr.* **76**: 143009q, 1972).

Mosier, B. 1970. Protecting and increasing the reflecting power of aluminium reflectors. *Ger. Offen.* 2,028,370 (*Chem. Abstr.* **75**: 120828w, 1971).

Mudki, J.P. and Warty, S.S. 1976a. A modified guar gum in reactive dye printing. I. Man made text. *India* **19**: 578–585 (*Chem. Abstr.* **86**: 156893e, 1977).

Mudki, J.P. and Warty, S.S. 1976b. Modified guar gum in reactive dye printing. II. Man made text. *India* **19**: 619–628 (*Chem. Abstr.* **87**: 24623c, 1977).

Muenzner, R. and Harmuth-Hoene, A.E. 1978. Influence of a guar gum containing diet of the fecal flora of rats. *Nutr. Metab.* **22**: 368–373.

Murphy, E.L. 1973. The possible elimination of legume flatulence by genetic selection. *Proceedings of the Symposium on Nutritional Improvement of Food Legumes by Breeding.* Sponsored by PAG at FAO Italy. July 3–5, 1972, p. 273–276.

Narasaki, T. and Fujimoto, K. 1965. Separation of hemicelluloses by paper electrophoresis. *Chem. Abstr.* **63**: 18648f.

Nath, B.S. and Gaitonde, R.V. 1976. Effect of admixture of guar gum and starch mucilage on micromeritics of granules and tablets of lactose. *Indian J. Pharm.* **38**: 99–101.

Negrevergne, G. 1969. Pyrazolidinedione derivatives substituted on the 4 position with a phenol-HCHO-polyamine ion exchange resin or a sugar. *U.S. Patent* 3,487,046 (*Chem. Abstr.* **72**: 79414f, 1970).

Nevins, D.G., English, P.D. and Albersheim, P. 1967. The specific nature of plan cell wall polysaccharides. *Pl. Physiol.* **42**: 900–906.

Nordgren, R. 1967. Aminoethyl gum ethers. *U.S. Patent* 3,303,184 (*Chem. Abstr.* **67**: 83183u, 1967).

Nordgren, R. 1968. Acrylamide adducts of polygalactomannans. *Brit. Patent*

1,107,687 (*Chem. Abstr.* **69**: 11603z, 1968).

Norwalk, M.H. 1968. Molded ceramic articles. *U.S. Patent* 3,379,543 (*Chem. Abstr.* **69**: 4943g, 1968).

Nuernberg, E. 1969. Stable water soluble vitamin B_{12} preparations. *Ger. Offen.* 1,290,661 (*Chem. Abstr.* **70**: 118096j, 1969).

Nuernberg, E , Mueller, H., Nowak, H. and Luecker, P. 1971. Galactomannan micro encapsulatum of 5 methylsulfadiazine. *Ger. Offen.* 2,017,495 (*Chem. Abstr.* **76**: 17808m, 1972).

Nuernberg, E., Rettig, E. and Mueller, H. 1972. Sustained release drugs containing galactomannans. *Ger. Offen.* 2,130,545 (*Chem. Abstr.* **78**: 62171a, 1973).

Onuki, M. 1932. *Nippon Nogei Kagaku Kaishi.* **8**: 445–462 (*Chem. Abstr.* **26**: 4308, 1932).

Onuki, M. 1933. *Sci. Pap. Instt. Phys. Chem. Res. Jpn.* **20**: 201–244 (*Chem. Abstr.* **27**: 3454, 1933).

Opie, J.W. and Keen, J.L. 1966. Periodate oxidized polygalactomannan gum. *U.S. Patent* 3,228,928 (*Chem. Abstr.* **64**: 11430a, 1966).

Oveharov, K.E. and Koshelev, Y.P. 1974. Sugar content in corn seeds with different viability. *Fiziol. Rast.* **21**: 969–974 (*Chem. Abstr.* **82**: 14036w, 1975).

Palmer, K.J. and Ballantyne, M. 1950. The structure of (1) some pectin esters and (2) guar galactomannan. *J. Amer. Chem. Soc.* **72**: 736–741.

Peat, S., Bourne, E.J. and Nicholls, M.J. 1948. Starches of the wrinkled and smooth pea. *Nature* **161**: 206–207.

Petek, F. and Dong, T. 1961. *Enzymologia.* **23**: 133.

Petek, F., Villarroya, E. and Courtois, J.E. 1969. Purification and properties of α-galactosidase in germinated *Vicia sativa* seeds. *Europ. J. Biochem.* **8**: 395–402.

Plouvier, V. 1955. Pinitol in legumes Quercitol in *Pterocarpus lucens. Compt. rend. Paris* **241**: 1838–1840 (*Chem. Abstr.* **50**: 6604c, 1956).

Potter, A.L., Silveria, V., McCready, R.M. and Owens, H.S. 1953. Fractionation of starches from smooth and wrinkled seeded peas, molecular weights, end group assays and iodine affinities of the fraction. *J. Amer. Chem. Soc.* **75**: 1335–1338.

Plunguian, M. and Cornwell, C.E. 1976. Foamed cementitious compositions. *U.S. Patent* 3,989,534 (*Chem. Abstr.* **86**: 33552k, 1977).

Pridham, J.B. and Dey, P.M. 1974. The nature and function of higher plant α-galactosidases. In: *Plant Carbohydrate Biochemistry.* Pridham, ed. p. 83–96. Academic Press, London.

Pridham, J.B. and Hassid, W.Z. 1965. Biosynthesis of raffinose. *Pl. Physiol.* **40**: 984–986.

Quiller, M. and Bourdon, D. 1956. Glycosidic metabolism of soybean (*Soja hispida* var. vilnensis). Maltose, important fraction of stock glycosidic extract of the petioles, stems and roots. *C.R. hebd. Seanc. Acad. Sci. Paris.* **242**: 1054–1056 (*Chem. Abstr.* **50**: 11442b, 1957).

Radhakrishnamurty, B. and Srinivasan, V.R. 1959. Studies on peanut shells. III. Structure of groundnut shell hemicellulose B 1. *Proc. Ind. Acad. Sci.* **49A**, 98–103.

Radhakrishnamurty, B. and Srinivasan, V.R. 1960. Studies on groundnut shells. IV. Structural features of hemicelluloses B2, C1 and C2. *J. Sci. Ind. Res.* **19C**: 157–158.

Rafique, C.M. and Smith, F. 1950. The constitution of guar gum. *J. Amer. Chem. Soc.* **72**: 4634–4637.

Rao, P.S. and Srivastava, H.C. 1973. Tamarind. In: *Industrial Gums.* Whistler and BeMiller, eds. pp. 369–411. Academic Press, New York.

Rao, P.S., Ghose, T.P. and Krishna. 1946. Tamarind seed pectin. *J. Sci. Ind. Res.* **4**: 705–710.

Rao, P.V. and Belavady, B. 1978. Oligosaccharides in pulses: Varietal differences and effects of cooking and germination. *J. Agr. Food Chem.* **26**: 316–319.

Reese, E.T. and Shibata, Y. 1965. β-mannanases of Fungi. *Can. J. Microbiol.* **11**: 167–183.

Reichert, E.T. 1913. The differentiation and specificity of starches in relation to genera, species. Vols. I and II, Publ. 173, Carnegie Inst. Washington, USA.

Reid, J.S.G. and Meier, H. 1970. Formation of reserve galactomannan in the seed of *Trigonella foenum-graecum. Phytochemistry* **9**: 513–520.

Reid, J.S.G. and Meier, H. 1973. Enzymic activities and galactomannan mobilisation in germinating seeds of fenugreek (*Trigonella foenum graecum* L. Leguminosae). *Planta* (Berl.) **112**: 301–308.

Reinders, M.A. and Gotlieb, K.F. 1969. Mayonnaise like dressings and sauces. *Neth. Patent* 6,715,194 (*Chem. Abstr.* **71**: 122603k, 1969).

Reynolds, J.H. 1974. Immobilized α-galactosidase continuous flow reactor. *Biotechnol. Bioeng.* **16** (1): 135–147.

Richards, J.B., Evans, P.J. and Hemming, F.W. 1972. In: *Biochemistry of the Glycosidic Linkage.* Piras and Pontis, eds. p. 207. Academic Press, London.

Rol, F. 1973. Locust bean gum. In: *Industrial Gums.* Whistler and BeMiller, eds. pp. 323–337. Academic Press, New York, London.

Rothmans of Pall Mall. 1969. Tobacco sheet fabrication. *Brit. Patent* 1,157574 (*Chem. Abstr.* **71**: 88622d, 1969).

Rubenstein, I.H. 1967. Fat free stabilizer compositions. *U.S. Patent* 3,343,967 (*Chem. Abstr.* **68**: 2122u, 1968).

Sakr, A.M. and Elsabbagh, H.M. 1976. Effect of particle size distribution on the disintegrating efficiency of guar gum. *Pharm. Ind.* **38**: 732–734.

Sakr, A.M. and Elasbbagh, H.M. 1977. Evaluation of guar gum as a tablet additive. *Pharm. Ind.* **39**: 399–403.

Sands, L. and Nutter, P. 1935. The hemicelluloses extracted from mesquite wood after chlorination. *J. Biol. Chem.* **110**: 17–22.

Sannella, J.L. and Whistler, R.L. 1962. Isolation and characterization of soybean hull hemicellulose B. *Arch. Biochem. Biophys.* **98**: 116–119.

Sarrade-Locheur, J., 1971. Coatings preventing dust formation on fertilizer granules. *Ger. Offen.* 2,037,647 (*Chem. Abstr.* **74**: 86964e, 1971).

Sastry, P.S. and Kates, M. 1964. Hydrolysis of monogalactosyl and digalactosyl diglycerides by specific enzymes in runnerbean leaves. *Biochemistry* **3**(9): 1280–1287.

Sautarius, K.A. and Milde, H. 1977. Sugar compartmentation in frost hardy and partially dehardened cabbage leafcells. *Planta* **136**: 163–166.

Savur, G.R. and Sreenivasan, A. 1946. Tamarind seed 'pectin'. *Curr. Sci.* **15** 43–44.

Savur, G.R. and Sreenivasan, A. 1948. Isolation and characterization of tamarind seed (*Tamarindus indica*) polysaccharide. *J. Biol. Chem.* **172**: 501–509.

Savur, G.R. 1955. Tamarind seed polysaccharides. *J. Chem. Soc.* 2600.

Savur, G.R. 1956a. Tamarind "pectin" industry of India. *Chem. Ind.* (London), pp. 212–214.

Savur, G.R. 1956b. Use of pectin from tamarind seeds in the textile industry. *Melliand Textilber* **37**: 588–90 (*Chem. Abstr.* **60**: 15091g, 1956).

Saxena, V.K. 1965. Guar gum, a versatile product. *Res. Ind.* **10**: 101–106.

Schweizer, T.F., Horman, I. and Wursch, P. 1978. Low molecular weight carbohydrates from leguminous seeds; a new disaccharide galactopinitol. *J. Sci. Fd. Agric.* **29**: 148–154.

Sehgal, K., Nainawatee, H.S. and Lal, B.M. 1973. Galactomannan degrading enzyme from germinating *Cyamopsis tetragonoloba* seeds. *Biochem. Physiol. Pflanz.* **164**: 423–428.

Senser, M. and Kandler, O. 1967. Vorkommen und Verbreitung von Galactinol in Blattern hoherer Pflanzen. *Phytochemistry* **6**: 1533–1540.

Shallenberger, R.S. and Moyer, J.C. 1961. Relation between changes in glucose, fructose, galactose, sucrose and stachyose, and formation of starch in peas. *J. Agri. Food Chem.* **9**: 137.

Sioufi, A., Percherson, F. and Courtois, J.E. 1970. Nucleoside-diphosphate-oses et metabolism glucidique au cours de la germination chez Le fenugrec. *Phytochemistry* **9**: 991–999.

Smith, F. and Montgomery, R. 1959. *The Chemistry of Plant Gums and Mucilages*. Reinhold, New York.

Societe de Prayon. 1972. Treatment of galvanized scrap iron by a wet method. *Belg. Patent* 773,906 (*Chem. Abstr.* **77**: 129544z, 1972).

Srivastava, H.C. and Singh, P.P. 1967. Structure of the polysaccharide from tamarind kernel. *Carbohydr. Res.* **4** (4): 326–342.

Steggerda, F.R. 1968. *Ann. N.Y. Acad. Sci.* **150**: 57.

Stewart, C.M. 1966. The chemistry of secondary growth in trees. *Divn. For. Prod.* C.S.I.R.O. Tech. paper No. 43.

Strominger, J.L., Y. Higashi, Sandermann, H., Stone, K.J. and Willonghley E. 1972. In: *Biochemistry of the Glycosidic Linkage*. Piras and Pontis, eds. p. 135. Academic Press, London.

Sundararajan, P.R. and Rao, V.S.R. 1970. Conformational studies of linear β-D-1, 4 mannan and galactan. *Biopolymers*, 9: 1239–1247.

Svolos, T. 1971. Hydrocolloids as icing stabilizers. *Baker's Dig.* 45: 57–61.

Synthelabo, S.A. 1971. Guar gum containing compositions for treating gastro intestinal disorders. *Fr. Demande* 2,073,254 (*Chem. Abstr.* 77: 39247a, 1972).

Tadros, W. and Kamel, M. 1952. A preliminary study of the main polysaccharide of *Lupinus termis* seeds. *J. Chem. Soc.* 4532–4533.

Tanner, W. 1969. Function of myoinositol glycosides in yeasts and higher plants. *Ann. N.Y. Acad. Sci.* 165: 726–742.

Tanner, W. Jung, P. and Linden, J.C. 1972. In: *Biochemistry of the Glycosidic Linkage*. Piras and Pontis, eds. p. 227. Academic Press, London.

Tanner, W. and Kandler, O. 1966. Biosynthesis of stachyose in *Phaseolus vulgaris. Pl. Physiol.* 41: 1540–1542.

Tanner, W. and Kandler, O. 1968. *Europ. J. Biochem.* 4: 233.

Tawakley, M.S. and Bhatnagar, R.K. 1953. Chemical examination of the fixed oil of *Tamarindus indica. Indian Soap J.* 19: 113–115.

Tomoda, M. and Kitamura, M. 1967. *Chem. Pharm. Bull.* 15: 101 (*Biol. Abstr.* 49,6695).

Trapasso, L.E. 1977. Polygalactomannan gum formate esters. *U.S. Patent* 4,011,393 (*Chem. Abstr.* 86: 157395f, 1977).

Trivedi, B.M. and Parikh, M.B. 1972. Guar gum as a substitute for commonly used suspending agents. *Res. Ind.* 17: 96–97.

Tu, Shu-Tung and Appleton, J. 1969. Preparation of a shoe upper material by a collagen reaggregation process with hydrocolloid inclusion. *J. Amer. Leather Chem. Ass.* 64: 598–613.

Venkatarman, R. and Reithel, F.J. 1958. Carbohydrates of the sapotaceae. I. The origin of the lactose in *Achros Sapota. Arch. Biochem. Biophys.* 75: 443–452.

Vereecken, J. and Winand, R. 1976. Influence of polyacrylamides on the quality of copper deposits from acidic copper sulphate solutions. *Surf. Technol.* 4: 227–235.

Villarroya, H. and Petek, F. 1976. Purification and properties of β-mannanase from Alfalfa seeds. *Biochem. Biophys. Acta* 438: 200–211.

Vogel, T. and Schleiden, M.J. 1839. *Pogg Ann. Phys. Chem.* 46: 327.

Wallenfels, K. and Lehmann, J. 1957. The oligosaccharides of St. Johns bread (*Ceratonia siliqua*) isolation of primeverose and ceratose. *Chem. Ber.* 90: 1000–7.

Walker, R.E. 1965. Stabilization of viscous liquids. *U.S. Patent* 3,215,634 (*Chem. Abstr.* **64**: 3820e, 1966).

Wheeler, R.G. 1978. Pourable blasting agents. *African Patent* 7,702,059 (*Chem. Abstr.* **90**: 25630b, 1979).

Whistler, R.L. and Durso, D.F. 1951. The isolation and characterization of two crystalline disaccharides from partial acid hydrolysis of guaran. *J. Amer. Chem. Soc.* **73**: 4189–4190.

Whistler, R.L., Eoff, W.E. and Doty, D.M. 1950. Enzymatic hydrolysis of guaran. *J. Amer. Chem. Soc.* **72**: 4938–4949.

Whistler, R.L. and Gaillard, B.D.E. 1961. Comparison of Xylans from several annual plants. *Arch. Biochem. Biophys.* **93**: 332–334.

Whistler, R.L. and Smart, C.L. 1953. *Polysaccharide Chemistry*. Academic Press, New York and London.

Whistler, R.L. and Smith, C.G. 1952. A crystalline mannotriose from the enzymatic hydrolysis of guaran. *J. Amer. Chem. Soc.* **74**: 3795–3796.

Whistler, R.L. and Stein, J.S. 1951. A crystalline mannobiose from the enzymatic hydrolysis of guaran. *J. Amer. Chem. Soc.* **73**: 4187–4188.

Whitsitt, N.F., Garland, W.D., Baxter, J.K. and Setser, W.G. 1970. Friction reducing resin coatings. *S. African Patent* 6,904,960 (*Chem. Abstr.* **73**: 36220t, 1970).

Yamatani, N. 1978. Cholesterol free liquid egg. *Japan Kokai Patent* 7,832,161 (*Chem. Abstr.* **89**: 74458q, 1978).

Yancik, J.J., Schulze, R.E. and Rydlund, P.H. 1972. Blasting agents containing guar gum. *U.S. Patent* 3,640,784 (*Chem. Abstr.* **76**: 101828, 1972).

Yueh, M.H. and Schilling, E.D. 1971. Calcium salts of carboxy methyl galactomannan. *Ger. Offen.* 2,104,743 (*Chem. Abstr.* **76**: 35433q, 1972).

2
Legume Lipids

D.K. SALUNKHE, S.K. SATHE and N.R. REDDY

2.1. Introduction

Lipids are a group of heterogeneous compounds which are classified together because of their solubility in organic solvents, i.e. chloroform, ethyl ether, petroleum ether or benzene. This solubility differentiates them from other constituents such as proteins, carbohydrates, and nucleic acids in seeds. They include free fatty acids, mono-, di-, and tri-glycerides, phospholipids, sterols, sterol esters, glycolipids (cerebrosides), and lipoproteins; and are esters of acyl-glycerols or glycerides. They are abundantly present in certain legumes such as soybean and peanuts. The food legumes which include soybeans, peanuts, peas, and beans provide important sources of proteins, essential fatty acids and calories. Soybean is primarily utilized in the U.S.A. and other parts of the world as an oil seed. Among oil seeds, soybean ranked first in the total production in 1976 (Lischenko 1979). Soybean oil dominates the supply of vegetable oils consumed in the U.S.A. and other parts of the world, and most of the oil is used in the preparation of food products. In the U.S.A., the use of soybean oil in food products increased from 1.5 billion lbs in the early 1950s to 6.2 billion lbs in 1976 and with a projection of 10.2 billion lbs in 1985 (Kromer 1975). The per capita consumption of soybean oil also increased from 16 lbs per year in 1960 to 31 lbs per year in 1972 with an estimated consumption of about 41 lbs per year in 1985 (Kromer 1975). Another legume, peanut (groundnut), ranks fourth among oil seeds in production and is also one of the world's primary vegetable oil sources. The candies, confectionaries, peanut butter, and salted and roasted nuts producing industries are the major consumers of peanuts in the U.S.A., while in other parts of the world, peanuts are used for the production of oil, salted, roasted, and deep-fat fried nuts, and protein concentrates. Other food legumes are used as a supplemental food because of their variable nutritive value and amino acid pattern (Patwardhan 1962).

51

2.1.1 Lipid Storage Bodies

In mature legumes, a predominant portion of lipids is stored in oil bodies or spherosomes or lipid-containing vesicles in cotyledons, which differ in size and relative abundance in different cells depending on the species. In soybeans, lipids are deposited in spherosomes, which have been identified by electron microscopy (Saio and Watanabe 1968). Fig. 2.1 shows an electron photo-micrograph of a soybean cotyledon in which the structural elements are identified. The spherosomes in soybean cotyledons are interspersed between the protein bodies and are about 0.2–0.5 μ in diameter. The spherosomes are also principal sites of lipid storage in peanuts and are isolated, characterized, and identified by electron microscopy (Jacks et al. 1967; Yatsu and Jacks 1972). An electron photo-micrograph of isolated spherosomes from peanuts is presented in Figs. 2.2a and b. The spherosomes of peanuts are particles of about 1.0–2.0 μ in diameter, bound by a limiting membrane. Yatsu et al. (1971) reported that the repeatedly washed, isolated spherosomes of peanuts contained 99.55 per cent total lipids, 0.09 per cent phospholipids, and 0.22 per cent protein. Reports of Mollenhauer and Totten (1971 a, b) claim that there are two distinct classes of lipid-containing vesicles (simple and composite lipid vesicles) in pea and bush bean cotyledons (Fig. 2.3). The simple vesicles by their name are of nonassociating type, generally 0.5–3.0 μm in diameter and contain predominantly linolenic acid-rich triglycerides with phospholipids as minor components in a molar ratio of 40 : 1 (Allen et al. 1971). These simple vesicles would resemble the oil bodies or spherosomes of peanut and soybean. The other group comprises the composite vesicles which associate strongly among themselves as well as with smooth membranes (sheets) and appear in groups along with membranes. Composite lipid vesicles are smaller in diameter (0.1–1.0 μm) and contain a molar ratio of triglycerides to phospholipins of about 10 : 1 (Allen et al. 1971). These composite vesicles are associated with plastids or plasma membranes during the dormancy of seed and do not seem to occur generally among plant species. Appelqvist (1975) has discussed extensively about oil bodies, its ultra structure, chemical composition, and biosynthetic activity of different oil seeds such as castor, rape seed, etc.

2.1.2. Legumes as a Source of Essential Fatty Acids

Economically, the most important food legumes, soybeans and peanuts, have higher amounts of the common essential fatty acids in their oils. The lipid content of all food legumes except soybean, peanut, and chick pea, used for food; is between 1–3.6 per cent depending upon the species (Aykroyd and Doughty, 1964; Takayama et al. 1965). Soybean, peanut and chick pea have a lipid content of about 21.3, 48.0, and 5.0 per cent respectively (Exler et al.

1977). Linoleic and linolenic acids are the most important essential fatty acids required for growth, physiological functions, and maintenance which cannot be synthesized by the human body and we have to depend on dietary sources for their adequate supply. Legume seed lipids show a significant amount of variability in their fatty acid composition. Groundnut, chick pea, red gram, pea, broad bean, and lentil have oleic and linoleic acids as major fatty acids; while soybeans, black gram, horse gram, and green gram additionally have linolenic acid as the major fatty acid. Snap beans, lima beans, black eyed peas, pinto beans, kidney beans, California small white beans, field beans, and cow-peas have linoleic and linolenic acids as major fatty acids. Many lipids from legume seeds characteristically contain substantial amounts of saturated fatty acids, especially palmitic acid (Mahadevappa and Raina 1978; Baker et al. 1961; Korytnyk and Metzler 1963).

2.1.3. CURRENT STATUS OF LEGUME LIPIDS

The storage lipids of the oil rich legume seeds form a major source of dietary fat and food. As a result of their importance in the food industry, much has been known relatively only on the lipid composition, chemistry, flavour, and off-flavour developments, and their technological implications in foods of dry oil rich seeds such as soybeans and peanuts. Lipids from green peas are also investigated to some extent. Other food legume lipids have not been studied in any great detail because of their low lipid content and limited or no use as oil. The literature on the biochemical, nutritional, and toxicological aspects of lipids from other legumes is scanty when compared with the published reports of seed lipids from soybeans and peanuts.

This chapter deals in detail primarily with the legume lipids from soy-beans, peanuts, and green peas in terms of their involvement in the development of flavours and off-flavours, removal of off-flavours, and technological implications in foods and their interactions with other constituents to some extent. The nutritional and toxicological implications of legume lipids from soybeans, peanuts and some of the other food legumes are also discussed here.

2.2. Chemistry of Legume Lipids

2.2.1. LIPID CONTENT, DISTRIBUTION, AND COMPOSITION OF LEGUMES

The total lipid content and major fatty acids of different legumes with common and scientific names are presented in Table 2.1. The beans with the same scientific name are grouped together under the same common name for total lipid content, regardless of the country of origin. The total lipid content of all legumes, except peanut, soybean, and winged beans, varies from 1.00 to

TABLE 2.1. Lipid content of different legume seeds

Legume species (Common name)	Total lipid content %	Major fatty acids	Reference
1	2	3	4
Groundnut		oleic acid	Fristrom et al. (1975)
(*Arachis hypogaea* L.)	49.7	linoleic acid	
Runner type	50.7	oleic acid	Exler et al. (1977)
		linoleic acid	
Virginia type	47.6	oleic acid	Exler et al. (1977)
		linoleic acid	
Spanish type	49.2	oleic acid	Exler et al. (1977)
		linoleic acid	
Soybean	21.3	oleic acid	Exler et al. (1977)
(*Glycine max* L.)		linoleic acid	
Winged bean	15.0–16.8	—	Harding, et al. (1978)
(*Psophocarpus tetrago-nolobus* L.)			Smart (1976)
Chickpea	4.99	oleic acid	Exler et al. (1977)
(*Cicer arietinum* L.)		linoleic acid	
Great northern bean	3.00	oleic acid	Sessa and Rackis (1977)
(*Phaseolus vulgaris* L.)		linoleic acid	
Garden pea	2.40	oleic acid	Exler et al. (1977)
(*Pisum sativum* L.)		linoleic acid	
Horse gram	2.20	linoleic acid	Mahadevappa and Raina
(*Dolichos biflorus* L.)			(1978)
Red gram	2.19	linoleic acid	Mahadevappa and Raina
(*Cajanus cajan* L.)			(1978)
Cowpea	2.05	linoleic acid	Mahadevappa and Raina
(*Vigna unguiculata* L.)		linolenic acid	(1978)
Kidney beans	1.90	linoleic acid	Korytnyk and Metzler (1963)
(*Phaseolus vulgaris* L.)		linolenic acid	
Pinto beans	1.85	linoleic acid	Korytnyk and Metzler (1963)
(*Phaseolus vulgaris* L.)		linolenic acid	
California small	1.70	linoleic acid	Korytnyk and Metzler (1963)
white beans		linolenic acid	
(*Phaseolus vulgaris* L.)			
Field bean	1.68	linoleic acid	Mahadevappa and Raina
(*Dolichos lablab* L.)			(1978)
Black gram	1.64	linolenic acid	Mahadevappa and Raina
(*Phaseolus mungo* L.)			(1978)
Broad bean	1.60	oleic acid	Exler et al. (1977)
(*Vicia faba* L.)		linoleic acid	
Black eye pea	1.50	linoleic acid	Korytnyk and Metzler (1963)
(*Vigna sinensis* L.)		linolenic acid	
Common bean	1.48	linoleic acid	Exler et al. (1977)
(*Phaseolus vulgaris* L.)		linolenic acid	
Lima bean	1.41	linoleic acid	Exler et al. (1977)
(*Phaseolus lunatus* L.)			

1	2	3	4
Lentil (*Lens culinaris* L.)	1.17	oleic acid linoleic acid	Exler et al. (1977)
Lathyrus bean (*Lathyrus sativus* L.)	1.00	linoleic acid	Choudhury and Rahman (1973)
Green gram (*Phaseolus radiatus* L.)	2.14	oleic acid linoleic acid linolenic acid	Baker et al. (1961)
Jack bean (*Canavalia ensiformis* L.)	2.60	—	Wolff and Kwolek (1971)
Lupin seeds (*Lupinus sp.*)	7.20	oleic acid linoleic acid	Wolff and Kwolek (1971)

7.20 per cent depending upon the species. Peanuts, soybeans, and winged beans have a total lipid content of 49.7, 21.3 and 16.8 per cent respectively. The total lipid contents in Table 2.1 for the common bean (*Phaseolus vulgaris* L.) also include the data from other beans with the same scientific name. Several peanut varieties are grouped into three main types: Runner, Virginia, and Spanish according to Woodroof's (1973) classification and have varied amounts of total lipids. In most of the legumes, either of the unsaturated fatty-acid, i.e. linoleic acid and linolenic acid is present as a major fatty acid. In others, two or more unsaturated fatty acids are present as major fatty acids. For example, in soybeans and peanuts, oleic and linoleic acids are the major fatty acids. The total lipid contents given in Table 2.1 are for dry mature beans. The data on the total lipid contents of different legumes are not considered to be absolute, and may vary depending upon the variety, origin, location, climatic, seasonal and environmental conditions, and the type of soil on which they are grown (Worthington et al. 1972). With a few exceptions, most of the legumes are low in the total lipids content.

In legumes, the total lipids consist of several classes of lipids such as neutral lipids, phospholipids, and glycolipids. Their distribution in legume seeds is different and varies with the species and varieties. Neutral lipids are the predominant class of lipids in most of the legume seeds, however, phospholipids and glycolipids are also present in appreciable amounts (Table 2.2). Soybean lipids contain 89 per cent of neutral lipids as a major fraction with 10 and 2 per cent phospholipids and glycolipids respectively. Soybean phospholipids are a major source for the production of commercial lecithins. The distribution of neutral lipids, phospholipids, and glycolipids in other legumes are entirely different. The neutral lipids and phospholipids in other legumes vary from 32.0 to 51.0 per cent, and from 23.0 to 38.0 per cent of total lipids respectively. Glycolipids constitute the least part of total lipids, i.e. 8.0 to 12.0 per cent. Phospholipids and glycolipids are the essential components of the seed membrane. Neutral lipids primarily consist of triglycerides, accom-

TABLE 2.2. Lipids and their distribution in some of the legume seeds

Legume	Neutral lipids (%)	Phospholipids (%)	Glycolipids (%)
Soybean	88.10	9.80	1.60
Cowpea	46.88	36.82	8.98
Field bean	47.62	36.07	9.82
Red gram	51.05	34.37	8.84
Horse gram	47.74	23.73	11.41
Black gram	49.76	34.84	9.17
Pinto bean	45.21[a]	28.08	—
Lima bean	40.60[a]	24.83	—
Broad bean	35.67[a]	29.49	—
Black eye bean	41.72[a]	29.14	—
Garden pea	35.02[a]	37.74	—
Great northern bean	32.00[a]	34.00	—
Small red bean	40.40[a]	34.11	—

Data calculated from the following references: Takayama et al. 1965; Privett et al. 1973; Mahadevappa and Raina 1978.

[a]Represents only the triglycerides content.

panied by smaller proportions of free fatty acids, sterols, and sterol esters. Miyazawa et al. (1975) found that the pea seeds contained ten different kinds of neutral lipids, among which triglycerides, free sterol, and sterol esters were the major components, and monoglyceride, diglyceride, free fatty acids, wax, and certain pigments were the minor ones. The main component fatty acids in the neutral lipids, phospholipids, and glycolipids are palmitic, oleic, linoleic, and linolenic acids in many of the legumes studied (Takayama et al. 1965; Miyazawa et al. 1975; Mahadevappa and Raina 1978).

The fatty acids composition of legumes is presented in Table 2.3. The distribution of fatty acids shows variability amongst different legumes. Palmitic, oleic, linoleic, and linolenic acids are the principal component fatty acids in several legumes studied. Soybean and peanut lipids are rich in oleic and linoleic acids and range from 20 to 50.6 per cent and 25 to 50.2 per cent of total lipids respectively. High contents of either linoleic or linolenic acid are especially associated with legumes containing insignificant amounts of lipids. For example, black gram contains high amounts of linolenic acid, i.e. 47.5 per cent of the total lipids. Legume lipids also characteristically contain substantial amounts of saturated fatty acids, notably palmitic acid which ranges from 9.22 to 32.5 per cent of the total lipids (Table 2.3).

In the U.S.A. three main types of peanuts (Runner, Spanish, and Virginia) are produced in different locations (Woodroof 1973). About 50 per cent of mostly Spanish and Runner types are produced in Southeast U.S.A. About 30 per cent of the crop produced in the Virginia-North Carolina area is predominantly of the Virginia type, and the remaining 20 per cent, mostly of the Runner type, is grown in Southwest U.S.A. As can be seen in Tables 2.1 and

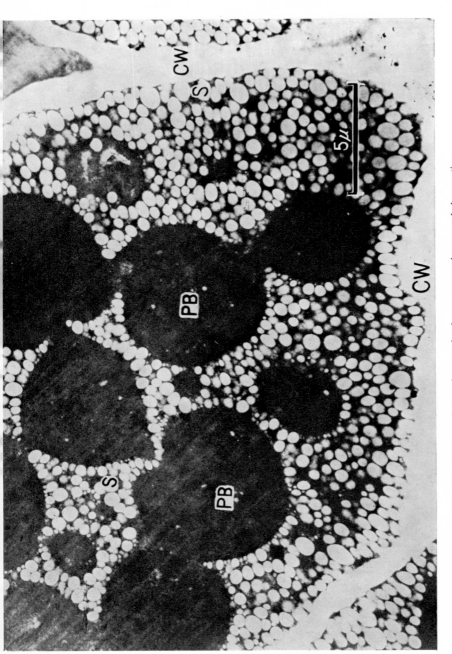

Fig. 2.1. Electron photo-micrograph of a mature soybean cotyledon section.
Seed was soaked in water overnight, fixed in osmium tetroxide, and stained with uranyl acetate and lead citrate.
(PB) protein bodies, (S) spherosomes, and (CW) cell wall are identified.

[*Source:* Saio and Watanabe (1968)]

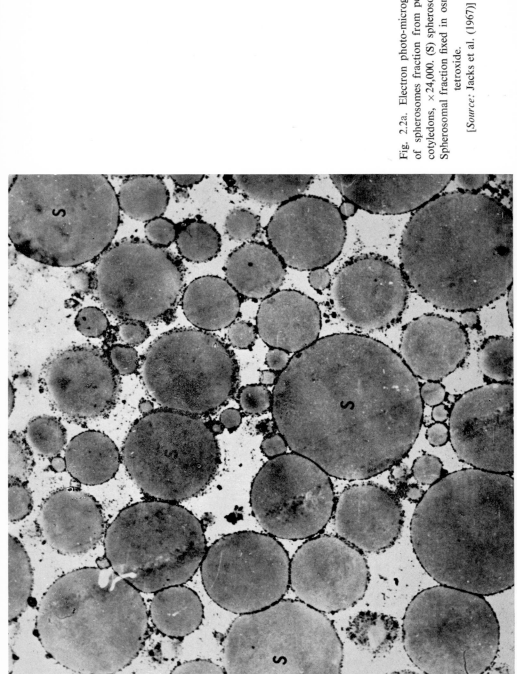

Fig. 2.2a. Electron photo-micrograph of spherosomes fraction from peanut cotyledons, ×24,000. (S) spherosomes. Spherosomal fraction fixed in osmium tetroxide.

[*Source*: Jacks et al. (1967)]

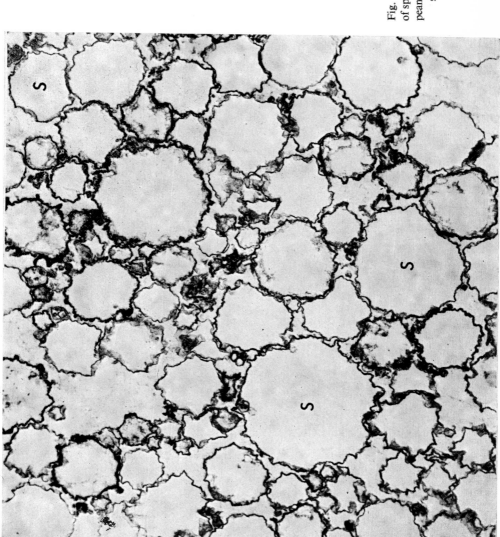

Fig. 2.2b. Electron photo-micrograph of spherosomes fraction, isolated from peanut cotyledons and fixed in potassium permanganate, ×16,000.
(S) spherosomes.
[*Source:* Jacks et al. (1967)]

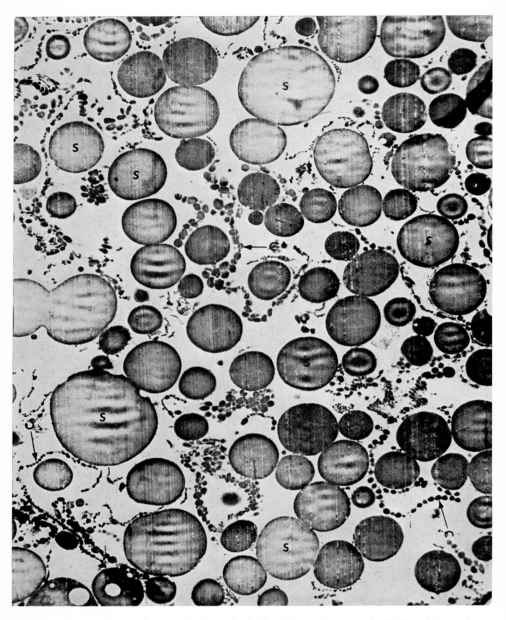

Fig. 2.3. Electron photo-micrograph of simple lipid vesicles and composite lipid vesicles isolated from bean cotyledons. (S) simple lipid vesicles, (C) composite lipid vesicles.

[*Source:* Mollenhauer and Totten (1971b)]

2.3, these three types have different contents of total lipid and fatty acids. The time of harvest and maturity also plays a role in the fatty acid composition of the seed. Usually, the amounts of stearic acid and oleic acid are greater in mature seeds and the amounts of linoleic and other fatty acids are lower (Young et al. 1972).

As with peanuts, the total lipid and fatty acids content in soybeans is also influenced by maturity. Privett et al. (1973) studied in detail the lipid composition of developing soybeans. During seed maturation, the total lipid content increases to more than 20 per cent. Although all lipid classes increase in absolute amounts, the neutral lipid content increases much faster than the phospholipid or glycolipid content. At maturity, neutral lipids are about 92.0 per cent of total lipids, phospholipid 6.1 per cent, and glycolipid 1.9 per cent (Singh and Privett 1970; Privett et al. 1973). Changes in the fatty acids content during the maturation of soybean seeds occur up to a certain time between twenty-five and sixty days after flowering and then level off (Rubel et al. 1972). The amounts of fatty acids such as palmitic, stearic, and linolenic decrease, while the fatty acids such as oleic and linoleic increase. The linolenic acid content increases sharply at the early stages of seed development and then decreases to 5.8 per cent at maturity (Rubel et al. 1972). Soybean oil is reported to be responsible for the production of off-flavours in foods and poor flavour stability, because of the presence of linolenic acid of about 5.8 per cent of the total lipids.

Several foliar fungicides and herbicides are also known to cause small changes in the fatty acids composition in seeds (Penner and Meggitt 1970; Worthington and Smith 1973).

Legume lipids contain appreciable amounts of free fatty acids and, in general, have high oxidation potentials. The susceptibility of legume lipids to oxidation leads to several changes in lipids as well as other food constituents. The oxidation of legume lipids could be further classified as enzymic (due to the inherent presence of lipoxygenase), and nonenzymic (primarily oxidative), both of which ultimately lead to the production of several hydroperoxides which in turn undergo decomposition to yield several products such as aldehydes, ketones, esters, acids, etc. It is these secondary products, particularly aldehydes, which are reactive, and may produce numerous undesirable compounds by reacting with other components present in the system. Of particular importance are the reactions between the decomposition products and proteins and amino acids, carbohydrates, vitamins, and minerals. There may be other interactions involved such as lipid-lipid, or lipid-pigments (chlorophylls, carotenoids) whose nutritional, technological, or biochemical importance at present is not understood at all in case of legumes. We will elucidate in this section on lipid-proteins and amino acids, lipid-carbohydrates, lipid-minerals, lipid-vitamins interactions in brief with particular emphasis on the chemistry involved.

TABLE 2.3. Fatty acids composition of important legume seeds[a]

Legumes	Saturated fatty acids[b] (%)				Total saturated fatty acids (%)	Unsaturated fatty acids[c] (%)				Total unsaturated fatty acids (%)
	16:0	18:0	20:0	22:0		18:1	18:2	18:3	20:1	
Groundnut	10.70	2.64	1.17	—	14.51	46.08	28.97	1.11	—	76.16
Groundnut (Runner type)	9.35	2.47	1.12	—	12.94	49.70	27.42	—	1.12	78.24
Groundnut (Virginia type)	9.37	2.92	1.07	—	13.36	50.63	25.63	—	1.09	77.35
Groundnut (Spanish type)	11.30	2.85	1.30	—	15.45	41.67	32.72	—	0.96	75.35
Soybean	10.80	3.62	—	—	14.42	20.80	50.23	7.65	—	78.68
Chickpea	9.22	1.20	—	—	10.42	21.84	43.29	2.00	—	67.13
Common bean	10.81	1.35	—	—	12.16	7.43	20.95	37.16	—	65.54
Broad bean	11.25	1.88	0.63	—	13.13	20.00	40.63	3.13	—	63.76
Lentil	12.82	0.85	0.85	—	14.52	16.24	36.75	8.55	0.85	62.39
Lima bean	19.86	2.13	—	—	21.99	9.22	31.21	14.89	—	55.32
Garden pea (dry)	12.92	2.08	—	—	15.00	15.42	36.25	6.67	—	58.34
Garden pea (raw)	12.35	2.47	—	—	14.82	16.05	37.04	6.17	—	59.26
Cowpea	23.50	5.60	0.60	2.20	31.90	8.40	34.00	25.70	—	68.10
Field bean	20.20	4.60	—	—	24.80	6.50	56.00	12.70	—	75.20
Red gram	20.50	6.90	0.80	—	28.20	10.50	56.30	5.00	—	71.80
Horse gram	27.10	1.70	—	—	28.80	13.00	44.55	13.65	—	71.20
Black gram	17.80	5.90	—	—	23.70	17.25	11.60	47.45	—	76.30
Green gram	14.10	4.30	—	9.30	27.70	20.80	16.30	35.70	—	72.80
Lathyrus pea	25.00	2.00	—	—	27.00	1.00	67.00	3.00	—	71.00
Black eye bean	32.50	4.60	—	2.50	39.60	7.20	31.20	22.00	—	60.40

Pinto bean	14.70	1.00	—	—	15.70	7.00	28.10	49.20	—	84.30
Kidney bean	13.40	0.74	—	—	14.14	8.30	26.90	50.60	—	85.80
California small white bean	12.20	0.65	—	—	12.85	9.70	23.20	54.30	—	87.20
Lupin seed	17.10	1.90	2.70	0.25	21.95	35.70	28.00	13.30	0.30	77.30

[a] Average data calculated from the following references: Choudhury and Rahman 1973; Korytnyk and Metzler 1963; Baker et al. 1961; Fristrom et al. 1975; Exler et al. 1977; Mahadevappa and Raina 1978.

[b] 16:0 = palmitic; 18:0 = stearic; 20:0 = arachidic; 22:0 = behenic.

[c] 18:1 = oleic; 18:2 = linoleic; 18:3 = linolenic; 20:1 = eicosenoic.

2.2.2. CHEMISTRY OF LIPID-PROTEIN AND LIPID-AMINO ACIDS INTERACTIONS

In the last 15 years or so lipid-protein and lipid-amino acids interactions have been studied in great detail primarily in model systems. Roubal and Tappel (1966a) studied the damage to adenine triphosphate (ATP) by peroxidizing lipids in model systems. They attributed the ATP damage to the production of free radicals during the peroxidation of lipids and discussed the similarities between the ionizing radiation damage to nucleotides. The free radicals produced during the peroxidation of lipids induce the polymerization of proteins. Lipids get entrapped, although in moderate amounts, in the polymerized protein matrix. These changes in the protein can affect the properties of proteins adversely. The free radical mechanism responsible for such changes was proposed by Roubal and Tappel (1966b) using model systems. They suggested the following mechanism:

$$\text{Initiation:} \quad \left.\begin{array}{c} \text{LOO.} \\ \text{or} \\ \text{LO.} \end{array}\right\} + 2\text{PH (Protein)} \rightarrow \begin{array}{c} \text{LOOH} + \text{P} \\ \text{or} \\ \text{LOH} + \text{P} \end{array}$$

$$\text{Polymerization:} \quad \text{P.} + \text{P} \rightarrow \text{P} - \text{P}$$
$$\text{P} - \text{P.} + \text{P} \rightarrow \text{P} - \text{P} - \text{P}$$
$$\text{P} - \text{P} - \text{P.} + \text{P} \rightarrow \text{P} - \text{P} - \text{P} - \text{P, etc.}$$

where, LOO. or LO. represent the lipid peroxide and P represents protein. Later, Roubal (1970) investigated the characteristics of the free radicals derived from oxidizing lipids and explored the possibility of participation of these free radicals in events leading to biological damage *in vitro*. He implied that in dry systems, even though radicals can be trapped in a matrix, radicals may be expected to react eventually. The production of free radicals leads to the damage of proteins, enzymes and amino acids. The damage is reported to be similar to that by radiation but less effective (1/10th) than that by radiation (Roubal and Tappel 1966b). They also found that methionine, histidine, cystine, and lysine were the most labile amino acids to damage and identified H_2S and cystine to be the major products of lipid peroxidation-cysteine interaction. This is particularly important for legumes and legume-based foods, as sulphur containing amino acids are the limiting amino acids in them. Apparently, nutritional studies on these aspects are not reported in legume systems. Labuza et al. (1971) studied the oxidation of methyl linoleate in protein and cellulose systems at different water activities and found that the rate of lipid oxidation increased with increasing water content. They attributed this to the increased mobilization of reactants and catalysts. They also showed that chelating agents such as ethylenediaminetetracetic acid, and citric acid reduced lipid oxidation significantly. Karel (1973) has reviewed protein-lipid interactions, and has described several aspects, primarily in

relation to meat and milk proteins. It appears that there is no single type
of bonding by which lipids interact with proteins. It seems that the electro-
static, salt bridge, hydrogen, van der Waals, covalent, and hydrophobic bonds
are involved in such interactions depending upon the food system and the
type of the lipid under consideration. The typical bond energies (Karel 1973)
for different lipid-protein bonds are represented in Table 2.4.

TABLE 2.4. Bond energies for different lipid-protein bonds

Type of bond	Energy (kcal/mole)	Dependence on distance(s)
Covalent	30–100	Max. attraction at 1–2 Å
Electrostatic	10–20	Energy $\propto s^{-1}$
Hydrogen	1.5–6	Max. attraction at 2–3 Å
van der Waals	0.5–2	Energy $\propto s^{-6}$

The contribution of water and other food components is also important
in understanding the lipid-protein interactions. Protein-water and lipid-water
interactions undoubtedly play an important role in deciding what type and
extent of interactions can take place between lipids and proteins. The ther-
modynamic consideration of such systems is beyond the scope of this chapter.

Recently, Gardner (1979a) has reviewed extensively lipid hydroperoxide
reactivity with proteins and amino acids. Much of the following discussion stems from this review. The chemical changes which occur due to the
interaction of lipid hydroperoxides and proteins include protein-protein
crosslinks, protein scission, protein-lipid adducts, and amino acid damage.
The secondary decomposition products arising from the decomposition of
hydroperoxides are capable of reacting with proteins and amino acids through
covalent bond formation. Aldehydes, particularly malondialdehyde seems to
be the most important in this regard because of its ability to form Schiff
base adducts with amino groups. Exposure of proteins to peroxidized lipids
results in significant proportion of lipid complexation with protein through
hydrophobic association and/or hydrogen bonds (Narayan and Kummerow
1958, 1963; Narayan et al. 1964). Several workers have attempted to characterize the types of reactions, mechanisms, and methodologies involved in such
studies but we will not elucidate on those aspects.

The homolytic bimolecular substitution as suggested by Roubal and Tappel (1966b) has been criticized recently (Karel et al. 1975), on the basis that
because of the high activation energy, the bimolecular homolytic substitution
is unlikely to occur at sp^3-hybridized 4-co-ordinate carbon atom (probably
with the exception of aromatic amino acids). They proposed that protein-
protein cross-linking occurs through the termination type reactions, thereby

TABLE 2.5. Damage to amino acid residues in proteins exposed to peroxidized lipids

Protein	Damage to amino acid residues (% loss)	Type of lipid (and conditions)
Cytochrome c	(A) His (59), Ser (55), Pro (53), Val (49), Arg (42), Met (38), Cys (35)	peroxidizing linolenic acid (aqueous, 37°C, 5 h)
	(B) Tyr (60–75), Try (48), Cys (37), His (31–35), Met (18–37)	linoleic acid hydroperoxide
Globulin	Lys (59), His (52), Tyr (51), Met (38), Cys (33)	peroxidizing ethyl arachidonate (aqueous, 37°C)
Bovine serum albumin	Gly (83), Cys (64), His (54), Ala (50), Val (48), Met (48)	peroxidizing ethyl arachidonate (aqueous, 37°C)
Hemoglobin	Tyr (91), Met (59), Lys (59), His (58)	peroxidizing ethyl arachidonate (aqueous, 37°C)
Ovalbumin	Met (80), His (38), Thr (28), Pro (28), Gly (28)	peroxidizing ethyl arachidonate (aqueous, 37°C)
Catalase	Lys (42), Ser (22), Val (21), Met (20), His (18)	peroxidizing ethyl arachidonate (aqueous, 37°C)
Ribonuclease A	Lys (35), His (35), Tyr (16)	peroxidizing ethyl arachidonate (aqueous, 37°C, 22 h)
Ribonuclease	Met (99), Tyr (62), His (54), Lys (51), Cys (40)	linoleic acid hydroperoxide (aqueous, 37°C, 40 min)
Trypsin	Met (83), His (12)	linoleic acid hydroperoxide (aqueous, 37°C, 40 min)
Pepsin	Met (99), Arg (13), Glu (13)	linoleic acid hydroperoxide (aqueous, 37°C, 40 min)
Lysozyme	Try (56), His (42), Lys (17), Met (14), Arg (9)	linoleic acid hydroperoxide (aqueous, 37°C, 8 days)
Ovalbumin	Lys (50), Met (42), Leu (22), His (21), Val (21)	peroxidized ethyl linoleate (80% relative humidity, 60°C, 4 days)
Casein	Lys (50), Met (47), Ile (30), Phe (30), Arg (29), Asp (29), Gly (29), His (28), Thr (27), Ala (27), Tyr (29)	peroxidized ethyl linoleate (80% relative humidity, 60°C, 4 days)
Ovalbumin	Met (17), Ser (10), Lys (9), Ala (8), Leu (8)	peroxidized ethyl linoleate (aqueous, 55°C, 24 h)
Casein	Lys (10), Thr (10), Val (10), Ala (9), Tyr (8), Phe (8), Ser (8), Arg (8), Asp (8)	peroxidized ethyl linoleate (aqueous, 55°C, 24 h)

Source: Gardner (1979a).

limiting the amount of cross-linking (as collision of two protein radicals has to take place). Very little is known about how a lipid binds to proteins. It appears that protein scission is favoured over protein-protein cross-linking particularly in the dehydrated state (Zirlin and Karel 1969). Amino acids seem to be readily attacked by peroxidized lipids in both aqueous and dry systems. The extent and type of amino acid residues damaged in proteins due to exposure to peroxidized lipids are listed in Table 2.5. Histidine, cysteine/

cystine, methionine, lysine, and tyrosine are the most susceptible residues. Gardner (1979a) implies that tryptophan is one of the most labile amino acid residues in addition to the above list. The degradation products arising through amino acid destruction by peroxidized lipids are embodied in Table 2.6. Malondialdehyde (secondary product of lipid peroxidation) has received considerable attention and seems to react with free amino groups of proteins,

TABLE 2.6. Products obtained after exposing amino acids to peroxidized lipids

Amino acid	Products	Type of lipid
Histidine	(A) imidazole lactic acid; imidazole acetic acid; two histidine-aldehyde Schiff base adducts	peroxidizing methyl linoleate or methyl linoleate hydroperoxide
	(B) histamine; valine; aspartic acid; ethylamine	peroxidizing methyl linoleate
Cysteine	(A) cystine; H_2S; alanine	peroxidizing ethyl arachidonate
	(B) cystine; cysteic acid; cystine disulfoxide	peroxidizing linoleic acid
Methionine	methionine sulfoxide	peroxidizing methyl linoleate
Lysine	diaminopentane; aspartic acid; glycine; alanine; 1, 10-diamino-1, 10-dicarboxy-decane; α-aminoadipic acid; pipecolic acid	peroxidizing methyl linoleate

Source: Gardner (1979a).

particularly ε-amino group of lysine. Other aldehydes (alkanals, alkenals, and alkdienals) formed during the peroxidation of lipids form a Schiff base with amino groups but are not important in the cross-linking of proteins in contrast to malondialdehyde. This is attributed to the occurrence of malondialdehyde in gem-dihydroxy configuration as opposed to the free carbonyl state, i.e. unhydrated, of most of the straight chain alkanals and alkenals.

Nonenzymic browning results from the reaction between the aldehydes and amines. In case of lipid auto-oxidation, the aldehydes are produced which react with amino groups of proteins and aldol polycondensation finally leads to the formation of pigments (usually brown). The reaction can be represented, in general, as follows:

$$RCH_2CHO + R'NH_2 \rightarrow RCH_2CH = NR'$$
$$\downarrow \leftarrow\!\!-R''CH_2CHO$$
$$RCCH = NR'$$
$$Repeat \leftarrow\!-\!-\!-\!-\!-\!-\!-\!\overset{||}{}$$
$$CHCH_2R''$$

Depending upon the nature of reactants and the extent of aldol condensation, the amino group may even cleave which explains why the nitrogen content of brown pigments varies over a range of values. Thus, aldol polycondensation of aldehydes with aldehyde-amino Schiff bases appears to be accepted as the root cause of nonenzymic browning during the auto-oxidation of lipids (Janicek and Pokorny 1971; Tai et al. 1974; Davidek and Jirousova 1975).

The biological importance of the reactions of peroxidizing lipids is still controversial. The biologically important results *in vivo* arising from peroxidizing lipids may include: damage to the lipoprotein membranes of subcellular particles, formation of aging pigments, cross-linking of several polymers in aging living beings, inactivation of certain enzymes either due to conformational changes or destruction of certain key amino acids at the active site or both (Matsushita 1975), malfunctioning of membranes, etc. A recent review (Gennis and Jonas 1977) has discussed the importance of lipid-protein interactions in details, particularly with respect to their importance in cells and cell membranes.

Another important consequence of lipid-protein interactions is the change in the functional properties of proteins. Very little is known about this aspect. Due to the binding of lipids, lipid peroxides or free fatty acids, etc. the change in the energy states of proteins may lead to the denaturation and subsequent loss of solubility of proteins. This change in the solubility of proteins in turn can affect several functional properties of proteins including emulsion capacity, emulsion stability, foaming capacity and stability, hydration, viscosity, gelation, oil and water holding capacity, textural characteristics, etc. Denaturation induced due to the binding of free fatty acids has been demonstrated (Bull and Breese 1967). In addition to the loss of functionality, these reactions may bring about undesirable changes in flavour, colour, and texture which are important technologically from the consumer acceptance viewpoint.

2.2.3. CHEMISTRY OF LIPID-CARBOHYDRATES INTERACTIONS

The interactions of lipids with carbohydrates can be classified broadly as (a) physical and (b) chemical. The interactions may depend on several parameters such as the type of carbohydrates, the type of lipids, and the environmental factors such as, temperature, pH, etc. of the system under consideration. Meagre to no information exists on legume lipids and carbohydrate interactions.

Carbohydrates are known to participate in the retention of volatiles. However, it is yet not well established whether this is due to the direct binding process or the mass transport phenomena involvement (especially during processing). Simple sugars which are soluble in water are capable of increasing the vapour pressure for number of food components and are likely to accentuate the flavour intensity, particularly of those components arising out of oxidized lipids which are water soluble. Amylose is thought to be primarily responsible for the formation of occlusion complexes between lipids and carbohydrates. Under a certain set of conditions the unbranched molecules coil up in a helix conformation with about 6 to 8 glucose moieties per turn. The hydrogen bonds between adjacent glucose units stabilize the helix structure. Oxygen atoms (inside the helix) of the glycosidic linkages give rise to a dense electron cloud due to their stacking in 3-dimensional arrangement. Any electron system which might get trapped inside the helix during the coiling of amylose may interact with this electron cloud leading to the binding of that system to the amylose fraction.

The dimensions of the amylose helix and the molecules which might get entrapped during coiling, both are important in such situations. The mechanisms and thermodynamics of such systems remain far from understood. Osman-Ismail (1972) studied flavour retention using several starches and amyloses as a trapping system for many volatiles. The volatiles of interest (from the legume technology viewpoint) included in the study were methanol, ethanol, 1-butanol, 1-pentanol, 1-hexanol, 1-octanol, and 1-decanol; 1-octanal, decanal; n-butyric, n-caproic, n-caprylic, n-capric, n-lauric, palmitic, lauric, stearic, oleic, and linoleic acids; and methyl ethyl ketone. He found that an inclusion formation was a function of temperature. The application of temperature gradient (which brings about the structural rearrangement of starch molecules) was necessary to form inclusion complexes. The inclusion complex formation took place over a range of temperatures (23 to 85°C) and the temperature of the inclusion complex formation was, in turn, the function of the type of starch as well as the type of volatile. He did not find any formation of an inclusion complex with pure amylose due to quick retrogradation of amylose, thus not giving enough time for the entrapment of the volatile in the matrix. Kinetic studies on binding lead him to conclude that the complexation can occur, both inside and outside the helix (surface or other sites on the macromolecules). He also studied the relationships between the molecular structure and binding of the volatiles and found that in case of C_6, C_8 and C_{10} alcohols, there was a decrease of stability but increase in glucose moieties involved per mole of the ligand (the volatile) while in case of C_{18} fatty acids with different degrees of unsaturation, there was an increase in the binding stability and increase in the glucose moieties involved in binding per ligand molecule with unsaturation. He concluded that double bond (unsaturation) favoured the stability of the complex but increased the steric instability.

Another important study in lipid-carbohydrate interactions (Hydar and Hadziyev 1973) investigated the effects of different matrices (carbohydrates and proteins) on the rates of oxidation in pea lipids which are discussed in the technology section.

From the foregoing discussion the following implications may be important in legume technology: (a) The oxidation products arising from legume lipids may form occlusion complexes with carbohydrates (particularly in the case of legume flours and protein concentrates, where carbohydrates are present in appreciable amounts); (b) The lipids thus trapped in carbohydrate portions may be those which are responsible for off-flavours; and (c) The peroxidized lipids entrapped can further interact with proteins, amino acids, vitamins, and minerals resulting in undesirable changes in legume flours and protein concentrates, particularly as regards the nutritionality and functionality.

2.2.4. CHEMISTRY OF LIPID-MINERALS INTERACTIONS

Minerals, such as Fe^{++}, Fe^{+++}, Cu^{++}, Mn^{++}, Ni^{++}, etc. are known to catalyze the auto-oxidation of lipids. Since oxidation occurs primarily in the hydrocarbon portion of lipids, the individual oxidative processes are generally the same, irrespective of the lipid's origin. The heavy metals (generally derived from equipment, during processing) reduce the induction period of auto-oxidation (time during which no measurable oxidation occurs). They can also affect the rates of chain initiation, propagation, termination, and hydroperoxide decomposition—all the steps of auto-oxidation. Metals may be present in lipids (such as Fe^{++}) in the naturally occurring lipids which have the same effect on auto-oxidation as by the added metals.

Ingold (1962) has given an excellent account of how metals can act in the auto-oxidation of lipids. The important aspects discussed therein seem to be very much valid even today and hence we will account them here briefly: (a) Metal salts are capable of stabilizing peroxy radicals and may decrease the chain propagation and should have an inhibitory effect on auto-oxidation unless the chain generation outweighs the rate of propagation; (b) In the absence of heavy metals, hydroperoxides can decompose by either unimolecular or bimolecular process to yield two free radicals. In the presence of heavy metals, the following additional decomposition reactions are also possible:

$$M^{n+} + ROOH \rightarrow M^{(n+1)+} + RO. + OH^- \qquad (i)$$
$$M^{(n+1)+} + ROOH \rightarrow M^{n+} + RO_2 + H^+ \qquad (ii)$$

where M = metal and ROOH = hydroperoxide.

Reaction (i) is a reductive decomposition and is known to occur at extremely rapid rates in polar solvents as compared to in non-polar solvents. The above reactions are rapid compared to thermal reactions alone primarily due to the low activation energy. Reaction (ii) is oxidative metal catalyzed decomposition of hydroperoxide and has lower rate than reaction (i) mainly because the energy of activation is much greater for reaction (ii) as compared to reaction (i); (c) Since free radicals result from one electron transfer reactions between heavy metal catalyst and a molecule, reverse reaction is also possible by which metals can inhibit formation of free radicals by one electron transfer reaction. Such inhibitory effect of metals in auto-oxidation of hydrocarbons, aldehydes (Ingold 1961; Denisov and Emanuel 1960) and lipids (Banks et al. 1961) has been demonstrated under a certain set of conditions.

The chelation of metal ions by fatty acids may also be possible in legume based foods. It remains to be seen whether such reactions have nutritional implications such as, effect on bioavailability of minerals.

2.2.5. CHEMISTRY OF LIPID-VITAMINS INTERACTIONS

Legumes contain several classes of lipids and can be broadly classified as polar and non-polar lipids. Practically nothing is known about how the vitamins interact with intact and oxidized lipids of legumes. The fat soluble vitamins A, D, E, and K and the water soluble vitamins like vitamins B and C may interact with lipids (intact or oxidized) depending upon the food system under consideration.

The probable interactions may be primarily oxido-reductive in nature as peroxidized lipids and their subsequent secondary products such as aldehydes, peroxides, ketones, etc. are the most likely compounds to be involved in such interactions. Vitamins A and D may undergo chemical changes due to interactions (presumably due to diene systems which are reactive) resulting in reduction or total loss of their biological activity. Vitamin C (a reducing agent) may be converted to dehydroascorbic acid, while vitamin E may act as an antioxidant and thereby retard the oxidation of lipids. Systematic efforts are needed to understand whether such interactions actually do occur in legume systems, and if they do occur, then to what extent and what are the nutritional implications of such interactions, remains to be seen.

2.3. Technology of Legume Lipids

2.3.1. IMPORTANCE OF LEGUME LIPIDS DURING PROCESSING

Amongst legumes, soybeans and peanuts are well studied but most of the other beans remain practically unattended. We will concentrate on lipids giving a special emphasis on their technological implications in this section.

Although dry beans have a good storage stability, off-aromas and off-flavours occasionally develop resulting in their rejection for the purposes of human food. Eley (1968) indicated that the major factor limiting the use of soybean protein products is flavour. Wilding (1970) and Cowan et al. (1973) supported this observation. The presence of significant amounts of polyunsaturated free fatty acids (PUFA) and enzymes such as lipoxygenases in legumes is presumably responsible for the development of off-flavours and off-aroma compounds at the various stages of food processing. Forss (1969) presented an excellent account of the role of lipids in flavours. He opined that "the best known contribution of lipids to flavour is as precursors". In this regard, the role of lipids during the processing of legumes is of primary importance, as the production of off-flavours during handling and processing leads to the rejection of legumes for human consumption. Recently, Rackis et al. (1979) have also reviewed the problems associated with the utilization of vegetable proteins in relation to the off-flavours and have indicated that "sensory evaluations are still needed to define more precisely flavour characteristics of other vegetable protein resources". Several reports clearly show that flavour is the most important single factor limiting the use of soy proteins in foods (Report of the National Soybean Research Coordinating Committee 1974; *Protein Resources and Technology: Status and Research Needs,* 1978; Myer and Williams 1976; Johnson 1976; Maga 1973). However, similar information on other legumes with probable exception of peanuts is by far negligible to none.

2.3.2. Lipid Derived Flavours

The word flavour is interpreted here as an overall sensual response experienced when food is taken in the mouth implying thereby an overall taste, odour, texture, and other oral sensations. It is also important to recognize the fact that lipids are not only important precursors of many volatile flavours, but they also contribute to other flavours as well. Intact lipids or low volatile breakdown products of lipids contribute to flavour largely through mouth stimulation (Forss 1969). The understanding of both desirable and undesirable flavours is equally important in developing a sound technology for the effective utilization of legumes. The problem of recognition and identification of individual components responsible for flavours becomes more involved as lipids and their degradation products are also capable of interacting with other food components such as proteins, carbohydrates, minerals, pigments and vitamins. In addition, lipids modify the taste and flavour of other compounds in foods and are capable of influencing the physical state of the food which affects the movement of compounds to the taste and odour receptor sites. When an oil is taken in the mouth, mixing of the oil with saliva results in the distribution of any flavour compounds present between the oil and aqueous saliva. The coating of the tongue with oil may decrease or even prevent the perception of water soluble taste compounds and might, thereby, magnify the effects of oil soluble flavours as well as its own flavour. The understanding of partition functions and the related thermodynamics thus may prove to be critical. The information on physical chemistry aspects in legume food systems is needed. Another important consideration is the knowledge of eating habits of the populus to which such types of products are to be offered. For example, the natural beany flavour of *Leguminosae* seeds may be offensive in most parts of the western hemisphere while it may be perfectly acceptable and desirable in certain Asian countries including China, India, Pakistan and Bangladesh. Flavour descriptions for legume products (predominantly flours, protein concentrates, protein isolates, and their end-products, i.e. bakery products, dairy products, etc.) are quite variant but the most frequently described flavours include beany, bitter, and nutty. Other flavour notes attributed to legume products consist of green, grassy, sweet, toasted, musty, stale, spoiled, raw beany, cardboard, chalky, etc. As evidenced in several reports (Rackis et al. 1979; Cowan et al. 1973) the flavour descriptions are partially dependent on the handling and type of processing involved prior to evaluation. The data relating to these descriptions for model and actual food systems are represented in Tables 2.7–2.10. Practically no information is available on the desirable

TABLE 2.7. Effects of steam on flavour score given by taste panel
for different soy flours

Flour sample	Treatment	Flavour score[a]	Remarks
Full fat	No steaming	1.5	Raw flour, beany, bitter, green
Full fat	3 min steaming[b]	4.5	Beany, bitter, nutty sweet, toasted
Full fat	10 min steaming	6.0	Beany, nutty, bitter
Full fat	20 min steaming	6.3	Beany, nutty, bitter
Full fat	40 min steaming	6.1	Beany, nutty, bitter
Hexane defatted	Washed with dilute acid at pH 4.6	3.6	—
Hexane defatted	80% methanol extraction	6.3	—
Hexane defatted	80% ethanol extraction	7.3	—
Hexane defatted	80% ethanol extraction plus 20 min steaming	8.0	Highest score

[a]Scoring 1–10 where 10 is bland and 1 is a strong flavour or odour.
[b]Live steam treatment in preheated autoclave at atmospheric pressure.
Source: Cowan et al. (1973).

TABLE 2.8. Flavour of full-fat soy flour: effects of steaming

Steaming, min	Flavour score[a]	Organoleptic flavour descriptions
0	1.5	Beany, bitter, green
3	4.5	Beany, bitter, nutty, sweet, toasted
10	6.0	Beany, nutty, bitter, toasted, sweet
20	6.3	Beany, nutty, bitter, toasted, sweet
40	6.1	Beany, nutty, bitter, toasted, sweet

[a]See Table 2.10 for scoring.
Source: Rackis et al. (1979).

flavour characteristics and the compounds responsible for them. Much of the efforts seem to be concentrated in isolating, identifying, and characterizing the components responsible for the production of undesirable off-flavours.

TABLE 2.9. Odour and flavour of some commercial soy flours, protein concentrates, and protein isolates

Sample	Odour		Flavour	
	Score[1]	Description	Score[1]	Description
Flours				
A	5.8a[2]	NP[3]	4.2a	Bitter, beany
C	7.0b	Beans	5.5b	Beany, bitter
F	7.4b	Toasted	6.6b	Toasted, beany, bitter
Concentrates				
A	6.4a	CW[3]	5.6a	Bitter, beany
C	6.9ab	NP	6.3ab	Beany
E	7.4b	NP	7.0b	Beany
Isolates				
A	7.7a	Musty, corn meal	5.9a	Beany, cardboard
F	7.4a	Flour	6.4a	Beany, flour, nutty

[1]See Table 2.7 for scoring.
[2]Within each group, i.e. flours, concentrates or isolates, scores with letters in common are not significantly different at the 0.05% level.
[3]NP, none predominant; CW, odour similar to oat cereal-singed wool in water.
Source: Cowan et al. (1973).

TABLE 2.10. Organoleptic evaluation of soy protein products

Products	Odour		Flavour	
	Score[a]	Description	Score[a]	Description
Soy flours A–G	5.8–7.5	NP[b], beany, corn meal vanilla, CW[c]	4.2–6.7	Beany, green beany, bitter, raw beany, toasted
Raw flours[d]	5.8	Beany	4.1	Raw beany, beany, bitter, green beany
Concentrates A–E	6.4–7.4	Beany, CW, NP, musty, stale, toasted, corn meal	5.6–7.0	Beany, bitter, astringent
Isolates A–F	6.8–7.7	CW, musty, beany, corn meal, spoiled	5.9–6.4	Beany, bitter, chalky, cardboard, astringent, toasted, nutty, cereal

[a]10, bland; 9–7, weak; 6–5, moderate; 3–4, strong; 1–2, very strong.
[b]NP=none predominant, several responses.
[c]CW=odour similar to combination of high protein oat cereal and singed wool in water.
[d]Laboratory prepared raw, defatted soy flour.
Source: Rackis et al. (1979).

It has been demonstrated that the lipid oxidizing potential of *Leguminosae* is high (Rhee and Watts 1966). These investigators interpreted the thiobarbituric acid (TBA) reactive materials in the extracts of several plants as an index of "lipid oxidation potential" and also demonstrated the wide distribution of lipoxidase in fruits and vegetables (Table 2.11). In another study (Pinsky et al. 1971) the lipid oxidizing activity in different legumes has been shown to be high. Thus, the presence of substantial quantities of polyunsaturated free fatty acids and lipoxygenase (linoleate/oxygen oxidoreductase, E.C. 1.13. 11.12) seem to be of an important concern in the understanding of lipid flavours in general and off-flavours in particular during the handling and processing of legumes. The hydroperoxidation of PUFA containing 1, 4-pentadiene system is catalyzed by lipoxygenase. The mechanism of lipoxygenase catalysis resulting in the production of hydroperoxides and their subsequent products has been reviewed (Eskin et al. 1977). Its role in the deterioration of the food quality and formation of objectionable flavours is of concern in legumes. It is important to recognize, that although the lipoxygenase and peroxidase activities are important, other factors may become more important in certain cases such as in roasted peanuts where the initiation of off-flavour formation due to the oxidation of lipids is attributed to metallo proteins (St. Angelo et al. 1977). Nonlipid precursors responsible for off-flavours in legumes have also been reported (Buttery et al. 1975). They showed that geosmin, an oxygenated hydrocarbon, to be responsible for the musty, mouldy, earthy odour of off-flavoured dry white navy beans. Characteristic unpleasant cooked odour—the major barrier to consumer acceptance of high temperature processed soy products—has been shown to be the result of thermal decarboxylation of the phenolic (p-coumaric and ferulic) acids.

Most lipids are hydrophobic, nonpolar materials which exist naturally as liquids (oils) or solids (fats) and may consist of one or more of the following classes: free fatty acids, mono-, di-, and triglycerides, phospholipids, sterols, plasmogens, and lipoproteins. During the degradation of lipids, lipase contributes to the first step which releases free fatty acids from triglycerides. However, it seems that its contribution to off-flavours is negligible as compared to that by lipoxygenases and peroxidases. The enzymatic formation of volatile aldehydes and alcohols responsible for off-flavours, derived from unsaturated fatty acids and roles of lipoxygenase and alcohol oxidoreductase have been reported by several workers (Weurman 1961; Eriksson 1975; Grosch 1967; Gonzales et al. 1972; Jadhav et al. 1972; Tressl and Drawert 1973; Gardner 1975; Stone et al. 1975; Galliard et al. 1976; Grosch et al. 1976; Singleton et al. 1976). Catalytic properties of lipoxygenase and hematin compounds in relation to food have been comprehensively reviewed (Watts 1954; Tappel 1961 and 1962; Love and Pearson 1971). "Hematin compounds in animal tissues are important lipid oxidation catalysts where they occur in high concentrations. The corresponding knowledge for plant tissues is lacking, probably due to the

TABLE 2.11. Lipid-oxidizing activity of vegetable and fruit extracts

	Latin name		Units per 10 g fresh tissues	Units per g protein
	Family	Genus and species		
Garden pea	Leguminosae	*Pisum sativum*	8.53	13.55
Blackeye pea	Leguminosae	*Vigna sinensis*	5.72	6.36
Pole bean	Leguminosae	*Phaseolus vulgaris*	5.48	28.84
Potato	Solanaceae	*Solanum tuberosum*	2.98	14.19
Lima bean, immatured	Leguminosae	*Phaseolus limensis*	1.20	1.43
Egg-plant	Solanaceae	*Solanum melongena*	1.11	9.25
Tender bean	Leguminosae	*Phaseolus vulgaris*	1.07	5.63
Sweet potato	Convolvulaceae	*Ipomea batatas*	0.96	5.65
Asparagus	Liliaceae	*Asparagus officinalis*	0.85	3.49
Hot pepper, whole	Solanaceae	*Capsicum annuum*	0.79	6.08
Squash, yellow	Cucurbitaceae	*Cucurbita pepo* var. melopepo	0.66	5.50
Green pepper pod	Solanaceae	*Capsicum annuum* var. grossum	0.62	5.17
Radish	Cruciferae	*Raphanus sativus*	0.61	6.10
Rutabaga	Cruciferae	*Brassica napobrassica*	0.55	5.00
Corn, immatured	Gramineae	*Zea mays*	0.52	1.49
Cabbage	Cruciferae	*Brassica oleracea* var. capitata	0.51	3.92
Carrot	Umbelliferae	*Daucus carota* var. sativa	0.47	4.27
Green bean	Leguminosae	*Phaseolus vulgaris*	0.44	2.32
Onion, yellow skin	Liliaceae	*Allium cepa*	0.44	2.93
Cauliflower	Cruciferae	*Brassica oleracea* var. botrytis	0.43	1.59
Apple, golden delicious	Rosaceae	*Malus sylvestris*	0.37	18.50
Apple, red delicious	Rosaceae	*Malus sylvestris*	0.36	18.00
Cucumber	Cucurbitaceae	*Cucumis sativus*	0.36	4.00
Pear	Rosaceae	*Pyrus communis*	0.34	4.89
Banana	Musaceae	*Musa paradisaca*	0.29	2.64
Tomato	Solanaceae	*Lycopersicon esculentum*	0.29	2.64

Celery	Umbelliferae	*Apium graveolens*	0.25	2.78
Rhubarb	Polygonaceae	*Rheum rhaponticum*	0.13	2.17
Squash, acorn	Cucurbitaceae	*Cucurbita maxima*	0.12	0.23
Turnip root	Cruciferae	*Brassica rapa*	0.10	1.00
Zucchini	Cucurbitaceae	*Cucurbita pepo* var. *medullosa*	0.05	0.10
Orange	Rutaceae	*Citrus sinensis*	0.02	0.29
Turnip green	Cruciferae	*Brassica rapa*	0.02	0.23
Green Crowder pea	Leguminosae	*Vigna species*	0.01	0.01

Source: Rhee and Watts (1966).

fact that lipoxygenase is far superior to hemoproteins as a lipid oxidation catalyst in enzymically active tissue" (Eriksson 1975).

The contribution to flavour may arise from both volatile and nonvolatile lipids. Lipids of low volatility (above C_{10}) are tasteless in terms of sour, sweet, bitter, and salty, partly because they are insoluble in water. A fifth sensation— the metallic taste particularly in many fatty foods—is attributed primarily to 1-Octen-3-one. However, "candle-like" flavour has been ascribed to these low volatile fatty acids. Any release of glycerol from saturated fats will impart a sweet taste. Kramer (1968) acknowledges that the feeling of the mouth is regarded by many people as a contribution to flavour but suggests that a more precise definition is required. Chewiness, grittiness, oiliness, mealiness, stickiness, fibrosity, etc. are essentially sensed by muscular forces during the process of mastication. The role of lipids in "mouth feel" is perhaps in their contribution towards texture. Practically nothing is known about legume lipids in this regard. Lipids do affect the flavour thresholds. Again, the perception level depends upon the physical nature of the medium in which the compounds are dispersed. Generally, the flavour potential is much stronger in an aqueous (lipophobic) than in an oily (lipophilic) medium. Other factors such as the polarity of the flavour compound (Kinsella et al. 1967) bonding to solvent molecules, etc. complicate the situation. Generally, the substances of low polarity (long chain fatty acids) may be expected to have low flavour thresholds in an aqueous medium and high thresholds in oil, while more polar substances (short chain fatty acids) might present lower flavour thresholds in oils and higher thresholds in water.

2.3.3. Lipid Derived Off-Flavours

The understanding of both enzymatic and nonenzymatic degradation of lipids is important for the recognition, isolation, and characterization of off-flavours in legumes. The primary factors are thus:
  (i) Nature and composition of lipids of legume under consideration;
  (ii) Handling, processing, and storage conditions employed (atmosphere, pH of the food system, oxygen, and others);
  (iii) Lipoxygenase and peroxidase activities;
  (iv) Interactions involved, of the lipids and their degradation products with other food components in a particular food system;
  (v) Presence or absence of catalysts; and
  (vi) Role of lipolytic and oxidative enzymes.

We will now consider the lipid degradation in brief and elucidate the types of compounds responsible for off-flavours in legumes and legume products. Although the necessary ingredients, i.e. polyunsaturated fatty acids, O_2, and catalysts are present, the living tissue is remarkably relatively immune to lipid peroxidation. This has been attributed to the precise compartmentalization of cellular organelles giving protection to the cell from self-destruction (Gardner 1979a). The enzymatic breakdown of lipids in plants can be conveniently divided into hydrolytic and oxidative processes both of which result in a net loss of lipids. Legume lipids are particularly susceptible to oxidative degradation and nonenzymatic reactions. During hydrolytic breakdown, fatty acids from triglycerides are released and are further degraded by lipoxygenases. Desnuelle and Savary (1963) have listed many substrates for plant "lipases" which include methyl, ethyl, p-nitrophenyl and naphthyl esters of fatty acids, the water soluble triacetyl glycerol, short- as well as long-chain triglycerides, and natural oils. It is interesting to note here that the term "lipase" has been used in the literature nebulously for any enzyme that liberates free fatty acids from acyl lipids. A true lipase (triglycerol acyl hydrolase, E.C. 3.1.1.3) may be defined as an enzyme system that hydrolyses emulsified triacylglycerols at an oil-water interface (Desnuelle and Savary 1963). The information on the properties of true lipases is limited. Specificity studies on several seed lipases (Williams and Bowden 1973) appear to conclude that seed lipases, in general, are non-specific for the position of acyl ester bonds in triglyceride hydrolysis. Greater confusion exists in the classification and nomenclature of lipolytic acyl hydrolases acting on polar lipids. These enzymes deacylate glycerolipids with the following general formula:

$$
\begin{array}{c}
\qquad\qquad\qquad\qquad\overset{\text{O}}{\overset{\|}{}} \\
\qquad\qquad CH_2-O-\overset{\text{O}}{\overset{\|}{C}}-R_2 \\
\overset{\text{O}}{\overset{\|}{R_1-C}}-O-\overset{|}{CH} \\
\qquad\qquad CH_2-O-X
\end{array}
$$

where R_1 and R_2 are fatty acyl residues and X represents a phosphate ester in phospholipids or a glycosidic moiety in glycolipids. The lack of systematic studies on substrate specificities on different hydrolytic activities described as phospholipases, galactolipases, sulpholipases, monoglyceride lipases, lipases, and esterases warns caution during the interpretation of the reported literature as it is possible that the same enzyme activity might have been given different names by different authors. Table 2.12 contains the information which illustrates this point. As can be seen, the purified lipolytic acyl hydrolase from potato tubers shows a differential specificity towards different substrates. Acyl transferase activities (the enzymatic transfer of fatty acids from endogenous donors, phospholipids and galactolipids, to acceptors like short-chain alcohols, e.g. methanol, ethanol, etc.) resulting in the production of methyl and ethyl esters in plants have been demonstrated (Galliard and Dennis 1974). Contribution of acyl transferase activities to legume flavours (desirable or undesirable) remains to be understood. The presence of other types of enzymes capable of lipid degradation such as fatty acid thiolester hydrolase,

TABLE 2.12. Substrate specificity of purified lipolytic acyl hydrolase from potato tubers

Lipid class	Substrate	Enzyme activity (μ equiv, fatty acid released/min/mg of protein)
monoglyceride	monolein	39.0
lyso-phospholipid	lyso-phosphatidylcholine	28.0
monogalactolipid	monogalactosyldiglyceride	12.0
methyl ester	methyl oleate	10.8
diglyceride	diolein	8.2
digalactolipid	digalactosyldiglyceride	6.8
phospholipid	phosphatidylcholine	5.1
triglyceride	triolein	<0.1
wax ester	octadecyl palmitate	<0.1
sterol ester	cholesteryl oleate	<0.1
("esterase" substrates)	p-nitrophenyl palmitate	15.5
	p-nitrophenyl acetate	0.14

Source: Galliard (1975).

phospholipase D (phosphatidylcholine phosphatidohydrolase; E.C. 3.1.4.4), phospholipase C (E.C. 3.1.4.3.)—which splits phospholipids to diglycerides and phosphate esters is known but their roles are far from being understood either in physiology or development of off-flavours. We will not discuss their biochemistry here as it is beyond the scope of this section.

Amongst the several enzymes involved in the oxidative degradation of lipids, lipoxygenases are the most important ones from the off-flavour production viewpoint. Lipoxygenases isolated from different sources differ in several respects, such as substrate specificity, pH optima, effects of inhibitors and, most significantly, in the isomeric form of the products. Several isoenzymic forms with different properties in the same source are also demonstrated, e.g. in soybean (Christopher and Axlerod 1971; Christopher et al. 1972; Grosch et al. 1972; Verhue and Francke 1972) and in peas (Eriksson and Svensson 1970; Arens et al. 1973; Anstis and Friend 1974a, b). All lipoxygenase enzymes specifically attack fatty acids with cis, cis-1, 4-pentadiene structures. Unsaturation at ω-6 (e.g. −12 position in C_{18} acids) has been reported to be essential for the lipoxygenase activity in case of soybeans (Hamberg and Samuelsson 1965). From linoleic or α-linolenic acids (the predominant endogenous substrates from plants) optically active derivatives of 9-D- or 13-L-hydro-peroxydiene are formed. The proportions of 9-D- and 13-L-hydro-peroxy derivatives of linoleic acid with different lipoxygenase preparations are represented in Table 2.13 which also serves to illustrate the effects of the other parameters (pH, temperature, and oxygen tension) on lipoxygenase activities.

TABLE 2.13. Influence of enzyme source and reaction conditions on the ratio of 9– to 13–hydroperoxide derivatives of linoleic acid produced by lipoxygenase enzyme preparations

Enzyme source	Reaction conditions			Ratio of 9 : 13 hydroperoxides
	pH	gas phase	temp. °C	
Soybean mixed isoenzymes	9.0	O_2	0	0–30 : 70–100
Soybean mixed isoenzymes	5.5	air	25	54 : 46
Soybean purified isoenzyme I	9.0	O_2	0	10 : 90
Soybean purified isoenzyme I	9.0	air	25	52 : 48
Soybean purified isoenzyme II	9.0	air	25	55 : 45
Soybean purified isoenzyme II	7.0	air	25	70 : 30
Soybean purified "acid isoenzyme"	7.0	O_2	ambient	ca. 50 : 50
Soybean purified "acid isoenzyme"	9.0	O_2	0	5 : 95
Corn germ	6.5	O_2	25	83–93 : 7–17
Corn germ	9.0	O_2	0	15 : 85
Potato tuber	5.5	air	25	95 : 5
Potato tuber	9.0	air	25	no reaction
Alfalfa seed	6.8	O_2	ambient	50 : 50
Flax seed	6.9	air	24	80 : 20
Dimorphotheca sinuata seed	6.9	air	23	0 : 100
Pea seed (purified enzyme)	6.75	1% O_2 100% O_2	0	55 : 45
Oat seed	7.0	O_2	ambient	ca. 90 : 10
Barley seed	6.8	air	ambient	ca. 80 : 20

Source: Galliard (1975).

The detailed mechanism of lipoxygenase activity is not yet elucidated. However, the involvement of the stereospecific abstraction of H at the C-11 methylene group of linoleic acid has been demonstrated using soybean enzyme (Hamberg and Samuelsson 1967). The removal of L_s-hydrogen results in 13-L-hydroperoxide formation. Egmond et al. (1972) further showed that, in the case of corn, the removal of D_R-hydrogen resulted in the formation of 9-D hydroperoxide isomer formation (Fig. 2.4). DeGroot et al. (1973) indicated that free radical intermediates are involved in lipoxygenase reaction. The involvement of iron (Chan 1973; Roza and Francke 1973; Pistorius and Axlerod 1974) and the tryptophan residues of the enzyme (Finazzi-Agro et al. 1973; Nagami 1973) at the active site have been reported. Galliard (1975) proposed a hypothetical mechanism compatible with the then known information (Fig. 2.5). However, it seems that still the mechanism is not completely understood. Lipoxygenase-catalyzed-co-oxidation of pigments is also well documented (Arens et al. 1973; Weber et al. 1973a, b; Anstis and Friend 1974b; Orthoefer and Dugan 1973). However their contribution towards off-flavour developments, if any, remains unknown.

The oxidative deterioration of food lipids involves, primarily, auto-oxida-

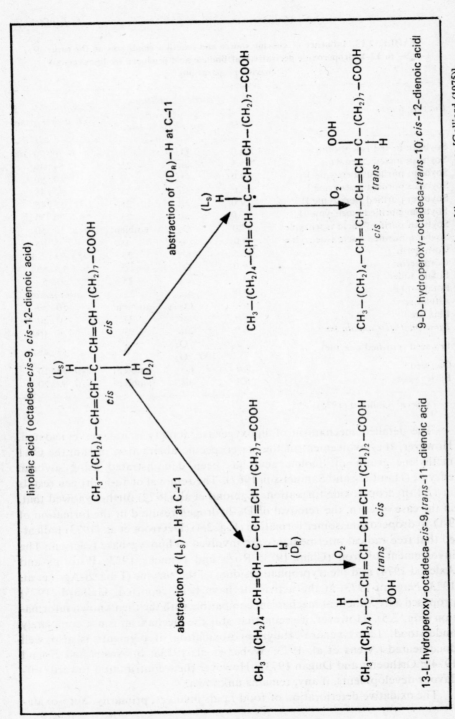

Fig. 2.4. Formation of 13-L and 9-D-hydroperoxy derivatives of linoleic acid by the action of lipoxygenases [Galliard (1975)].

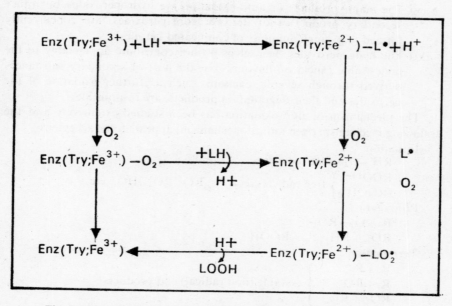

Fig. 2.5. Hypothetical scheme for lipoxygenase action [Galliard (1975)].

tive reactions. These reactions are further accompanied by secondary reactions which may include oxidative and non-oxidative reactions or interactions with other food components as well. Auto-oxidation, in general, involves the reaction of any material with molecular oxygen. The interesting feature of autocatalytic auto-oxidation of fats is that the reaction rate increases with time due to the formation of products which themselves catalyse the reaction. The important general features of autocatalytic auto-oxidation are:

(i) The reaction primarily involves unsaturated acyl groups and the hydro-peroxide group appears in alpha position relative to a double bond. Depending upon the amount of unsaturation in the acyl group and other factors, the formation of hydroperoxides may or may not involve a shift of the double bond.

(ii) Rates are greatly dependent upon the degree of unsaturation amongst common fatty acids.

(iii) The presence or absence of catalysts (trace metals) and light radiation (notably in the infra-red region) have markedly accelerating effects while a multitude of substances even in small concentrations (particularly several phenolic substances) have anti-oxidant effects.

(iv) Temperature has a marked effect on auto-oxidation; however, the manner in which temperature has effect, differs from more simple types of chemical reactions.

(v) The major products of auto-oxidation are hydroperoxides but other secondary products (not derived from peroxides) are concurrently formed (e.g. small amounts of conjugated ketones).

(vi) The hydroperoxides themselves do not contribute appreciably to the undesirable rancid off-flavours, but the host of secondary substances derived through several reactions, such as further oxidation of the peroxides and their degradation products, are responsible.

The mechanism of auto-oxidation has been studied extensively and the following mechanism (free radical mechanism) appears to be accepted.

Initiation:

$$RH + O_2 \rightarrow \text{free radicals}$$

$$\left.\begin{array}{l} ROOH \\ (ROOH)_2 \end{array}\right\} \text{free radicals (e.g. R., RO., RO_2., HO., etc.)}$$

Propagation:

$$R. + O_2 \rightarrow RO_2.$$

$$RO_2. + RH \rightarrow R. + ROOH$$

Termination:

$$\left.\begin{array}{l} R. + R. \\ R. + RO_2. \\ RO_2. + RO_2. \end{array}\right\} \rightarrow \text{stable (non radical) end products}$$

Another important aspect of lipid degradation is thermally induced reactions. Studies involving the effect of heat on lipids can be broadly divided into three categories: (a) Studies in which heat was applied under strictly non-oxidative conditions; (b) Studies where oxygen or air and heat both were present; and (c) Thermal studies where no such distinction was made. Several reactions including dehydration, decarboxylation, hydrolysis of ester bond, double bond conjugation, polymerization, dehydrocyclization, aromatization, dehydrogenation and degradation by carbon-carbon cleavage are known to occur under the influence of heat in the absence of oxygen. The formation of free fatty acids by heating in the absence of moisture has been demonstrated (Hurd and Blunck, 1938). Formation of acrolein, di-n-nonyl ketone, symmetrical ketones (di-n-butyl, di-n-amyl, and di-n-hexyl), decanoic acid, and capric acid from tricaprin under different conditions of heat in the absence of moisture has been demonstrated (Crossley et al. 1962). The effect of heat on lipids in the presence of moisture resulting in free fatty acids formation is known (Buziassy and Nawar 1968). The formation of free fatty acids from triglycerides, both in the presence and absence of moisture under the influence of heat is implicated. In the thermal hydrolysis of triglycerides, the preferential release of short chain and the unsaturated free fatty acids but no positional specificity, has been observed. Mechanisms involved in the formation of lactones, methyl ketones, hydrocarbons and mono- and dicarboxylic methyl esters in heated fats have been reviewed (Nawar 1969).

The compounds derived from lipids and their degradation products are numerous and can be broadly classified as follows:

(i) Free fatty acids.

(ii) Carbonyls—These compounds are known now for their intense characteristic flavours and are probably the most important ones from the legume derived off-flavours viewpoint.

Aldehydes belonging to three groups: Cl-10 n-alkanals; C3–12, n-2-alkenals; and C5–12, n-2, 4-alkadienals formed from oxidized unsaturated fatty acids (particularly oleic, linoleic, linolenic, and arachidonic) which are mainly responsible for oxidized flavours described as painty, oily, tallowy, or cardboardy.

Ketones derived from oxidized lipids, e.g. 1-Octen-3-one (metallic flavour) 1-penten-3-one (oily flavour).

(iii) Delta and gamma lactones. The former are important in dairy products while the latter in fruits. Their contribution in legumes is less understood.

(iv) Saturated alcohols which might play a minor role in flavours. Compounds like 1-penten-3-ol isolated from milk fat and meat have an oily, grassy flavour and may be of importance in legume technology as well.

(v) Aliphatic hydrocarbons, generally, have weak flavours.

(vi) Amines which are generally derived from proteins and amino acids might be formed from lipid portion (lecithin) through bacterial activities and can impart strong repulsive odours.

Apart from the above discussed classes of compounds derived predominantly from lipids, other classes of compounds such as esters, pyrazines, heterocyclic compounds, notably thio compounds have been reported particularly in processed legumes and legume products. The reported compounds from several legumes and legume products are presented in Table 2.14 which illustrate the effects of processing on legume lipids to some extent.

2.3.4. Technological Aspects of Off-Flavours

Off-flavours represent the barrier between the manufacturer and the consumer acceptance. This has been already experienced in the case of soybeans and their products. Off-flavours, particularly in soy flours, soy protein concentrates, and soy protein isolates have prevented them from being exploited to their potential for human consumption. It seems that in the case of soy products, carbonyl compounds are the ones responsible for the distinctively undesirable green-beany, grassy flavours. Of these, ethyl vinyl ketones, n-hexanal, 3-cis-hexenal, and n-pentyl furan have been identified as the most important ones for objectionable flavours. Data on other legumes are meagre and more systematic studies are needed in this area. The identification of compound(s) responsible for the characteristic aroma of the foodstuff can be a complicated exercise (Parliment and Scarpellino 1977). This is because of the

TABLE 2.14. Volatile compounds identified in legumes

Legume	Processing condition(s)	Volatile compounds identified
1	2	3
Green peas (*Pisum sativum* L.)	Unblanched, frozen	Methanol, Acetaldehyde, Ethanol, *trans, trans*-Hexa-2,4-dienal, Propanol, *trans*-Hex-2-enal, Butanol, Butan-2-ol, Pentanal, 2-Methylpropanol, Octanal, Pent-1-en-3-ol, Nonanal, *trans*-Pent-2-enol, Butan-2,3-dione, *cis*-Pent-2-enol, Acetic acid, Pentanol, γ-Hexanolactone, Pentan-2-ol, *trans, cis*-Hepta-2 4-dienal, Pentan-3-ol, *trans, trans*-Hepta-2,4-dienal, 2-Methylbutanol, *trans*-Hept-2-enal, 3-Methylbutanol, Heptanal, 2-Methylbutan-2-ol, *trans*-Oct-2-enal, *trans*-Hex-2-enol, *trans, cis*-Nona-2,4-dienal, *trans*-Hex-3-enol, *trans, trans*-Nona-2,4-dienal, *cis*-Hex-3-enol, *trans, cis*-Nona-2,6-dienal, Hexanol, *trans*-Non-2-enal, 4-Methylpentan-2-ol, *trans, cis*-Deca-2,4,-dienal, *trans*-Hept-2-enol, *trans, trans*-Deca-2,4-dienal, Heptanol, Benzaldehyde, Heptan-2-ol, Phenylacetaldehyde, Oct-1-en-3-ol, *trans*-Oct-2-enol, Octanol, Acetone, Octan-2-ol, *trans, cis*-Octa-3,5-dien-2-one, Octan-3-ol, *trans, trans*-Octa-3,5-dien-2-one, *cis*-Non-3-enol, A third octa-3,5-dien-2-one, Ethyl acetate, Oct-1-en-3-one, Ethyl hexanoate, *trans*-Oct-3-en-2-one, Methyl heptanoate, Octan-3-one, Octan-3-one, Ethyl heptanoate, 6-Methylheptan-2-one, Methyl octanoate Nonan-2-one, Propyl hexanoate, Undecan-2-one, Ethyl octanoate, Acetophenone, Methyl benzoate, *o*-Methylacetophenone, Ethyl benzoate, 3,5,5-Trimethylcyclohex-2-en-1-one (isophorone), Butyl benzoate, Ethyl phenylacetate, Ethyl cinnamate, Undecane, Dodecane, Tridecane, Tetradecane, Pentadecane, C$_{11}$-C$_{15}$ singly branched saturated, C$_{11}$-C$_{15}$ straight chain mono-unsaturated, Dimethylethylbenzenes (3), *m*- or *p*-Diethylbenzene, A methylindene, A methylindan, A dimethylindan, Naphthalene, 1-Methylnaphthalene, 2-Methylnaphthalene, Dimethylnaphthalenes (3), Diphenyl, Nonanal, Benzeneacetonitrile, Methyl formate, 1,2-Dimethoxybenzene; Ethyl formate, 2,3-Benzothiophene, Methyl 2-methylbutanoate, Methyl-2,3-benzothiophenes, Methyl 3-methylbutanoate, An ethylbenzothiophene, Ethyl 2-methylbutanoate, Dimethylbenzothiophenes (2), Ethyl 3-methylbutanoate, Diethyl-acetal β-Cyclocitral, 2-(or 4-) Methylpyridine, Linalool,-Terpineol, 3-Isopropyl-2-methoxypyrazine, 1,8-Cineole, 3-*sec*-Butyl-2-methoxypyrazine, -Ionone, 3-Isobutyl-2-methoxy-pyrazine, β-Ionone

Peas	Unblanched, frozen, stored (2 yr, −17.8°C)	Propane-2-one, Ethanal, Propanal, Hexanal, Pent-2-enal, Hex-2-enal, Hept-2-enal Oct-2-enal, Non-3-enal, Hept-2,4-di-enal, Non-2,4-dienal, Dec-2,4-dienal
Runner beans (*Phaseolus coccineus* L.)	Fresh	Methanethiol, Acetaldehyde, Dimethyl sulfide, Propionaldehyde, Acetone, n-Butaraldehyde, Ethyl methyl ketone, Methyl alcohol, Ethyl alcohol, Diethyl ketone, But-2-en-1-al, Dimethyl disulfide, n-Hexanal, Allyl alcohol, Dipropyl ketone, Trans-pent-2-en-1-al, Allyl cyanide, Trans-but-2-en-1-ol, Methyl propyl disulfide, Cis-pent-3-en-1-ol, Butyl methyl disulfide or propyl isothiocyanate, Cis-hex-3-en-1-ol, Butyl isothiocyanate, Trans-hept-3-en-1-ol
	Frozen	All above compounds with following exceptions. Acrolein, Trans-hex-2-en-1-al, and Dipropyl disulfide were additionally present, while Trans-hept-3-en-1-ol, Butyl isothiocyanate, Cis-hex-3-en-1-ol, Cis-pent-2-en-1-ol, Trans-but-2-en-1-ol, were absent.
Bush snap beans (*Phaseolus vulgaris* L.)	Fresh, frozen, canned	Acetaldehyde (traces), ethanol, 3-pentanone, n-hexanol, cis-3-hexanol, trans-2-hexanol (traces), 1-octen-3-ol, linalool
Dry red beans (*Phaseolus vulgaris* L.)	Vacuum distilled oil	Oct-cis, 5-en-2-one, Oct-3,5-dien-2-one, 2-Methylbutanol, 3-Methyl butanol, Hexanol, Hex-cis, 3-enol, Octanol, Octan-3-ol, Oct-1-en-3-ol, Oct-cis, 5-en-2-ol, Nonanol, 2,5-Dimethyl pyrazine, Benzaldehyde, 2-Phenylacetaldehyde, Benzene, Toluene; Xylene, p-Cymene, Trimethylbenzene, Naphthalene, 1-and 2-methylnaphthalene, Dimethyl-naphthalene, Lineonene, γ-Terpinene, α-Terpineol.
	Atmospheric steam volatile oil	In addition to the volatiles listed for vacuum distilled oil following volatiles were identified, additionally, 2-Pentyl furan, 2-Methyl-3-keto-tetrahydrofuran, furfural, 2,6-Dimethyl pyrazine, 2,3-Dimethyl pyrazine, Pyridine, 2-Methyl pyridine, 2-Methyl-5-ethylpyridine, 2-Acetylpyridine, 2,4,5-Trimethyl-thiazole, 2,5-Dimethyl-4-ethylthiazole, 2,4-Dimethyl-5-ethylthiazole, 2-Isopropyl-4,5-dimethylthiazole, 2-Isobutyl-4,5-dimethylthiazole, 2-Acetylthiazole, 2,4,5-Trimethyl-2-thiazoline, Thialdine (2,4,6-Trimethylperhydro-1,3,5-dithiazine), 3,5-Dimethyl-1,2,4-trithiolane (two isomers), p-Vinylguaiacol (2-Methoxy-4-vinyl-phenol), and Benzyl alcohol.
Lima beans (*Phaseolus lunatis* L.), Common beans (*Phase-*	Headspace analysis	Methanol, Ethanol, 2-Propanol, 1-Propanol, 2-Butanol, 1-Butanol, 2-Methylpropanol, Acetaldehyde, Acetone, Methylpropanal, 2-Butanone, Methylbutanal, 2-Pentanone, 1-Pentanol, Hexanal, 1-Hexanol, Ethylbenzene, o-Xylene, Styrene, Benzaldehyde, (*Contd.*)

1	2	3
olus vulgaris L.), Lentils (*Lens culinaris*), Mung beans (*Phaseolus aureus* L.), Split peas (*Pisum sativus* L.)		Dodecane, Phenylacetaldehyde, Acetophenone, Diethyl ether, Dimethyl sulfide, Carbon disulfide, Methyl acetate, Chloroform, Benzene, Pentanal, Cumene, Dichlorobenzene, Acetonitrile, Toluene, Trichlorobenzene, Naphthalene, Methylnaphthalene.
Peanuts (*Arachis hypogaea* L.)	Roasting (Basic fraction)	Pyridine, Pyrazine, 2-Methylpyrazine, 2,5-Dimethylpyrazine, 2,6-Dimethylpyrazine, 2-Ethyl-pyrazine, 2-3-Dimethylpyrazine, 2-Ethyl-6-methylpyrazine, 2-Ethyl-5-methylpyrazine, 2-Ethyl-3-methylpyrazine, Trimethylpyrazine, 2,5-Dimethyl-3-ethylpyrazine, 2,3-Dimethyl-5-ethylpyrazine, 2,6-Dimethyl-3-ethylpyrazine, 2,6-Diethyl-3-methylpyrazine or 2-3-Diethyl-3-methylpyrazine, 2-Ethyl-3,5,6-trimethylpyrazine, Methyl-2,3-cyclopentanopyrazine, 2-Isopropenylpyrazine, Methylisopropenylpyrazine.
	Raw	Pentane, Methyl formate, Octane, Acetaldehyde, 2-Butanone, Acetone, Methanol, Ethanol, Pentanal, Hexanal.
	High temperature cured off-flavour	Formaldehyde, Acetaldehyde, Ethanol, Acetone, Isobutyraldehyde, Ethyl acetate, Butaraldehyde, Isovaleraldehyde, 2-Methyl valeraldehyde, Methyl butyl ketone, Hexaldehyde, 2-Methyl-1-butanol or 3-Methyl-1-butanol, Furfural
Soybeans (*Glycine max* L.)	Oil	Butane, Butene, Pentane, Hexane, n-Butane, Butenal, n-Pentanal, 2-Pentenal, Octane, 2-Octene, Octadiene, 2-Hexenal, 2-Heptenal, 2,4-Heptadienal, Octenal, Benzyl alcohol, Nonanal, 2-Decenal, 2,4-Decadienal.
	Raw	Dimethylamine, Acetaldehyde, Acetone, Methanol, Ethanol, n-Hexanal, 2-Pentanol, Pentanol acetate, Isopentanol, n-Pentanol, n-Hexanol, n-Heptanol, Acetic acid, Hydrogen sulfide. Ethanol, 2-Propanone, Glyoxal, Methyl glyoxal, Glyoxylic acid, Pyruvic acid, α-Ketoisocaproic acid, α-Ketoglutaric acid, Benzaldehyde, Protocacechuic aldehyde, Propionic acid, Isovaleric acid, n-Valeric acid, Isocaproic acid, n-Caproic acid, n-Caprylic acid, n-Nonanoic acid, n-Capric acid, Monomethylamine, Dimethylamine, Piperidine, n-Pentanol acetate, Ammonium hydroxide. 3-cis-Hexenal, n-Pentyl furan, Cis-2-(1-pentenyl) furan, Trans-2-(1-pentenyl) furan, Ethyl vinyl ketone.
	Defatted	Methanal, Ethanal, n-Hexanal, 2-Heptanal, 2,4-Decadienal, 2-Propanone, 2-Pentanone,

Soy protein	2-Hexanone, Syringic acid, Vanillic acid, Ferulic acid, Gentisic acid, Salicylic acid, p-Coumaric acid, p-Hydroxybenzoic acid, Isochlorogenic acid, Chlorogenic acid, Methanol, Ethanol, Isopentanol, n-Pentanol, n-Hexanol, n-Heptanol, Acetone, Methanal, Ethanal, Isopentanol, n-Pentanol, n-Hexanol, n-Heptanol, Ethanal, Propanal, n-Hexanal, 2-Propanone
Deep fat fried	2-Methylpropanal, Acetone, Ethanol, 2-Methyl-2-butanol, 2,3-Pentanedione, Dimethyl disulfide, Hexanal, 1-Penten-3-ol, 2-Heptanone, Pyrazine, Pentanol, 2-Methylpyrazine, 2-Octanone, 2,5-Dimethylpyrazine, 2,3-Dimethylpyrazine, 2-Ethyl-5-methylpyrazine, Trimethyl pyrazine, 1-Octen-3-ol, 2-Ethyl-3,6-dimethyl pyrazine, Furfural, 2-Furfuryl methyl ketone, Benzaldehyde, 5-Methyl furfural, Phenylacetaldehyde, Furfuryl alcohol, 2,4-Decadienal, Guaiacol, Acetopyrrole, Pyrrol-2-Carboxaldehyde, 1-Methylpyrrole-2-carboxaldehyde, 4-Vinyl guaiacol, 2,3-Dihydrobenzofuran 5-(pentenyl)-2-furaldehyde, Chloromethylfuraldehyde
Oxidized soy phospholipids	

Sources: Murray et al. 1976; MacLeod and MacLeod 1970; Whitfield and Shipton 1966; Buttery et al. 1975; Toya et al. 1974; Lovegren et al. 1979; Shu and Waller 1971; Pattee et al. 1965; Pattee et al. 1969; Jackson and Giacherio 1977; Cowan et al. 1973; Smith and Circle 1972; Rackis et al. 1979; Wilkens and Lin 1970; Sessa and Plattner 1979.

complexity of interactions involved within and between the food component(s). The problem becomes a jeopardy particularly in the case of protein concentrates and isolates when treatments aimed at removal of off-flavours bring about undesirable changes in proteins leading towards adverse effects on protein functionality and nutritionality. Recently Gardner (1979a) has reviewed extensively the chemical and biochemical aspects of lipid hydroperoxide reactivity with protein and amino acids. The detailed description will be avoided here, however, a brief description would bring about the nature and complexity of the situation one can encounter.

Potentially reactive hydroperoxides and their degradation products arising as a result of lipid oxidation are capable of the deterioration of proteins and amino acids. The consequences of exposure of proteins to peroxidized lipids include the formation of lipid-protein complexes purely through physical forces, chemical changes due to the interaction of lipid hydroperoxide and protein such as protein-protein cross links, protein scission, protein-lipid adducts, and amino acid damage. Secondary products of hydroperoxide decomposition (particularly aldehydes) readily damage proteins and amino acids through the formation of covalent bonds. Malondialdehyde capable of cross-linking proteins via Schiff base formation appears to be important amongst these aldehydes.

The important aspect which is not studied to a significant extent is the possible interaction of oxidized lipids (including the secondary products) and carbohydrates (one of the major constituents of legumes) and their technological importance in legume processing. It remains to be seen how the legume processing can be improved, particularly the commercially unexploited varieties in view of the preceding discussions.

2.3.5. CHANGES IN LIPIDS DURING PROCESSING

At present, the commercial processing of legumes is limited. Soybeans and peanuts are probably the only exceptions, which are processed extensively for the purpose of extraction of oil and/or production of flours, protein concentrates, and protein isolates. The next commercially most important legume in the U.S.A. is probably green peas, which are frozen. Almost all other legumes in the U.S.A. are dried and sold as dry beans. Outside the U.S.A. however, the situation is quite different where legumes are consumed in appreciable quantities and are an inseparable part of the daily diet. In countries such as India, Pakistan, Bangladesh, China, and Africa, legumes are processed, prior to consumption, at home. Such processing includes soaking, cooking, germination, fermentation (particularly in combination with some cereal such as rice in the case of *Idli* in India) etc. Scattered studies have been reported on several individual legumes in a particular processing with respect to changes in lipids.

Rackis et al. (1970) demonstrated that little oxidation of lipids took place during the preparation of raw defatted soybean flakes and was of no practical consequence as regards the off-flavours. It is noteworthy that they found that the beany, bitter, and grassy flavour constituents were present in the last 0.02 per cent lipids which could not be extracted by pentane-hexane but were soluble in aqueous ethanol and hexane-alcohol azeotropes employed to remove them. The storage stability of peanuts and peanut products has been reviewed recently (Schewfelt and Young 1979). These authors implied that cultivation, harvesting, and moisture content are the most critical factors in storage stability. The oxidative changes in lipids are the ones responsible for bitter tastes and off-flavours. Sumner et al. (1979) studied the storage stability of pea flour, protein concentrate and starch. They found that odour of pea flour and protein concentrate stored at the highest moisture levels (13.6 per cent) and temperature (30°C) underwent significant changes. The fresh pea odour at the beginning changed to a musty, fishy odour after 52 weeks. They did not find any objectionable odour development in samples stored at or less than 30°C and 10 per cent moisture. They attributed these off-flavour developments to the enzymatic modifications of the lipids. Although they implied involvement of lipoxygenase, they did not elaborate on responsible modifications caused by lipoxygenase. They remarked, however, "the unpleasant odours were volatile and disappeared rapidly," but did not report what treatment was given, if any, to effect this disappearance. They also analyzed the changes in the total and free fatty acids during storage and found that linolenic acid content in both total and free fatty acids decreased sharply in lipids from the stored pea flour but to a lesser extent in lipids from the pea protein concentrate while a reduction in linoleic acid in free fatty acids was slight in pea flour during storage. They attributed these decreases to oxidation and possible polymerization.

Hinchcliffe et al. (1977) studied the flavour of faba beans as affected by heat and storage. They found that the faba bean flour developed a bitter flavour after one year's storage at ambient temperature. Toya et al. (1974) reported the effects of processing and maturity on volatile components in bush snap beans (*Phaseolus vulgaris* L.) and found that in the concentration of the 17 volatile components in canned, frozen, and fresh green bean pods studied, all the components were greatly reduced or lost in frozen pods as compared to those in fresh ones. In canned pods most of the "higher boiling" compounds were found to decrease, while the concentration of the "lower boiling" components greatly increased. Of particular interest was the finding that the compound 1-Octen-3-ol which they found later to be characteristic of bush snap beans flavour (Toya et al. 1974), increased with time after thawing of the frozen samples while very little changes were observed in concentration of other compounds. They neither identified the precursors of these volatiles nor studied the involvement of lipids. Murray et al. (1976) carried out a detailed study of off-flavour compounds in unblanched peas. They identified several classes of compounds in peas as well as pea shells including alcohols, carbonyls, esters, hydrocarbons, terpenes, and 3-alkyl-2-methoxypyrazines. They concluded that most of the strong odours were associated with alkanals, alka-2-enals, alka-2, 4- and 2, 6-dienals, octa-3, 5-dien-2-ones, 3-alkyl-2-methoxypyrazines, and hexanol and remarked "no compound, which, by itself, could account for the 'haylike' off-flavour of peas which was evident". They also reported that all the compounds identified in that study were the result of the degradation of the unsaturated pea lipids. They implicated that the most likely compounds responsible for the off-flavours in peas were the mono- and di-unsaturated carbonyls and saturated and mono-unsaturated alcohols as a result of lipoxygenase and alcohol dehydrogenase activity. Other studies on pea lipids degradation have also been reported and we will include here only few of them to represent several processing and/or handling conditions and their effect on lipid oxidation or lipid derived off-flavours. Rhee and Watts (1966) studied the lipid oxidation in frozen vegetables in relation to the flavour change. They concluded that the rancidity was not the cause of off-flavours in blackeye peas (*Vigna sinensis* L.) and blanching inactivated the lip-oxidase activity which could not be regenerated during the frozen storage of garden peas (*Pisum sativum* L.). Bengtsson and Bosund (1966) studied the lipid hydrolysis in unblanched frozen peas. They investigated the effects of several parameters such as temperature, oxygen, and pea maturity on the rate of hydrolysis of total lipids as well as of individual classes of lipids within the temperature range of -5 to $-20°C$. Their findings include: a) Q_{10} value for the formation of free fatty acids to be about 2.5 with corresponding value for off-flavour development to be about 3.0, b) increased rate of formation of free fatty acids with increase in temperature, c) no significant difference in hydrolysis rate of neutral fat and phospholipids, d) inositol phosphatides to be more resistant towards hydrolysis than lecithin and cephalin phosphatides, e) apparent preferential hydrolysis of PUFA, and f) no net change in any single acid in combined lipids. Whitfield and Shipton (1966) reported the carbonyls of unblanched stored frozen peas using thin-layer partition chromatography, infra-red, ultraviolet, and visible spectra of 2, 4-dinitrophenyl hydrazones. They found ethanal to be the major carbonyl responsible for flavour of stored peas (96 per cent of the total volatiles). Other identified compounds included propanal, hexanal, propan-2-one, pent-2-enal, hex-2-enal, hept-2-enal, oct-2-enal, non-3-enal, and hept-, non-, and dec-2, 4 dienals and attributed their origin to lipid oxidation.

Lee and Mattick (1961) also investigated the changes occurring in the free fatty acids of peas stored for 1 year at $-17.8°C$ in the raw and enzyme-inactivated conditions. They found that there were substantial losses in the free fatty acids in the phospholipid fraction of raw peas as opposed to in the enzyme-inactivated ones. The neutral fat of the same samples showed losses for all the unsaturated fatty acids in raw peas as compared to enzyme-inacti-vated peas. The net gain in palmitic acid was thought to be at the cost of three C_{18} unsaturated fatty acids. Recently Gardner and Sessa (1977) reviewed the degradation of fatty acid hydroperoxides by cereals and legumes. They stat-ed, "Recent work with soybeans and other legumes indicates an absence of linoleic acid hydroperoxide isomerase (LAHI)", and proposed that in soy and perhaps in other legumes the enzymes acting on hydroperoxides are per-oxidases and other metalloproteins like lipoxygenase and cytochromes. They further concluded that there was an unusually good correlation in the pro-ducts isolated from the legumes, soy and pea versus from the chemical models.

Takayama et al. (1965) studied the lipid composition of several legumes including several type of beans and Alaska peas in an attempt to find a cor-relation with cooking time but did not get any significant correlation between the lipid content and the cooking time. Hydar and Hadziyev (1973) have studied the pea lipids' oxidation on protein and carbohydrate matrices and es-sentially concluded that lipid polarity, rather than its unsaturation, predomi-nantly controlled the lipid oxidation. They further stated that the rate of oxidation was influenced by the physical and chemical interactions of lipids and the matrices (carbohydrates and proteins). Besides these, they have re-ported other interactions emerging from the same study and interesting fea-tures are summarized here: (a) The amylopectin-neutral lipid system retarded the lipid oxidation; (b) Amylose did not form inclusion complexes with neu-tral lipids; (c) On the contrary, amylose did form inclusion complexes with polar lipids as suggested by the decrease in the rate of oxidation of polar lipids on amylose matrix; (d) Pectin matrix enhanced rate of oxidation of polar lipids but retarded the same for neutral lipids; (e) Effect of protein ma-trix on neutral lipids oxidation was relatively small as compared to on polar lipids; and (f) Globulins promoted lipid oxidation while albumins retarded it. The finding (f) is in contrast to the previous report (Togashi et al. 1961). Since albumins and particularly globulins are the major proteins in legumes, inves-tigations on the role of albumins and globulins in lipid oxidation would pro-vide valuable information.

The bulk of the information on off-flavour technology comes from the studies primarily in soybeans. Very little or no information is available on effects of processings on legume lipids. We will include few examples here to illustrate the reported changes in lipids.

Rabari et al. (1961) reported the changes in lipids during the germina-tion in peanut seeds and showed that the lipids were converted into carbohy-drates (soluble and insoluble glucosides). They found that saturated fatty acids were metabolized at a greater rate than unsaturated fatty acids. Beuchat and Worthington (1974) investigated the changes in the lipid content of fermented peanuts and found that the free fatty acid fraction of fermented peanuts was lower in linoleic and higher in saturated acids in total lipids over the unfermented counterpart. They indicated on the basis of this observation, that lipase activity was specific for 1, 2 positions of triglycerides and also noted that no preferential utilization of free fatty acids was observable during the course of fermentation. They suggested that the increase in fatty acids might increase the digestibility of peanuts thereby increasing their nutritional value.

Numerous studies have been reported on effects of processing on soybean lipids. Wilkens and Lin (1970) isolated, fractionated, and identified the volatile components of deep-fat fried soybeans and found several carbonyls, aromatic compounds, pyrazine, and pyrrole derivatives in volatiles. Of particular significance was 2, 4-decadienal which they described to arise from the auto-oxidation of linoleic acid of soybean oil as well as corn oil (used for deep fat frying) which has been previously shown to impart desirable deep fatty flavour to a food (Mookerjee et al. 1965). Large quantities of carbonyls and alcohols found in raw soybeans (Mattick and Hand 1969), a majority of which have been described to impart undesirable "raw" or "beany" flavours were present in small amounts in the volatiles of deep fat fried soybeans with the exceptions of 2, 4-decadienal and 1-octen-3 ol in their investigation. Other important volatiles which were identified included aldehydes such as phenyl acetaldehyde, 4-vinyl guaiacol (from ferulic acid's thermal degradation), furfural, 5-methyl furfural, 2-furfuryl methyl ketone, furfuryl alcohol (from carbohydrate degradations), several pyrazine and pyrrole derivatives, and ammonia (probably from proteins). Honig et al. (1976) studied the effect of toasting and solvent extractions (hexane : ethanol, water, calcium hydroxide) in an attempt to improve the nutritionality and flavour of defatted soy flakes, protein concentrates, and protein isolates. They found that toasting and azeotropic extraction improved the flavour scores of soybean flour protein concentrates due to the reduction in grassy/beany, bitter, and astringent flavour components. Toasted flakes, together with azeotropic extraction, did not increase the flavour score of protein isolates. They also found, in the case of protein isolates, that grassy/beany flavour intensity had reduced. However, a distinct bitter flavour was retained. The bitter flavour (taste) in soybeans is attributed to the auto-oxidation degradation products of fatty acids attached to phosphatidylcholine in soybeans (Kalbrener et al. 1971; Kalbrener et al. 1974; Sessa et al. 1969; Sessa et al. 1974; Sessa et al. 1976; Sessa et al. 1977). Cowan et al. (1973) have reviewed the problems associated with soy protein products as regards the off-flavour development in details. Other reviews of interest which deal with off-flavour problems associated with soybean products in particular and vegetable products (related to legumes primarily) in general include those by Maga (1973); Wolf (1975); Nawar (1969); Forss (1969); Eriksson (1975); Rackis et al. (1979); and Gardner (1979 a). Recently Gardner (1979b) has reviewed lipid enzymes in detail. This review is particularly important in understanding the chemistry and biochemistry of the oxidation of lipids as it has taken into account the latest developments in the recent past in this area. Attempts are also being made to understand the chemistry of volatile products arising from severe heat processing of vegetable oils in model systems (Thompson et al. 1978).

2.3.6. REMOVAL OF OFF-FLAVOUR

The removal of off-flavour necessitates the knowledge of the formation and physico-chemical properties of the compounds responsible. The level of the removal of off-flavours will be partially dependent on the end use of the product as well. The situation can be very complex and demanding in terms of technological skills particularly when the same starting material has to be processed into several fractions for diverse end uses. This type of situation can be envisaged in case of soybeans, peanuts, peas, etc. The raw materials (e.g. soybeans) are processed for oil, flour (full fat and defatted), protein concentrates, protein isolates, etc. where each fraction may undergo different end uses such as oil for frying, protein concentrate and isolates for fortification, extension, or nutrification of other flours such as wheat flour, or may be used as functional ingredients in different foods. In such a situation, the problem can be "sticky". For example, in an attempt to remove the residual fats from protein concentrates or isolates, drastic conditions such as heating at high temperatures or use of denaturing solvents might bring about the irreversible denaturation of proteins resulting in multitudes of changes in their physical and chemical properties which in turn might affect adversely their functional and/or nutritional properties. Thus, depending upon the final use intended, the severity of the treatments to effect the removal of off-flavours will be governed, at least partially. Identification, genesis, and characteristics of the responsible compounds for the undesirable off-flavours certainly constitute the first step toward solution. The knowledge and recognition of degradation products and their interactions with other food constituents is equally important.

The present understanding of the problem is still in a stage of analyzing several factors, and elucidating the underlying mechanisms responsible for the off-flavour developments in legume lipids and legume products. It seems, however, lipolysis and autolytic lipid auto-oxidation resulting in the production of hydroperoxides and their decomposition products, notably carbonyls, are of prime importance in the off-flavour developments. Attempts to alleviate and control these problems can be categorized into two main classes: (a) Physical treatments to remove the off-flavours and/or the substrates (lipids) responsible for their development. Such treatments include primarily the application of heat, dry or moist (steam); controlled storage (reduction in heat thereby reducing the rates of reactions); the control of mechanical treatments (particularly milling); blanching, roasting, toasting, etc.; (b) Chemical and biochemical approach which would include the genetic control of lipids synthesized in the legumes, adjustment in chemical parameters to prevent/ reduce the rates of reactions concerned such as the adjustment in pH, chemical extraction (use of defatting solvents such as ethanol, hexane, butanols, etc.), treatments aimed at the inactivation of enzymes (particularly lipoxygenases) with the help of chemicals and hot treatments like blanching, etc.

Out of these, physical treatments appear to be extensively studied, at least in the case of soybeans and peanuts. These include the application of heat in the form of roasting and steaming to "distill off" the off-flavours although the success of such treatments is limited in spite of several claims made by many workers. The problem associated with this approach is that it cannot remove the non-volatile compounds responsible for the off-flavours, particularly those compounds which are capable of interacting with other food components covalently and are not readily volatile. The other approach is to extract the substrates and/or products involved in off-flavour developments using suitable solvents. The solvents which have been reported primarily include alcohols, hydrocarbons, and their mixtures with or without the application of heat. This approach also suffers the same drawbacks as for physical treatments, in that solubility of the compounds responsible for the off-flavour development governs the success of the process. The residual solvents in view of the possible toxicological effects will definitely pose further limitations on this approach. The inactivation of lipases and lipoxygenases undoubtedly helps in the reduction of off-flavours. The complete understanding of lipoxygenases, especially the mechanism of action and their physiological role in legume biochemistry is yet to be achieved. The approach involving both physical and chemical treatments in the pre- and post-off-flavour developments might be more useful. Another approach, apparently not yet tried, is to chemically inactivate or neutralize or hydrolyze the components responsible for the off-flavour developments, as the masking of off-flavours has been proved to be a difficult proposition. More rigorous and dedicated efforts are needed, as evidenced, to recognize, analyze, identify, and understand all the phases of off-flavour developments associated with legumes and legume products in the near future in order to develop a sound technology for the effective and efficient utilization of these economical oil and protein rich resources.

2.3.7. TECHNOLOGICAL IMPLICATIONS

The effective utilization of legumes as a nutritious but low-cost food for human consumption depends upon the consumer acceptance. Beany, grassy, haylike, bitter, and astringent flavours associated with legumes and legume products is the major barrier today between the manufacturer of such products and the consumers. The knowledge of off-flavour development in such products is essentially inadequate and hence the utilization of legume products for human consumption has not yet been realized to their real potential. Today's technology is unsuccessful in solving these off-flavour problems associated with such products. In view of the increasing cost of living, the newer food resources will have to be utilized efficiently and effectively. Legumes offer in them one such inexpensive nutritious source for human food. The problem of off-flavour is complex and will have to be solved through systematic efforts in the near future in order to effectively utilize legumes to their full potential.

2.4. Nutritional and Health Importance of Legume Lipids

The significance of essential fatty acids in nutrition was known long ago. In 1929, Burr and Burr described a deficiency syndrome of fatty acids in rats by feeding a fat-free diet. The signs of deficiency included scaliness of skin, reduced growth, necrosis of tail, degeneration of kidneys, failure to reproduce, and a typical eye condition. This deficiency syndrome was traced back to the deficiency of a single fatty acid: linoleic acid (Burr and Burr 1930), and possibly α-linolenic acid.

Specific deficiency symptoms of α-linolenic acid have not been found in humans or rats, but recent studies with rats indicate that α-linolenic acid and its derived 20- and 22-carbon chain unsaturated fatty acids have a specific role in the development and function of the brain (Lamptey and Walker 1976) and retina (Benolken et al. 1973). Due to its physiological role, α-linolenic acid has been suggested to be an essential fatty acid. However, more research is needed in clarifying the role of α-linolenic acid and its derived fatty acids in human nutrition. Animals and humans cannot synthesize linoleic and α-linolenic acids. The metabolic degradation of linoleic acid and α-linolenic acid in humans and/or animals produces 20- and 22-carbon chain unsaturated fatty acids such as arachidonic and docos-ahexaenoic acids. These two acids are required for normal growth, cell structure, functions of all tissues, and prostaglandin synthesis (Lehninger 1975; FAO WHO/Report 1977). The studies conducted by Mohrhauer and Holman (1963) indicated that the linoleic and arachidonic acids are required for normal growth and physiological functions such as normal tissue and lipids synthesis, in rats. The unsaturated fatty acids are used for the esterification of cholesterol and subsequently reduce the cholesterol content in the serum and liver of rats (Alfin-Slater 1957). In the absence of essential fatty acids such as linoleic and α-linolenic acids, the cholesterol is esterified with more saturated fatty acids and tends to accumulate in the blood of the artery, reduce the metabolism of cholesterol, and result in a hardening of the inner walls of the artery—Atherosclerosis (Fig. 2.6). The role of lipids in atherosclerosis has been reviewed by Kummerow (1975). Linoleic acid, in addition to being an essential fatty acid, is also hypocholesterimic, while palmitic acid is hypercholesterimic (Grande et al. 1970). Despite the effect of fatty acids on obesity, coronary heart disease, and myocardial infraction, Americans consume high amounts of calories (40–60 per cent) from animal and other fats. In the USA and other parts of the world, consumers are shifting from animal fat products to vegetable oil products such as soybean, peanut, and other vegetable oils because of their ability to lower the blood cholesterol level (Kromer 1975). The lipids of legumes have substantial amounts of polyunsaturated fatty acids (Takayama et al. 1965; Sessa and Rackis 1977; Exler et al. 1977). The polyunsaturated fatty acids from a legume also have the ability to lower the serum cholesterol level compared to animal and milk fats (Keys et al. 1960; Dairy Council Digest 1975). Black gram, horse gram and red gram (Devi and Kurup 1970; Devi and Kurup 1972), chick pea (Mathur et al. 1964; Jaya and Venkataraman 1979; Jaya et al. 1979), and green gram (Jaya and Venkataraman 1979; Jaya et al. 1979) have been shown to lower the cholesterol levels in serum, liver, and the aorta of rats. This lowering has been attributed, in part, to the high content of polyunsaturated fatty acids such as linoleic and linolenic acids, which are present in higher amounts in legumes (Mahadevappa and Raina 1978; Exler et al. 1977; Baker et al. 1961). Studies with rabbits by Kritchevsky et al. (1973) indicate that the peanut oil is atherogenic.

During frying and heating of oils, many oxidative and hydrolytic changes occur which decrease their functional, sensory and nutritional qualities (Melnick 1957; Rice et al. 1960). For example, a liquid fat high in polyunsaturated fatty acids will deteriorate more rapidly than a hardened animal fat. Alfin-Slater et al. (1959) studied the nutritional evaluation of some heated fats such as cottonseed oil, soybean oil and lard by using rats. Soybean oil was heated *in vacuuo* for either 70 or 100 min at 321°C and was fed to male and female rats at a level of 15 per cent in the diet over a complete life span. They measured nutritional indices such as growth, including food digestibility, reproduction and lactation, longevity, tissue cholesterol, and total lipid content in rats and found no evidence of the impaired nutrition or harmfulness. Isaacks et al. (1963) reported that the diets containing up to 30 per cent of soybean oil, resulted in increased growth and feed efficiency in the chick. Thomasson et al. (1966) found no abnormal changes in rats after feeding them with large amounts of hydrogenated soybean and linseed oils during a 12-week period.

Crampton et al. (1951) demonstrated that severely heat treated soybean and linseed oils at levels of 10 or 20 per cent in rat diets, reduced rates of weight gain and decreased caloric intake. The damages were proportional to the duration of the heat treatment. They also observed that the appetite was depressed when thermally polymerized oils were fed and the faeces of the animals were dark and sticky, hair coats were oily and matted, whereas the controls were sleek and clean. Reporter and Harris (1961) found a diet of

18% oxidized soybean oil to cause enlarged kidneys, growth depression, and severe diarrhoea in rats.

Nolen et al. (1967) investigated the effect of heat and aeration of frying on the nutritional properties of fats. Partially hydrogenated soybean oil was used for frying under the practical restaurant type frying conditions until it was unfit for human consumption. These thermally oxidized fats were fed to groups of 50 male and 50 female rats at a level of 15 per cent in the diet for two years. They found no adverse effects pathologically or clinically except for a slight decrease in the absorption of fried fats compared to un-heated control fats. They also isolated distillable non-urea adductable fractions from the fried soybean oil which proved to be toxic when large doses were administered to weaning rats. The fried oils themselves produced no appreciable deleterious effects on animals when consumed in small quantities. The harmlessness of properly heated fats has got further support from Nolen's (1973) study, who fed dogs with diets containing 15 per cent fresh partially hydrogenated soybean oil or 15 per cent of a partially hydrogenated soybean oil previously used for commercial deep-fat frying. The male dogs fed on the diets with used fried fat grew about the same as those fed commercial dog feed, but both groups had a reduced growth compared to dogs fed on the diets with fresh fat. They found no apparent differences in the growth of female dogs fed on all these diets. The reduced growth rate for male dogs was attributed to the lower absorption of used fried fat compared to fresh. Clinical observations on dogs revealed no adverse effects and there were no abnormalities on histopathological examination of the organs. Artman (1969) concluded in his review that the fats heated in normal cooking processes are not harmful to humans.

2.5. Toxicology of Legume Lipids

The naturally occurring toxicants in legume lipids are either present at low levels or are absent. Most of the toxicities of fats result from heating at low or high temperatures in the presence of air. During deep-fat frying, oxidative and thermal decompositions in oils may take place with the formation of volatile and non-volatile decomposition products (Chang et al. 1978). Some of the volatile and non-volatile compounds are in excessive amounts and are harmful to human health.

Raju et al. (1965) studied the nutritive value of heated peanut oil. They reported that the peanut oil heated for 8 hours at 270°C produced growth depression, fatty livers, elevated levels of glucose and cholesterol in blood, and disruption in the digestion and absorption of carbohydrates in rats.

The heating of a fat in the presence of air results in three alterations in fats: (1) Peroxides of unsaturated fatty acids are formed; (2) Peroxides decompose to carbonyl and hydroxy acids; and (3) Partially oxidized fats polymerize (non-urea adduct forming fraction). This non-urea adduct forming fraction contains mainly cyclized monomeric acids which have been shown to cause acute toxicity at a level of 2.5 per cent in the diet in weaning rats (Crampton et al. 1951; Kummerow 1962). Intravenous injection of peroxides has also been shown to cause an acute toxicity in rats and chickens (Kummerow 1962). O'Brien and Frazer (1966) studied the effects of lipid peroxides on the biochemical components of the cell. They reported that lipid hydroperoxides cause severe damage to cytochrome C by interacting with it bringing about significant interferences in the mitochondrial electron transport system.

Nolen (1972) investigated the reproductive and teratological effects in rats by feeding them with fresh hydrogenated or deep fried hydrogenated soybean oil at a level of 15 per cent in the diet until third pregnancy. He sacrificed all the female rats during the third pregnancy to examine for skeletal, soft tissue, and embryonic abnormalities on the reproductive parameters or any teratogenic effects due to either fresh hydrogenated or deep fried soybean oil for 56 hours and found no deleterious effects.

Andrews et al. (1960) investigated the toxicity of air oxidized soybean oil. They indicated that the growth depression in rats fed with oxidized soybean oil was proportional to the extent of oxidation from the peroxide numbers of 100 to 1200. Using cupric and ferric ions as catalysts, they oxidizced soybean oil by aeration at 60°C to peroxide numbers as high as 1200. When rats were fed with 1200 peroxide number fat at a level of 20 per cent in the diet caused immediate severe diarrhoea, sustained losses in weight and deaths of all animals within three weeks. The dilution of this product with fresh soybean oil to give mixtures with peroxide values of 400 and 800 lessened the severity of symptoms and prevented the deaths at a level of 20 per cent in the diet. The histopathological examination of rats fed with oxidized soybean oil yielded negative results. Much of the toxicity of severely heated vegetable oils has been associated with a non-urea adducting acid fraction which contains cyclized monomeric and dimeric acids (Shue et al. 1968).

Vidyasagar et al. (1974) investigated the nutritive and chemical changes associated with deep fat fried peanut oil at 180°, 220° and 260°C. They evaluated the deep fat fried peanut oils for chemical composition, *in vitro* digestibility, and polyunsaturated fatty acids content. Fresh refined peanut oil contains monomeric material and a very small quantity of polymers but no dimers. The heating of peanut oil causes an increase in the dimers and polymers content but a decrease in the monomers content. After six hours of frying at 180°C, the content of dimers was 11.8 per cent and that of polymers 10.0 per cent (Table 2.15). At 220°C the dimers and polymers contents were 12.5 per cent and 10.0 per cent, respectively. At 260°C, the increase in the content of dimers and polymers was the highest (i.e., 18.6 per cent and 12.2 per cent respectively) compared to other frying temperatures. The heating of

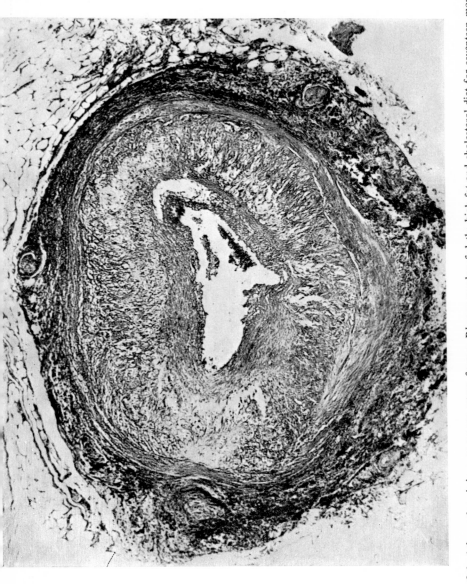

Fig. 2.6. Atherosclerosis in a coronary artery from a Rhesus monkey fed the high fat and cholesterol diet for seventeen months. Note closure of coronary artery by the deposition of lipids.

(Courtesy of Dr. M.L. Armstrong, The University of Iowa, College of Medicine, Iowa City, Iowa)

TABLE 2.15. Percentage of monomers, dimers and polymers in refined groundnut oil
heated at 180, 220, and 260°C for frying purposes

Temp. of frying, °C	Period, hr	Monomers, %	Dimers, %	Polymers, %
180	0	96.7	0.0	3.3
	3	90.7	4.5	4.9
	6	79.2	11.8	10.0
220	3	84.2	9.8	6.0
	6	77.5	12.5	10.0
260	3	81.4	11.0	7.2
	6	69.2	18.6	12.2

Source: Vidyasagar et al. (1974).

oils also causes a decrease in the pancreatic digestibility of fats and polyunsaturated fatty acids compared to unheated fresh peanut oil (Vidyasagar et al. 1974).

2.6. Summary and Conclusions

The information on the lipids of legumes is presented in five sections: introduction, chemistry of legume lipids, technology of legume lipids, nutritional and health importance of legume lipids, and toxicology of legume lipids. The food legumes, including soybean and groundnut, have a total lipid content between 1 and 48 per cent. The legume lipids contain appreciable amounts of polyunsaturated fatty acids including essential fatty acids as well as some saturated fatty acids.

Lipid oxidation products such as carbonyls are capable of interacting with other food components such as proteins, carbohydrates, minerals, and vitamins. These interactions may contribute to undesirable changes in foods during processing and storage. These undesirable changes include a loss of nutritionality and functionality and an off-flavour development. The off-flavour development in processed legumes is the major barrier between the processors and consumers. Today's technology is not yet successful in solving the off-flavour problems associated with processed legumes. The elucidation of underlying mechanisms involved in the formation of off-flavours is important in the effective utilization of inexpensive nutritious legumes.

Legume lipids are effective in lowering the serum and liver cholesterol levels. Legume lipids do not contain natural toxicants but most of the lipid toxicities occur due to the heating of lipids at high temperatures.

Major information on legume lipids is derived from soybeans, peanuts, and peas. However, very little is known about the lipids of other legumes.

References

Alfin-Slater, R.B. 1957. Newer concepts of the role of essential fatty acids. *J. Am. Oil Chem. Soc.* **34**: 574–578.

Alfin-Slater, R.B., Auerbach, S. and Aftergood, L. 1959. Nutritional evaluation of some heated oils. *J. Am. Oil Chem. Soc.* **36**: 638–641.

Allen, C.F., Good, P., Mollenhauer, H.H. and Totten, C. 1971. Studies on seeds. IV. Lipid composition of bean cotyledon vesicles. *J. Cell. Biol.* **48**: 542–546.

Andrews, J.S., Griffith, W.H., Mead, J.F. and Stein, R.A. 1960. Toxicity of air-oxidized soyabean oil. *J. Nutr.* **70**: 199–210.

Anstis, P.J.P. and Friend, J. 1974a. Lipoxygenase in dark-grown *Pisum sativum*. *Phytochem.* **13**: 567–573.

Anstis, P.J.P. and Friend, J. 1974b. The isoenzyme distribution of etiolated pea seedling lipoxygenase. *Planta* **115**: 329–335.

Appleqvist, L. 1975. Biochemical and structural aspects of storage and membrane lipids in developing oil seeds. In: *Recent Advances in the Chemistry and Biochemistry of Plant Lipids*. T. Galliard and E.I. Mercer, eds. pp. 247–270. Academic Press, New York, NY.

Arens, D., Seilmeier, W., Weber, F., Kloos, G. and Grosch, W. 1973. Purification and properties of a carotene co-oxidizing lipoxygenase from peas. *Biochim. Biophys. Acta* **327**: 295–305.

Artman, N.R. 1969. The chemical and biological properties of heated and oxidized fats. In: *Advances in Lipid Research*. R. Paoletti and D. Kritchevsky, eds. Vol. 7, pp. 245–285. Academic Press, New York.

Aykroyd, W.R. and Doughty, J. 1964. Legumes in human nutrition. In: *Food and Agriculture Organization of the United Nations*. No. 19, p. 43. Rome, Italy.

Baker, B.E., Papaconstantinou, J.A., Cross, C.K. and Khan, N.A. 1961. Protein and lipid constitution of some Pakistani pulses. *J. Sci. Food Agr.* **12**: 205–207.

Banks, A., Eddie, E. and Smith, J.G.M. 1961. Reactions of cytochrome-C with methyl linoleate hydroperoxide. *Nature* **190**: 908–909.

Bengtsson, B. and Bosund, I. 1966. Lipid hydrolysis in unblanched frozen peas (*Pisum sativum*). *J. Food Sci.* **31**: 474–481.

Benolken, R.M., Anderson, R.E. and Wheeler, T.G. 1973. Membrane fatty acids associated with the electrical response in visual excitation. *Science* **182**: 1253–1254.

Beuchat, L.R. and Worthington, R.E. 1974. Changes in the lipid content of fermented peanuts. *J. Agr. Food Chem.* **22:** 509–512.

Bull, H.B. and Breese, K. 1967. Denaturation of proteins by fatty acids. *Arch. Biochem. Biophys.* **120:** 309–315.

Burr, G.O. and Burr, M.M. 1929. A new deficiency disease produced by the rigid exclusion of fat from the diet. *J. Biol. Chem.* **82:** 345–367.

Burr, G.O. and Burr, M.M. 1930. The nature of the role of fatty acids essential in nutrition. *J. Biol. Chem.* **86:** 587–621.

Buttery, R.G., Guadagni, D.G. and Ling, L.C. 1975. Geosmin, a musty off-flavour of dry beans. *J. Agr. Food Chem.* **24:** 420–421.

Buziassy, C. and Nawar, W.W. 1968. Specificity in thermal hydrolysis of triglycerides. *J. Food Sci.* **33:** 305–307.

Chan, H.W.S. 1973. Soybean lipoxygenase: An iron containing dioxygenase. *Biochim. Biophys. Acta* **327 (1):** 32–35.

Chang, S.S., Peterson, R.J. and Ho, C.T. 1978. Chemical reactions involved in the deep fat frying of foods. *J. Am. Oil Chem. Soc.* **55:** 718–727.

Choudhury, K. and Rahman, M.M. 1973. Fatty acids in different pulses produced and consumed in Bangladesh. *J. Sci. Food Agr.* **24:** 471–473.

Christopher, J. and Axelrod, B. 1971. On the differential positional specificities of peroxidation of linoleate shown by two isozymes of soyabean lipoxygenase. *Biochem. Biophys. Res. Commun.* **44:** 731–736.

Christopher, J.P., Pistorius, E.K. and Axelrod, B. 1972. Isolation of a third isozyme of soyabean lipoxygenase. *Biochim. Biophys. Acta* **284:** 54–62.

Cowan, J.C., Rackis, J.J. and Wolf, W.J. 1973. Soyabean protein flavour components: A review. *J. Am. Oil Chem. Soc.* **50:** 426A–444A.

Crampton, E.W., Common, R.H., Farmer, F.A., Wells, A.F. and Crawford, D. 1951. Studies to determine the nature of the damage to the nutritive value of some vegetable oils from heat treatment. III. The segregation of toxic and non-toxic material from the esters of heat-polymerized linseed oil by distillation and by urea adduct formation. *J. Nutr.* **49**: 333–346.

Crossley, A., Heyes, T.D. and Hudson, J.F. 1962. The effect of heat on pure triglycerides. *J. Am. Oil Chem. Soc.* **39:** 9–14.

Dairy Council Digest. 1975. The biological effects of polyunsaturated fatty acids. **46:** 3.

Davidek, J. and Jirousova, J. 1975. Formation of volatile compounds and brown products in the model system: n-hexanal-glycine. *Z. Lebensm. Unters. Forsch.* **159:** 153–159.

DeGroot, J.J.M.C., Garssen, G.J., Vligenthart, J.F.G. and Bolding, J. 1973. The detection of linoleic acid radicals in the anaerobic reaction of lipoxygenase. *Biochim. Biophys. Acta* **326:** 279–284.

Denisov, E.T. and Emanuel, N.M. 1960. Catalysis by metals of variable valency in reactions of liquid-phase oxidation. *Uspekhi Khim.* **29:** 1409–1438.

Desnuelle, P. and Savary, P. 1963. Specificities of lipases. *J. Lipid Res.* **4:** 369–384.

Devi, K.S. and Kurup, P.A. 1970. Effects of certain Indian pulses on the serum, liver, and aortic lipid levels in rats fed a hypercholesterolaemic diet. *Atherosclerosis* **11:** 479–484.

Devi, K.S. and Kurup, P.A. 1972. Hypolipidaemic activity of *Phaseolus mungo* L. (black gram) in rats fed a high-fat-high-cholesterol diet. *Atherosclerosis* **15:** 223–230.

Egmond, M.R., Vliegenthart, J.F.G., and Bolding, J. 1972. Stereospecificity of the hydrogen abstraction at carbon atom n-8 in the oxygenation of linoleic acid by lipoxygenases from corn germs and soyabeans. *Biochem. Biophys. Res. Commun.* **48:** 1055–1060.

Eley, C.P. 1968. Food uses of soy protein. In: *Marketing and Transportation Situation*, ERS-388, p. 27.

Eriksson, C.E. 1975. Aroma compounds derived from oxidized lipids: Some biochemical and analytical aspects. *J. Agr. Food Chem.* **23:** 126–128.

Eriksson, C.E. and Svensson, S.G. 1970. Lipoxygenase from peas, purification and properties of the enzymes. *Biochim. Biophys. Acta* **198:** 449–459.

Esskin, N.A.M., Grossman, S. and Pinsky, A. 1977. Biochemistry of lipoxygenase in relation to food quality. *Crit. Rev. Food Sci. Nutr.* **9** : 1–40.

Exler, J., Avera, R.M. and Weihrauch, J.L. 1977. Comprehensive evaluation of fatty acids in foods. XI. Leguminous seeds. *J. Am. Diet. Assoc.* **71:** 412–415.

FAO/WHO Report. 1977. Dietary fats and oils in human nutrition. *FAO*. Rome, Italy.

Finazzi-Agro, A., Arigliano, L., Veldink, G.A., Vliegenthart, J.F.G. and Bolding, J. 1973. The influence of oxygenase as the fluorescence of lipoxygenase. *Biochim. Biophys. Acta* **326:** 462–470.

Fisher, G.S., Legendre, M.G., Lovegren, N.V., Schuller, W.H. and Wells, J.A. 1979. Volatile constituents of Southern pea seed (*Vigna unguiculata* (L.) Walp.) *J. Agr. Food Chem.* **27:** 7–11.

Forss, D.A. 1969. Role of lipids in flavours. *J. Agr. Food Chem.* **17:** 681–685.

Fristrom, G., Stewart, B.C., Weihrauch, J.L. and Posati, L.P. 1975. Comprehensive evaluation of fatty acids in foods. IV. Nuts, peanuts and soups. *J. Am. Diet. Assoc.* **67:** 351–355.

Galliard, T. 1971. The enzymic deacylation of phospholipids and galactolipids in plants. *Biochem. J.* **121:** 379–390.

Galliard, T. 1975. Degradation of plant lipids by hydrolytic and oxidative enzymes. In: *Recent Advances in the Chemistry and Biochemistry of Plant Lipids*. T. Galliard and E.I. Mercer, eds. pp. 319–357. Academic Press, New York.

Galliard, T. and Dennis, S. 1974. Isoenzymes of lipolytic acyl hydrolase and esterase in potato tubers. *Phytochem.* **13**: 2463–2468.

Galliard, T., Philipps, D.R. and Reynolds, J. 1976. The formation of cis-3-nonenal, trans-2-nonenal and hexanal from linoleic acid hydroperoxide by a hydroperoxide cleavage enzyme system in cucumber (*Cucumis sativus*). *Biochim. Biophys. Acta* **441**: 181–192.

Gardner, H.W. 1975. Decomposition of linoleic acid hydroperoxide: Enzymic reactions compared with nonenzymic. *J. Agr. Food Chem.* **23**: 129–136.

Gardner, H.W. 1979a. Lipid hydroperoxide reactivity with proteins and amino acids: A review. *J. Agr. Food Chem.* **27**: 220–229.

Gardner, H.W. 1979b. Lipid enzymes: Lipases, lipoxygenases, and "hydroperoxidases". In: *Proceedings: Basic symposium on food lipids. Institute of Food Technologists 39th annual meeting.* St. Louis, MO.

Gardner, H.W. and Sessa, D.J. 1977. Degradation of fatty acids of hydroperoxide by cereals and a legume: A comparison. *Ann. Technol. Agric.* **26**: 151–159.

Gennis, R.B. and Jonas, A. 1977. Protein-lipid interactions. *Ann. Rev. Biophys. Bioeng.* **6** : 195–238.

Gonzales, J.G., Coggon, P. and Sanderson, G.W. 1972. Biochemistry of tea fermentation: Formation of t-2-hexenal from linoleic acid. *J. Food Sci.* **37**: 797–798.

Grande, F., Anderson, J.T. and Keys, A. 1970. Comparison of effect of palmitic and stearic acids in the diet on serum cholesterol in man. *Am. J. Clin. Nutr.* **23**: 1184–1193.

Grosch, W. 1967. Enzymatische bildung von neutralen carbonyl verbindungen aus den lipoiden der ersbse (*Pisum sativum* var Gottinga). *Z. Lebensm. Unters. Forsch.* **135**: 75–76.

Grosch, W., Hoener, B., Stan, H.J. and Schormueller, Jr. 1972. Preparation of a pure lipoxygenase/GLO (guaiacolinoleic hydroperoxide oxidoreductase) complex from soybeans and its fragmentation into subunits. *Fette. Seifen. Anstrichm.* **74 (1)**: 16–20.

Grosch, W., Laskawy, G. and Weber, F. 1976. Formation of volatile carbonyl compounds and co-oxidation of β-carotene by lipoxygenase from wheat, potato, flax and beans. *J. Agr. Food Chem.* **24**: 456–459.

Hamberg, M. and Samuelsson, B. 1965. On the specificity of the lipoxidase catalyzed oxygenation of unsaturated fatty acids. *Biochem. Biophys. Res. Commun.* **21**: 531–536.

Hamberg, M. and Samuelsson, B. 1967. On the specificity of the oxygenation of unsaturated fatty acids catalyzed by soyabean lipoxidase. *J. Biol. Chem.* **242**: 5329–5335.

Harding, J., Martin, F.W. and Kleiman, R. 1978. Seed protein and oil yields of the winged beans (*Psophocarpus tetragonolobus*) in Puerto Rico. *Trop. Agr.* (Trinidad) **55**: 307–314.

Hinchcliffe, C., McDaniel, M., Vaisey, M. and Eskin, N.A.M. 1977. The flavour of fababeans as affected by heat and storage. *Can. Inst. Food Sci. Technol. J.* **10**: 181–184.

Honig, D.H., Warner, K. and Rackis, J.J. 1976. Toasting and hexane: Ethanol extraction of defatted soy flakes. *J. Food Sci.* **41**: 642–646.

Hurd, C.D. and Blunck, F.H. 1938. The pyrolysis of esters. *J. Am. Chem. Soc.* **60**: 2419–2425.

Hydar, M. and Hadziyev, D. 1973. Pea lipids and their oxidation on carbohydrate and protein matrices. *J. Food Sci.* **38**: 772–778.

Ingold, K.V. 1961. Inhibition of the auto-oxidation of organic substances in the lipid phase. *Chem. Rev.* **61**: 563–589.

Ingold, K.V. 1962. Metal catalysis. In: *Symposium on Foods: Lipids and their Oxidation.* H.W. Schultz, E.A. Day, R.O. Sinnhuber, eds. pp. 93–121. The AVI Publishing Co. Inc., Westport, Connecticut.

Isaacks, R.E., Davies, R.E., Reiser, R. and Couch, J.R. 1963. Growth stimulating effects of high levels of vegetable oils. *J. Am. Oil Chem. Soc.* **40**: 747–749.

Jadhav, S., Singh, B. and Salunkhe, D.K. 1972. Metabolism of unsaturated fatty acids in tomato fruit: Linoleic and linolenic acids as precursors of hexanal. *Plant and Cell Physiol.* **13**: 449–459.

Jacks, T.J., Yatsu, L.Y. and Altschul, A.M. 1967. Isolation and characterization of peanut spherosome. *Plant Physiol.* **42**: 585–597.

Jackson, H.W. and Giacherio, D.J. 1977. Volatiles and oil quality. *J. Am. Oil Chem. Soc.* **54**: 458–460.

Janicek, G. and Pokorny, J. 1971. Nichtenzymatische traunung. *Z. Lebensm. Unters. Forsch.* **145**: 142–147.

Jaya, T.V. and Venkataraman, L.V. 1979. Germinated legumes and their influence on liver, serum cholesterol levels in rats: Influence of different components of chick pea and green gram on tissue cholesterol levels in rats. *Nutr. Rep. Intern.* **20**: 383–392.

Jaya, T.V., Venkataraman, L.V. and Krishnamurthy, K.S. 1979. Influence of germinated legumes on the levels of tissue cholesterol and liver enzymes of hypercholesterolemic rats: Influence of ungerminated and germinated chick pea and green gram (whole). *Nutr. Rep. Intern.* **20**: 371–381.

Johnson, D.W. 1976. World soybean research. In: *Proceedings of the World Soyabean Research Conference.* L.D. Hill, ed. pp. 1014–1017. The Interstate Printers and Publishers, Inc., Danville, IL.

Kalbrener, J.E., Eldridge. A.C., Moser, H.A. and Wolf, W.J. 1971. Sensory evaluation of commercial soy flours, concentrates and isolates. *Cereal Chem.* **48**: 595–600.

Kalbrener, J.E., Warner, K. and Eldridge, A.C. 1974. Flavors derived from linoleic and linolenic acid hydroperoxides. *Cereal Chem.* **51**: 406–416.

Karel, M. 1973. Symposium: Protein interactions in biosystems, protein-lipid interactions. *J. Food Sci.* **38**: 756–763.

Karel, M., Schaich, K. and Roy, R.B. 1975. Interaction of peroxidizing methyl linoleate with some proteins and amino acids. *J. Agr. Food Chem.* **23:** 159–163.

Keys, A., Anderson, J.T. and Grande, F. 1960. Diet type and blood lipids in man. *J. Nutr.* **70:** 257–266.

Kinsella, J.E., Patton, S. and Dimick, P.S. 1967. The flavor potential of milk fat. A review of its chemical nature and biochemical origin. *J. Am. Oil Chem. Soc.* **44:** 449–454.

Korytnyk, W. and Metzler, E.A. 1963. Composition of lipids of lima beans and certain other beans. *J. Sci. Food Agr.* **14:** 841–844.

Kramer, A. 1968. Basic principles of sensory evaluation. ASTM STP 433. p. 49. American Society for Testing and Materials, Philadelphia, PA.

Kritchevsky, D., Tepper, S.A., Vesselinovitch, D. and Wissler, R.W. 1973. Cholesterol vehicle in experimental atherosclerosis. 13: Randomized peanut oil. *Atherosclerosis* **17:** 225–237.

Kromer, G.W. 1975. Fats and oils: Natural and processed foods. In: *Nutrients in Processed Foods: Fats and Carbohydrates.* P.L. White, D.C. Fletcher and M. Ellis, eds. 39 p. American Medical Association Publishing Sciences Group Inc., Acton, MA.

Kummerow, F.A. 1962. Toxicity of heated fats. In: *Symposium on Foods: Lipids and their Oxidation.* H.W. Schultz, E.A. Day and R.O. Sinnhuber, eds. 294 p. AVI Publishing Co. Inc., Westport, Connecticut.

Kummerow, F.A. 1975. Lipids in atherosclerosis. *J. Food Sci.* **40:** 12–17.

Labuza, T.P., Heidelbaugh, N.D., Silver, M., and Karel, M. 1971. Oxidation at intermediate moisture contents. *J. Am. Oil Chem. Soc.* **48:** 86–90.

Lamptey, M.S. and Walker, B.L. 1976. A possible essential role for dietary linolenic acid in the development of the young rat. *J. Nutr.* **106:** 86–93.

Lee, F.A. and Mattick, L.R. 1961. Fatty acids of the lipids of vegetables. I. Peas (*Pisum sativum*). *J. Food Sci.* **26:** 273–275.

Lehninger, A.L. 1975. *Biochemistry* (Second edition). 279 p. Worth Publishers Inc., New York.

Lischenko, V.F. 1979. World production of food proteins: Situation, structure and trends. *J. Am. Oil Chem. Soc.* **56:** 178–180.

Love, J.D. and Pearson, A.M. 1971. Lipid oxidation in meat and meat products: A Review. *J. Am. Oil Chem. Soc.* **48:** 547–549.

Lovegren, N.V., Fisher, G.S., Legendre, M.G. and Shuller, W.H. 1979. Volatile constituents of dried legumes. *J. Agr. Food Chem.* **27:** 851–853.

MacLeod, A.J.J. and MacLeod, G. 1970. Chemistry of vegetable flavor. *Flavor Ind.* **1 (10):** 665–672.

Maga, J.A. 1973. A review of flavor investigations associated with the soy products, raw soybeans, defatted flakes and flours, and isolates. *J. Agr. Food Chem.* **21:** 864–868.

Mahadevappa, V.G. and Raina, P.L. 1978. Nature of some Indian legume lipids. *J. Agr. Food Chem.* **26:** 1241–1243.

Mathur, K.S., Singhal, S.S., and Sharma, R.D. 1964. Effects of Bengal gram in experimentally induced high levels of cholesterol in tissues and serum in albino rats. *J. Nutr.* **84:** 201–204.

Mattick, L.R. and Hand, D.B. 1969. Identification of a volatile component in soybeans that contributes to the raw beany flavor. *J. Agr. Food Chem.* **17:** 15–17.

Matsushita, S. 1975. Specific interactions of linloeic acid hydroperoxides and their secondary degraded products with enzyme proteins. *J. Agr. Food Chem.* **23:** 150–154.

Melnick, D. 1957. Nutritional quality of frying fats in commercial use. *J. Am. Oil Chem. Soc.* **34:** 578–582.

Miyazawa, T., Ito, S. and Fujino, Y. 1975. Fatty acid composition of glycerides and stereospecific analysis of triglyceride in pea seeds. *J. Nutr. Sci. Vitaminol.* **21:** 137–142.

Mohrhauer, H. and Holman, R.T. 1963. The effect of dose level of essential fatty acids upon fatty acid composition of the rat liver. *J. Lipids Res.* **4:** 151–159.

Mollenhauer, H.H. and Totten, C. 1971a. Studies on seeds. II. Origin and degradation of lipid vesicles in pea and bean cotyledons. *J. Cell Biol.* **48:** 395–405.

Mollenhauer, H.H. and Totten, C. 1971b. Studies on seeds. III. Isolation and structure of lipid containing vesicles. *J. Cell Biol.* **48:** 533–541.

Mookerjee, B.D., Deck, R.E. and Chang, S.S. 1965. Relationship between monocarbonyl compounds and flavor of potato chips. *J. Agr. Food Chem.* **13:** 131–134.

Murray, K.E., Shipton, J., Whitfield, F.B. and Last, J.H. 1976. The volatiles of off-flavored unblanched green peas (*Pisum sativum*). *J. Sci. Food Agric.* **27:** 1093–1107.

Myer, E.W. and Williams, L.D. 1976. World soybean research, *Proceedings of the world soyabean research conference.* L.D. Hill, ed. pp. 904–917. The Interstate Printers and Publishers, Inc., Danville, IL.

Nagami, K. 1973. The participation of typtophan residue in the binding of ferric iron to pyrocatechase. *Biochem. Biophys. Res. Commun.* **51:** 364–369.

Narayan, K.A. and Kummerow, F.A. 1958. Oxidized fatty acid-protein complexes. *J. Am. Oil Chem. Soc.* **35:** 52–56.

Narayan, K.A. and Kummerow, F.A. 1963. Factors influencing the formation of complexes between oxidized lipids and proteins. *J. Am. Oil Chem. Soc.* **40:** 339–342.

Narayan, K.A., Sugai, M. and Kummerow, F.A. 1964. Complex formation between oxidized lipids and egg albumin. *J. Am. Oil Chem. Soc.* **41:** 254–259.

National Agricultural Research Policy Advisory Committee. 1974. In: *National Soybean Research Needs: A report of the National Soybean Research Coordinating Committee.*

Nawar, W.W. 1969. Thermal degradation of lipids: A review. *J. Agr. Food Chem.* **17**: 18–21.

Nolen, G.A. 1972. Effects of fresh and used hydrogenated soybean oil on reproduction and teratology in rats. *J. Am. Oil Chem. Soc.* **49**: 688–693.

Nolen, G.A. 1973. A feeding study of a used, partially hydrogenated soybean oil, frying fat in dogs. *J. Nutr.* **103**: 1248–1255.

Nolen, G.A., Alexander, J.C. and Artman, N.R. 1967. Long-term rat feeding study with used frying fats. *J. Nutr.* **93**: 337–348.

O'Brien, P.J. and Frazer, A.C. 1966. The effect of lipid peroxides on the biochemical constituents of the cell. *Proc. Nutr. Soc.* **25**: 9–18.

Orthoefer, F.T. and Dugan, L.R. 1973. The coupled oxidation of chlorophyll with linoleic acid catalyzed by lipoxidase. *J. Sci. Food Agric.* **24**: 357–365.

Osman-Ismail, F. 1972. The formation of inclusion compounds of starches and starch fractions. Thesis, Swiss Federal Institute of Technology, Zurich, No. 4829.

Parliment, T.H. and Scarpellino, R. 1977. Organoleptic techniques in chromatographic food flavor analysis. *J. Agr. Food Chem.* **25**: 97–99.

Pattee, H.E., Beasley, E.O. and Singleton, J.A. 1965. Isolation and identification of volatile components from high temperature-cured off-flavor peanuts. *J. Food Sci.* **30**: 388–392.

Pattee, H.E., Singleton, J.A. and Cobb, W.Y. 1969. Volatile components of raw peanuts: Analysis by gas-liquid chromatography and mass spectrometry. *J. Food Sci.* **34**: 625–627.

Patwardhan, V.N. 1962. Pulses and beans in human nutrition. *Am. J. Clin. Nutr.* **11**: 12–30.

Penner, D. and Meggitt, W.F. 1970. Herbicide effects on soybean (*Glycine max* L. Merrill) seed lipids. *Crop Sci.* **10**: 553–555.

Pinsky, A., Grossman, S. and Trop, M. 1971. Lipoxygenase content and antioxidant activity of some fruits and vegetables. *J. Food Sci.* **36**: 571–572.

Pistorius, E.K. and Axlerod, B. 1974. Iron, an essential component of lipoxygenase. *J. Biol. Chem.* **239**: 3183–3186.

Privett, O.S., Dougherty, K.A., Erdahl, W.L. and Stolyhwo, A. 1973. Studies on the lipid composition of developing soybeans. *J. Am. Oil Chem. Soc.* **50**: 516–520.

Protein Resources and Techology: Status and Research Needs. 1978. M. Milner, N.S. Scrimshaw and D.I.C. Wang, eds. AVI Publishing Co. Inc., Westport, Connecticut.

Rabari, L.F., Patel, R.D. and Chohan, J.G. 1961. Studies on germinating peanut seeds. *J. Am. Oil Chem. Soc.* **38**: 4–5.

Rackis, J.J., Honig, D.H., Sessa, D.J. and Steggerda, F.R. 1970. Flavor and flatulence factors in soybean protein products. *J. Agr. Food Chem.* **18**: 977–982.

Rackis, J.J., Sessa, D.J. and Honig, D.H. 1979. Flavor problems of vegetable food proteins. *J. Am. Oil Chem. Soc.* **56**: 262–271.

Raju, N.V., Rao, M.N. and Rajagopalan, R. 1965. Nutritive value of heated vegetable oils. *J. Am. Oil Chem. Soc.* **42**: 774–776.

Reporter, M.C. and Harris, R.S. 1961. Effects of oxidised soyabean oil on the vitamin A nutrition of the rat. *J. Am. Oil Chem. Soc.* **38**: 47–51.

Rhee, K.S. and Watts, B.M. 1966. Evaluation of lipid oxidation in plant tissues. *J. Food Sci.* **31**: 664–668.

Rice, E.E., Poling, C.E., Mone, P.E. and Warner, W.D. 1960. A nutritive evaluation of over-heated fats. *J. Am. Oil Chem. Soc.* **37**: 607–613.

Roubal, W.T. and Tappel, A.L. 1966a. Damage to proteins, enzymes, and amino acids by peroxidizing lipids. *Arch. Biochem. Biophys.* **113**: 5–8.

Roubal, W.T. and Tappel, A.L. 1966b. Polymerization of proteins induced by free-radical lipid peroxidation. *Arch. Biochem. Biophys.* **113**: 150–155.

Roubal, W.T. 1970. Trapped radicals in dry lipid-protein system undergoing oxidation. *J. Am. Oil Chem. Soc.* **47**: 141–144.

Roza, M. and Francke, A. 1973. Soybean lipoxygenase: An iron containing enzyme. *Biochim. Biophys. Acta* **327**: 24–31.

Rubel, A., Rinne, R.W. and Canvin, D.T. 1972. Protein, oil, and fatty acid in developing soybean seeds. *Crop Sci.* **12**: 739–741.

Saio, K. and Watanabe, T. 1968. Observation of soybean foods under electron microscope. *J. Food Sci. and Technol.* (Japan) **15**: 290–296.

Schewfelt, A.L. and Young, C.T. 1979. Storage stability of peanut-based foods: A review. *J. Food Sci.* **42**: 1148–1152.

Sessa, D.J., Gardner, H.W., Kleiman, R. and Weisleder, D. 1977. Oxygenated fatty acid constituents of soybean phosphatidylcholines. *Lipids* **12**: 613–619.

Sessa, D.J., Honig, D.H., and Rackis, J.J. 1969. Lipid oxidation in full fat and defatted soybean flakes as related to soybean flavor. *Cereal Chem.* **46**: 675–686.

Sessa, D.J. and Plattner, R.D. 1979. Novel furaldehydes from oxidized soy phospholipids. *J. Agr. Food Chem.* **27**: 209–210.

Sessa, D.J. and Rackis, J.J. 1977. Lipid derived flavors of legume protein products. *J. Am. Oil Chem. Soc.* **54**: 468–473.

Sessa, D.J., Warner, K. and Honig, D.H. 1974. Soybean phosphatidylcholine develops bitter taste on autooxidation. *J. Food Sci.* **39**: 69–72.

Sessa, D.J., Warner, K. and Rackis, J.J. 1976. Oxidized phosphatidylcholines from defatted soybean flakes taste bitter. *J. Agr. Food Chem.* **24**: 16–21.

Shu, C.K. and Waller, G.R. 1971. Volatile components of roasted peanuts: Comparative analyses of the basic fraction. *J. Food Sci.* **36**: 579–583.

Shue, G.M., Douglass, C.D., Firestone, D., Friedman, L. and Sage, J.S. 1968. Acute physiological effects of feeding rats: Non-urea adducting fatty acids (urea-filtrate). *J. Nutr.* **94**: 171–177.

Singh, H. and Privett, O.S. 1970. Studies on glycolipids and phospholipids of immature soybeans. *Lipids* **5**: 692–697.

Singleton, J.A., Pattee, H.E. and Sanders, T.H. 1976. Production of flavor volatiles in enzyme and substrate enriched peanut homogenates. *J. Food Sci.* **41**: 148–151.

Smart, J. 1976. *Tropical Pulses.* Longman Group Limited, London, U.K.

Smith, A.K. and Circle, S.J. 1972. Protein products as food ingredients. In: *Soybeans: Chemistry and Technology*, Vol. 1. *Proteins.* A.K. Smith and S.J. Circle, eds. pp. 339–388. The AVI Publishing Co. Inc., Westport, Connecticut.

St. Angelo, A.J., Kuck, J.C. and Ory, R.L. 1977. Enzymes in food and beverage processing. R.L. Ory and A.J. St. Angelo, eds. pp. 229–243. *ACS Symposium series 47, American Chemical Society*, Washington, D.C.

Stone, E.J., Hall, R.M. and Kazeniac, S.J. 1975. Formation of aldehydes and alcohols in tomato fruit from U-^{14}C-labeled linolenic and linoleic acids. *J. Food Sci.* **40**: 1138–1141.

Sumner, A.K., Whalley, L.L., Blankenagel, G. and Youngs, C.G. 1979. Storage stability studies on pea flour, protein concentrate and starch. *Can. Inst. Food Sci. Technol. J.* **12**: 51–55.

Tai, P.T., Pokorny, J. and Janicek, G. 1974. Nonenzymic browning. X. Kinetics of the oxidative browning of phosphatidyl ethanolamine. *Z. Lebensm. Unters. Forsch.* **156**: 257–262.

Takayama, K.K., Muneta, P. and Wiese, A.C. 1965. Lipid composition of dry beans and its correlation with cooking time. *J. Agr. Food Chem.* **13**: 269–272.

Tappel, A.L. 1961. In: *Auto-Oxidation and Antioxidants.* Vol. 1, W.O. Lundberg, ed. p. 325. Interscience, New York.

Tappel, A.L. 1962. In: *Symposium on Foods: Lipids and their Oxidation.* H.W. Schultz, E.A. Day, and R.O. Sinnhuber, eds. p. 122. AVI Publishing Co., Inc., Westport, Connecticut.

Thomasson, H.J., Gottenbos, J.J., Kloeze, J. and Vles, R.O. 1966. Nutritional evaluation of hydrogenated fats. *Proc. Nutr. Soc.* **25**: 1–4.

Thompson, J.A., May, W.A., Paulose, M.N., Peterson, R.J. and Chang, S.S. 1978. Chemical reactions involved in the deep-fat frying of foods. VII. Identification of volatile decomposition products of trilinolein. *J. Am. Oil Chem. Soc.* **55**: 897–901.

Togashi, H.J., Henick, A.S. and Koch, R.B. 1961. The oxidation of lipids in thin films. *J. Food Sci.* **26**: 186–191.

Toya, D.K., Frazier, W.A., Morgan, M.E. and Baggett, J.R. 1974. The influence of processing and maturity on volatile components in bush snap beans (*Phaseolus vulgaris* L.). *J. Am. Soc. Hort. Sci.* **99**: 493–497.

Tressl, R. and Drawert, F. 1973. Biogenesis of banana volatiles. *J. Agr. Food Chem.* **21**: 560–565.

Verhue, W.M. and Francke, A. 1972. The heterogeneity of soybean lipoxygenase. *Biochim. Biophys. Acta* **284**: 43–53.

Vidyasagar, K., Arya, S.S., Premavalli, K.S., Parihar, D.B. and Nath, H. 1974. Chemical and nutritive changes in refined ground nut oil during deep fat frying. *J. Food Sci. and Technol.* **11**: 73–75.

Watts, B.M. 1954. Oxidative rancidity and discoloration in meat. *Advan. Food Res.* **5**: 1–52.

Weber, F., Arens, D. and Grosch, W. 1973a. Identifizierung von lipoxygenaseisoenzymen als carotinoxidasen. *Z. Lebensm. Unters. Forsch.* **152**: 152–154.

Weber, F., Laskawy, G. and Grosch, W. 1973b. Enzymatischer carotinabbau in erbsen, sojabohnen, weizen und leinsamen. *Z. Lebensm. Unters. Forsch.* **152**: 324–331.

Weurman, C. 1961. Gas liquid chromatographic studies on the enzymatic formation of volatile compounds in raspberries. *Food Technol.* **15**: 531–536.

Whitfield, F.B. and Shipton, J. 1966. Volatile carbonyls in stored unblanched frozen peas. *J. Food Sci.* **31**: 328–331.

Wilding, M.D. 1970. Oil seed proteins: Present utilization patterns. *J. Am. Oil Chem. Soc.* **47**: 398–401.

Wilkens, W.F. and Lin, F.M. 1970. Gas chromatographic and mass spectral analyses of soybean milk volatiles. *J. Agr. Food Chem.* **18**: 333–336.

Williams, P.M. and Bowden, B.N. 1973. Triglyceride metabolism in germinating *Andropogon gayanus* seeds. *Phytochem.* **12**: 2821–2827.

Wolf, W.J. 1975. Lipoxygenase and flavor of soybean protein products. *J. Agr. Food Chem.* **23**: 136–141.

Wolff, I.A. and Kwolek, W.F. 1971. Lipids of the leguminosae. In: *Chemotaxonomy of the Leguminosae*. J.B. Harborne, D. Boulter and B.L. Turner, eds. pp. 231–255. Academic Press, London, U.K.

Woodroof, J.C. 1973. *Peanuts: Production, Processing and Products*, second edition. AVI Publishing Co., Westport, Connecticut.

Worthington, R.E., Hammons, R.O. and Allison, J.R. 1972. Varietal differences and seasonal effects on fatty acid composition and stability of oil from 82 peanut genotypes. *J. Agr. Food Chem.* **20**: 727–730.

Worthington, R.E. and Smith, D.H. 1973. Effects of several foliar fungicides on the fatty acid composition and stability of peanut oil. *J. Agr. Food Chem.* **21**: 619–621.

Yatsu, L.Y., Jacks, T.J. and Hensarling, T.P. 1971. Isolation of spherosomes (oleosomes) from onion, cabbage, and cottonseed tissues. *Plant Physiol.* **48**: 675–682.

Yatsu, L.Y. and Jacks, T.J. 1972. Spherosome membranes: Half unit membranes. *Plant Physiol.* **49**: 937–943.

Young, C.T., Mason, M.E., Matlock, R.S. and Waller, G.R. 1972. Effect of maturity on the fatty acid composition of eight varieties of peanuts grown at Perkins, Oklahoma in 1968. *J. Am. Oil Chem. Soc.* **49**: 314–317.

Zirlin, A. and Karel, M. 1969. Oxidation effects in a freeze dried gelatinmethyl linoleate system. *J. Food Sci.* **34**: 160–164.

3

Storage Proteins of Legume Seeds

J. MOSSÉ and J.C. PERNOLLET

3.1. Introduction

Legume seeds have a high level of protein content ranging approximately from 20 to 40 per cent, as compared to other sources of plant proteins. Their amino acid pattern is close to the ideal aminogram—with the exception of sulphur amino acids—for human beings and animals. They contain at least two to three times more protein than cereal grains and the amount of lysine is also two to three times higher in their proteins than in those of cereal grains in which lysine is the most limiting essential amino acid. Therefore they are major sources of proteins and other nutrients in the diets of almost all countries around the world.

In the developing countries, there is a scarcity of animal proteins, or more accurately, of well-balanced proteins from the point of view of essential amino acids. In these countries, legume seeds have been labelled as the "meat for poor people". In contrast, the advanced or developed countries convert inefficiently tremendous quantities of plant proteins into meat with the help of animals. On the one hand, the increase of world population will not allow this use of plant proteins on a large scale for a long time, and on the other hand, new technology progressively enables the isolation and processing of seed proteins into new foods (Altschul 1978; Lillford 1978; Mossé and Fauconneau 1973; Tombs 1978). This finding is recent in the field of proteins, but it has been made for centuries with oil in the field of edible lipids and for years with sucrose and starch in the field of consumable glucids.

There are now innumerable considerations in the literature on the exceptional importance of legume seeds as a source of proteins from the nutritional and economic points of view (FAO 1964; PAG 1975; Pirie 1975; Orr 1978).

Beside the general reviews which will be mentioned in the following sections, several collective books or proceedings of symposia are dealing with biochemistry, cytology and genetics of legume seed proteins (Harborne et al.

1971; Milner 1975) or with the different related fields (Miege 1975; Pirie 1975; N.A.S. 1976; Norton 1978; Rubenstein et al. 1979; Leaver 1980).

As any analytical data on the crude protein and amino acid content of legume seeds are often scattered, the first section of the present chapter is focused on this field. The subsequent sections are almost exclusively devoted to the storage proteins of legume seeds: subunit composition and structure, accumulation and location and biosynthesis and evolution during germination. Although represented by a relatively small number of species, storage proteins account for around 70 per cent of legume seed nitrogen. Besides, there are still several thousand different enzymes (and enzyme inhibitors), regulatory, transporting, structural and recognition proteins (Boulter and Derbyshire 1978). Some of them pose exciting problems and form the subject of numerous studies which have been surveyed in several reviews. Such is the case of urease (Bailey and Boulter 1971), protease inhibitors (Liener and Kakade 1980) and phytohemagglutinins or lectins (Lis and Sharon 1973; Jaffé 1980).

One of the main aims of researches which are now prevailing on these storage proteins is to determine the number of genes coding for their subunits, as already emphasised by Millerd (1975), with the ultimate object to control and improve both quantitatively and qualitatively the legume protein production. Two converging ways seem able to successfully reach this goal. The first one involves the knowledge of the number, properties and structure of the implicated storage protein subunits and an understanding of their quaternary structure which is among the first steps of subcellular organisation (Sund and Weber 1966). But the complexity of the protein pattern is such that it becomes obvious that progresses will also and possibly sooner occur through molecular biology studies, i.e. the second way: purification of mRNA coding for storage proteins, preparation of cDNA, cloning of the latter and then the transcription and translation either *in vitro* or *in vivo*. The traditional methods joined to cytology have nevertheless brought clearer ideas about seed proteosynthesis over the past years, but have not been successful to completely elucidate the genetic aspects of seed proteins and of regulation steps of proteosynthesis. The location of reserve proteins within the cotyledon cell is now clear-cut since it has been conclusively found that the protein bodies are typical organelles for such a protein storage. The formation of these subcellular organelles, which is related to proteosynthetic processes, is still poorly known in spite of numerous studies (Briarty 1978; Yarwood 1978). Compared to the cereal protein body formation within the caryopsis endosperm, the situation is much more intricate and we will try to draw a theoretical scheme of this process from the numerous data of the literature. The mobilisation of seed storage proteins during germination is globally known, as well as protein body degradation, but the molecular mechanisms involved in protein degradation are still ignored: we will review the latest works on proteolysis, but no general mechanism or regulation can be put out.

3.2. Protein Content and Amino Acid Composition of the Main Legume Seeds

As far as we know, there are very few systematic studies and practically no exhaustive reviews specifically dealing with protein level and amino acid composition of legume seeds. In an analytical study of different seed samples of many plant families, Earle and Jones (1962) have determined the seed protein content of several hundred species of Leguminosae. The amino acid composition of seeds arising from many species, out of which some sixty leguminosae species, has been determined, except for cysteine and tryptophane, by Van Etten et al. (1963 and 1967). Busson (1965) has also performed remarkably accurate amino acid analyses of many African edible seeds, roots and leaves and he has studied the composition of the main legume seeds. The data on the amino acid composition of consumable seeds have then been collected by FAO (1970). It is initially interesting to briefly discuss the protein content and amino acid composition of the main legume seeds.

It is a fact that the protein content (Blixt 1977) as well as the amino acid composition change from one legume species to another, and also from one variety or cultivar to another. This mainly arises from genetic differences. But the protein content and composition are likely to vary according to the agronomical and physiological conditions of plant growing and development, as shortly tackled by Byers et al. (1978). And changes can as well be expected either inside of a pod or of the following pod location on the plant. So that it is needed to tentatively address the following questions: What do the words 'seed protein content' exactly mean? Concerning the possible variations of protein and amino acid content, what are the ranges of such change in each of the cases mentioned immediately above? In such variations, how is the amino acid composition related to the protein content? How are these kinds of relationships explained?

3.2.1. EVALUATION OF SEED PROTEIN CONTENT

In the present section, the protein contents are always evaluated by multiplying the nitrogen percentage (on a dry basis) by the coefficient 6.25. Such an evaluation is the only one usual in the field of nutrition science. Theoretically, the nitrogen content of proteins can vary between two extreme values: 8.6 per cent for polytyrosine and 35. 9 per cent for polyarginine. Practically, the most frequent value reached by the nitrogen content of all known proteins of any origin is around 16 per cent, which explains the generally adopted conversion factor of 6.25, in spite of its too often forgotten conventional characteristics. In fact, when used for seeds, it gives rather their content in crude protein which involves a complex mixture of very different nitrogen compounds. Besides proteins, there are free amino acids, osamines, complex lipids, puric and pyrimidic bases, nucleic acids, alkaloids, and so on. All these compounds come under "non protein nitrogen" (N.P.N.) compounds. There is a lot of studies looking for the chemical nature and botanical distribution of many N.P.N. compounds. Alkaloids have been reviewed by Mears and Mabry (1971), and non protein amino acids by Bell (1971). But the literature is very poor in the quantitative data concerning the different classes of N.P.N. components. Earle and Jones (1962) have carried out quantitative analyses of N.P.N. in many legume seeds. They extracted it by 0.8 trichloracetic acid. The ratio of N.P.N. to the total seed nitrogen is practically always in the region of 10 to 15 per cent. This shows that proteins are largely predominant and constitute 85 to 90 per cent of the total seed nitrogen.

Concerning the nitrogen to the protein conversion factor, a few authors use more realistic values than the classical 6.25. For instance, Cherry (1977) uses 5.46 for *Arachis hypogaea* seed proteins. Tkachuk (1969) has proposed more accurate factors for some cereal and oilseed meals, calculated from the quantitative amino acid data. He has shown that the resulting values vary from 5.4 to 5.8 (with 5.49 for *Glycine max* seeds). Therefore, it must be kept in mind that the use of 6.25 always gives an overestimate of the total seed protein content, corresponding to 10–15 per cent in excess.

Besides the nitrogen determinations (by Kjeldahl or Dumas methods), there are of course several other analytical techniques for measuring the seed protein content. They comprise colorimetric methods which are more rapid, are rather easy to be performed and are particularly useful for screening: for instance infra-red reflectometry; dye binding capacity (DBC), method using acid orange-12 dye; modified biuret method, and so on. We do not intend to discuss them. They have been compared several times, as well with cereal grains (Georgi et al. 1979) as with legume seeds (Singh and Jambunathan 1980). But in every case, nitrogen determination remains the only reference technique, because of the very clear definition of nitrogen quantity and of its additive feature. Therefore we will restrict the following discussion almost exclusively to those studies which involve nitrogen determinations.

3.2.2. PROTEIN CONTENT AND ITS VARIATIONS

The Leguminosae family is a large one. In it, the seeds show a very important variation in their protein content, when taking into account as many species as possible. This was shown by Earle and Jones (1962) who have performed systematic analyses of some 1,418 seed samples from over 900 species in 501 genera and 113 families. In their screening of the plant kingdom, 564 samples of seeds arose from some 320 different species in 86 genera of Leguminosae. Among these latter, the value for the protein content of seeds is one of the highest (32 per cent). It varies from 12 to 55 per cent, showing the widest range among the 113 studied families. This is partially due to the exceptional size of the family of Leguminosae (Heywood 1971) and possibly to the fact the number of screened samples of this family represented more than one-third of the total number of samples analysed by Earle and Jones (1962). The results of these authors show also that the oil content of Leguminosae seeds has one of the lowest average (around 5 per cent) but a wide range of variation spreading out from 1 up to 45 per cent. This is a consequence of the fact that only a few occasional species of the family have seeds which have a high oil content.

COMPARISON OF PROTEIN CONTENT OF THE SEEDS OF THE MAIN LEGUME GENERA

Figure 3.1 shows the most frequent protein levels found among cultivated legumes. The first column is concerned with the most common of these ones. They can be pooled in three groups, by the order of increasing protein content: *Cicer arietinum* is close to 22 per cent. *Arachis hypogaea, Pisum sativum, Vigna unguiculata, Phaseolus vulgaris, Lens culinaris* and *Vicia faba* range from 25 to 30 per cent. *Psophocarpus tetragonolobus, Glycine max, Lupinus albus* from 35 to 40 per cent. In the second column of Fig. 3.1 are mentioned a dozen of less common genera, ranging from 18 (*Voandzeia subterranea*) to 33 per cent (*Trigonella foenum-graecum*). The third column of Fig. 3.1 indicates the range of most frequent protein content of a few different species of three genera: *Arachis* of which 14 selected wild species have been studied by Cherry (1977), *Lupinus* (14 species) and *Phaseolus* (8 species other than *Phaseolus vulgaris*) after Baudet and Mossé (unpublished results). The *Arachis* species ranges from 19 to 35 per cent, *Phaseolus* from 20 to 30 per cent, while the *Lupinus* species appears as able to have the highest protein content of all cultivated legumes, with the protein level varying from 27 to 50 per cent. From a technological or nutritional point of view, it can be reminded that a few of these legumes also contain significant amounts of lipids besides protein. Roughly 45 per cent of oil is available in *Arachis hypogaea*, 20 per cent in *Glycine max*, 15 in *Psophocarpus tetragonolobus* and from 5 to 15 per cent in many *Lupinus* species. The presence of enough oil allows to get a cake the protein content of which is a function of the oil content of the seed.

INTRASPECIFIC VARIATION OF PROTEIN CONTENT OF SEEDS

For only a few legume species, the analyses of the crude protein content were made on several ten-seed samples, corresponding to different inbred lines, cultivars or varieties (Table 3.1). Such data often enable to draw a frequency curve, to calculate the average protein content for each species, and to get an idea of its possible range of variation. In most cases, the magnitude of variation reaches 10 to 15 percentage units. For instance, for *Cicer arietinum*

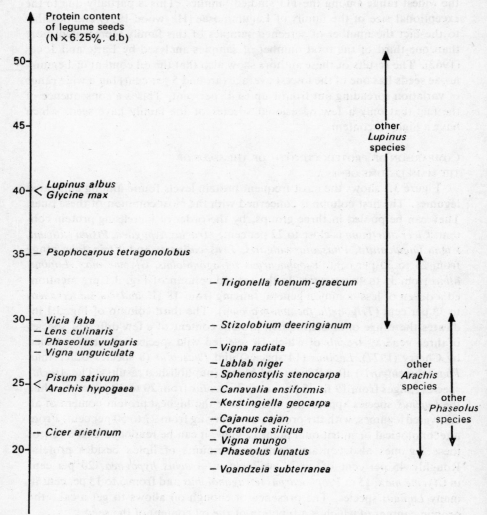

Fig. 3.1. Average protein content of main consumable legume seeds.

(Singh and Jambunathan 1980) the protein content of 150 lines varies from 15 to 29.6 per cent, with an average of 22.2 per cent. Among 87 samples of *Phaseolus vulgaris* seeds analysed by Earle and Jones (1962), 81 range from 22 to 32 per cent of crude protein, with an average close to 28 per cent, and so on.

In order to make valuable comparisons of the different genotypes within

TABLE. 3.1. Intervarietal ranges of seed protein content (N × 6.25% d.b.) inside of a few legume species

Species	Sample		Protein content		Reference
	Number	Nature	Min.	Max.	
Arachis hypogaea	21	(1)	23.5	33.3	Cherry (1977)
Cicer arietinum	150	(1)	14.9	29.6	Singh and Jambunathan (1980)
Phaseolus vulgaris	87	(3)	21.1	39.4	Earle and Jones (1962)
Pisum sativum	28	(1)	21.2	28.5	Bajaj *et al.* (1971)
Pisum sativum	57	(1)	24.3	32.9	Richter (1976)
Pisum sativum	50	(1)	21.5	37.5	Baudet *et al.* (1977)
Trigonella foenum-graecum	23	(3)	26.1	39.8	Earle and Jones (1962)
Vicia faba	12	(1)	26.4	34.7	Bhatty (1974)
Vicia faba	70	(2)	23.1	38.1	Mossé and Baudet (1977)
Vicia faba	33	(1)	22.9	38.5	Griffiths and Lanes (1978)
Vigna unguiculata	22	(1)	22.9	30.2	Bliss (1975)
Vigna unguiculata	21	(1)	22.9	34.6	Boulter et al. (1975)

(1) Cultivars, lines, varieties or collection accession.

(2) Cultivars or varieties with a few different samples of identical varieties cultivated in different conditions.

(3) Samples of unknown origin.

a species, the phenotypic variability should be known first. As far as we know, no systematic investigations have been made toward this. Nevertheless it is possible to have an idea of the same with the results obtained by Baudet et al. (1977) who performed systematic analyses on single seeds. A study of 50 single seeds randomly picked up within a seed sample of the same variety ("Minarette") harvested at the same place, the same year, shows that the protein content draws from 23 to 40.5 per cent. Moreover, when the material arises from two single plants, the protein content of single seeds ranges from 21.6 to 32 per cent for one plant, and from 23 to 34 per cent for the other (two seeds corresponding to extreme values being excluded in each case). In the same study, like in the work of Wolf (1975a and b), the individual seed weight was also taken into account. Several conclusions can be mentioned. The different seeds are very close for their individual weight and protein content, when located inside the same pod or in pods bound to the same nod. But the seed weight as well as the protein weight per seed diminish when the distance from the roots increases. And a linear relationship is found between the seed weight and the protein weight per seed. This relationship does not involve the seed position in the plant but appears as being related to the genetic characteristics of the varieties.

All these results point out the importance of working with samples as much well-defined and representative as possible. The knowledge of the seed weight distribution is essential. They also show the interest of measuring the weight of protein per seed. Contrary to the protein to dry matter percentage, the protein weight per seed does not depend on the non-proteic material content, like glucids. It is obvious that the increase of the glucid content of seed dry matter would result in a decrease in the protein content. We think that such considerations are not worthless for plant breeding when dealing with seed composition and quality.

3.2.3. AMINO ACID COMPOSITION OF LEGUME SEEDS

A survey of the amino acid composition of legume seeds is not easy. In spite of many analyses performed by Busson (1965) or by Van Etten et al. (1963 and 1967) and of the compositions pooled by FAO (1970), the data of literature remain insufficient. On the other hand, even when the same species, like *Phaseolus vulgaris* or *Vicia faba*, have been analysed by several different laboratories, the results are not always in good agreement. This can be explained at first by the diversity of problems posed to the chemist. A recent comparison of the composition of the same sample studied by different laboratories revealed considerable discrepancies (Anonymous 1978). Disagreements can also arise either from significant genetic differences between the cultivars of the same species, or from samples which, in spite of their genetic identity, correspond to distant protein contents. Ideal analyses should have to obey several conditions, a few of which have already been pointed out by FAO (1970):

1. Amino acids are analysed after 24 and 48 hour hydrolyses in 6 N HCl at 115°C and subsequent chromatographies on ion exchange columns are performed with automatic analysers. Separate analyses are carried out for methionine and cysteine (after hydrolyses of preliminarily oxidised meals), and also for tryptophane.

2. The amounts of 17 to 18 of the 20 proteic amino acids of the code of Nirenberg are given. When analysing a seed flour, it is practically impossible to differentiate glutamine from glutamic acid and one can only know the sum of the concentrations of both (Glx = Gln + Glu). It is just the same for asparagine and aspartic acid (Asx = Asn + Asp). Moreover, tryptophane is very hard to be analysed accurately. It requires special investigations, so that its content, which is one of the lowest, is often lacking.

3. The possible uncommon amino acids are taken into account, whatever they are: proteic ones like hydroxyproline, or free ones like canavanine.

4. The results indicate or allow to calculate either the total weight of the amino acids analysed, or the recovered nitrogen, i.e. the yield of the analysis.

5. They also mention the nitrogen (or crude protein) content of the sample analysed.
6. They specify the origin or nature or name of the cultivar (or variety) of the sample, its total weight and the average weight of its seeds.

AMINO ACID COMPOSITION OF THE MAIN LEGUME SPECIES

As far as possible, all the analytical problems mentioned above were taken into account in the selected data given in Table 3.2 which concerns the main cultivated legumes already cited in Fig. 3.1. Concerning Table 3.2, when several compositions were given by the same author, the results we have reported are averages. When several authors analysed the material of the same species, they are also referenced. If necessary, the data have been recalculated in grams of amino acid per 16 g of nitrogen.

With a few exceptions, the total weight of the recovered amino acids in g per 16 g N is 95 ± 5 g. Such yields are rather good. It is a fact that 100 g of protein correspond to 100 g of amino acid residues, i.e. to about 100–115 g of amino acids. But the presence of non amino acid nitrogen in the seeds and the use of a too high conversion factor like 6.25 are sufficient to give the total weights of the amino acids which cannot be higher than 95–100 g in the best analytical conditions.

For each amino acid, the highest and the lowest amount found among the 23 tabulated genera or species are shown, respectively in bold and italic type and the ratio (maximum/minimum) calculated. It can be seen that about half of the amino acids show ratios ranging from 1.5 to 1.9 (Gly, Ala, Val, Leu, Ile, Ser, Thr, His, Asx) while the other ones show ratios from 2.25 to 2.85 (Tyr, Phe, Pro, Met, Cys, Lys, Arg, Glx). For instance, the lysine amount can vary from 3.5 in *Arachis hypogaea* which is exceptionally low to 7.85 g/16 g N in *Vigna radiata*. Methionine varies from 0.7 (*Lupinus albus* and *Vicia faba*) to 1.9 (*Sphenostylis stenocarpa* and *Vigna unguiculata*) and cysteine from 0.85 (*Vigna radiata*) to 1.9 (*Lens culinaris*).

INTRASPECIFIC AND PHENOTYPIC VARIATION OF AMINO ACID COMPOSITION

The question of the possible variations of the amino acid composition of seeds among the varieties of the same species is also difficult for several reasons. The amino acid analyses are complex and time consuming, and they need to be repeated in order to reach a sufficient accuracy. The variations of the composition can arise as well from the genotypical differences between varieties as from phenotypical factors like protein content. Within a species, it is neither easy to work with samples of the same protein content for comparisons of possible genome effect nor often possible to get the seed samples of the same variety with very different protein contents.

These reasons explain that rather few literature data concern themselves with intraspecific comparisons. Busson (1965) was one of the first to analyse

TABLE 3.2. Amino acid composition

Species	Prot. %	Gly	Ala	Val	Leu	Ile	Ser	Thr	Tyr	Phe
Abrus precatorius	19.4	3.9	3.5	3.9	5.9	3.4	4.4	3.0	3.6	3.6
Arachis hypogaea	25	5.9	4.05	4.4	6.55	3.6	4.95	2.7	4.0	5.05
Cajanus cajan	22	3.5	4.2	3.9	7.2	3.3	4.75	3.5	2.9	9.05
Canavalia ensiformis	25	4.4	5.1	5.2	8.1	4.5	5.8	4.9	3.9	6.3
Ceratonia siliqua	21	5.3	4.1	4.4	6.5	3.5	5.0	3.6	3.5	3.2
Cicer arietinum	19.4	4.0	4.1	4.6	7.6	4.4	5.2	3.5	3.3	6.6
Glycine max	40	4.4	4.3	5.15	7.9	5.0	5.15	3.9	3.6	5.1
Kerstingiella geocarpa	23.3	4.5	4.5	6.3	7.7	4.5	5.9	3.8	3.5	5.8
Lablab niger[a]	26.4	4.3	4.5	5.1	8.2	4.4	5.5	3.7	3.6	4.8
Lens culinaris	29	3.9	4.0	4.2	7.1	3.75	4.9	3.45	2.2	4.6
Lupinus albus	40	3.85	3.1	3.9	6.9	4.2	5.25	3.55	4.2	3.6
Phaseolus lunatus	20.5	4.0	4.4	4.7	8.25	4.5	6.5	4.4	4.2	5.75
Phaseolus vulgaris	27	3.8	4.0	4.85	8.3	4.25	6.0	4.3	3.4	6.1
Pisum sativum	25	4.4	4.35	4.6	6.95	4.1	4.85	3.8	3.15	4.55
Psophocarpus tetragonolobus	35	4.1	4.05	3.9	8.6	3.8	5.7	4.15	5.65	5.05
Sphenostylis stenocarpa	26	4.3	4.3	5.6	7.7	4.4	6.2	4.1	4.3	5.3
Stizolobium deeringianum	29.4	4.2	3.4	4.9	7.0	4.7	4.1	4.0	4.8	4.4
Trigonella foenumgraecum	33	4.2	3.1	3.3	5.65	4.3	4.1	2.9	2.8	4.1
Vicia faba	30	4.3	4.15	4.55	7.55	4.1	5.2	3.55	3.2	4.15
Vigna mungo[b]	21	4.4	4.5	4.65	8.3	3.95	5.3	3.75	3.7	5.6
Vigna radiata[c]	28	3.75	4.25	5.1	8.35	4.4	5.1	3.4	3.3	5.9
Vigna unguiculata[d]	27	4.1	4.2	5.2	7.3	4.1	5.0	3.8	3.2	5.2
Voandzeia subterranea	18.5	4.0	4.5	5.3	7.8	4.4	5.6	3.5	3.5	5.6
Ratio (*max/min*)		1.7	1.7	1.9	1.5	1.5	1.6	1.8	2.5	2.85

[a] formerly *Dolichos lablab*. [b] in the past *Phaseolus mungo*.

References:
(1) Van Etten et al. (1967); (2) Busson (1965); (3) Baudet and Mossé unpublished (1978); (7) Boudet and Mossé (1980); (8) Vuyst et al. (1963); (9) Smith and Circle (1972); Evans and Boulter (1974); (14) Maneepun et al. (1974); (15) Palmer et al. (1973); (16) Sosulski (1979); (20) Evans and Boulter (1980); (21) Sauvaire et al. (1976); (22) Eppendor- (25) Coffman and Garcia (1977); (26) Otoul (1973); (27) Bliss (1975); (28) Boulter et al.

several cultivars of the same species. For instance, he has given the composition of twenty varieties of *Arachis hypogaea*, close enough to one another, of two samples of *Lablab niger*, nine varieties of *Glycine max*, and three of *Kerstingiella geocarpa*. *Lens culinaris* has been analysed by Bhatty et al. (1976) (six genotypes) and by McCurdy et al. (1978) (five varieties); *Lupinus luteus* and *Lupinus angustifolius* by Hove (1974); *Lupinus termis* by

of legume seeds (g/16g N)

Trp	Pro	Met	Cys	Lys	His	Arg	Asx	Glx	Total	Ref.	Other Ref.
	8.6	1.2		5.4	3.2	*4.8*	*8.2*	13.7	80.2	(1)	
1.05	4.7	1.4	1.45	*3.45*	2.2	11.6	11.7	18.9	97.6	(2)	(8)
	4.3	1.4	1.35	6.55	3.45	6.0	9.8	19.9	95	(3)	
1.2	4.7	1.7	1.2	5.9	3.0	4.9	10.3	11.5	92.6	(2)	
	4.0	1.0		5.6	2.5	**11.8**	9.0	**28.0**	101	(4)	
	4.3	1.4		7.2	2.3	8.8	11.7	16.0	95	(1)	
1.3	5.9	1.55	1.65	6.35	2.75	8.1	11.8	18.0	101.9	(2)	(9)
	5.4	1.4	1.0	6.6	2.8	6.5	11.5	17.4	99.1	(2)	
	5.4	0.9	1.1	6.2	2.8	7.0	11.5	16.3	95.3	(2)	
0.9	*3.5*	1.6	**1.9**	7.3	2.6	6.9	12.0	15.8	90.55	(5)	10)
0.7	3.95	*0.7*	1.7	4.5	*2.05*	9.95	10.3	21.5	93.85	(3)	(11, 12)
	4.3	1.5	1.55	6.75	3.15	5.4	**12.4**	13.9	96.55	(3)	(13, 14)
	3.5	1.15	1.1	6.7	2.7	6.7	**12.4**	15.9	95.05	(3)	(15, 16, 17)
0,8	4.0	0.95	1.7	7.2	2.35	9.55	11.5	17.1	95.9	(3)	(18, 19, 20)
	6.5	1.0	1.5	7.4	2.75	6.85	11.0	14.8	96.7	(6)	
1.3	4.6	**1.9**	1.5	6.8	**3.7**	6.2	10.8	13.4	96.4	(2)	(13)
	5.5	1.1		6.3	2.3	6.6	11.6	*11.8*	86.7	(1)	
0.8	3.95	0.8	1.1	5.5	2.15	8.5	9.2	15.2	81.65	(21)	
0.85	4.1	*0.7*	1.35	6.35	2.45	9.6	11.7	17.6	95.45	(7)	(22, 23, 24)
	3.9	1.85	1.05	7.25	2.8	6.35	12.0	16.1	95.4	(3)	
	4.05	1.3	*0.85*	**7.85**	2.9	7.35	11.8	17.6	97.25	(3)	(25)
	3.5	**1.9**	1.7	6.3	3.1	7.1	11.0	16.6	93.3	(2)	(26, 27, 28)
	5.05	1.75	1.05	6.4	3.0	6.3	11.8	16.8	96.25	(2)	
	2.45	2.7	2.25	2.3	1.8	2.45	1.5	2.35			

[c] in the past *Phaseolus aureus*.   [d] also named *Vigna sinensis*.

results; (4) Van Etten et al. (1963); (5) McCurdy et al. (1978); (6) Gillespie and Blagrove (10) Bhatty et al. (1976); (11) Duranti and Cerletti (1979); (12) Ballester et al. (1980); (13) Kelly and Bliss (1975); (17) Pusztai et al. (1979); (18) Eppendorfer (1974); (19) Holt and fer 1971); (23) Kaldy and Kasting (1974 and 1978); (24) Lafiandra et al. (1979); (1975).

Abdel Fattah et al. (1974); *Phaseolus lunatus* by Busson (1965) (four varieties), Evans and Boulter (1974) and Maneepun et al. (1974) and Otoul (1976); *Vicia faba* by Kaldy and Kasting (1974) (eight cultivars), Mossé and Baudet (1977) and Baudet and Mossé (1980) (seventy samples among which different varieties or cultivars), Lafiandra et al. (1979) (eighteen varieties); *Pisum sativum* by Holt and Sosulski (1979) (thirty-three samples or lines); *Vigna unguiculata* by Otoul (1973) (sixteen varieties) and Bliss (1975); and *Voandzeia subterranea* by Busson (1965) (six varieties).

In many of these works dealing with the seeds of different varieties, the protein content, when it is specified, does not vary much among the analysed samples. This partially explains that the differences in amino acid composition are often not significant. It is a fact that only mutations could produce large variations in the amino acid pattern as it was shown in the cereal regulatory mutants by Mertz et al. (1964), Nelson et al. (1965) and Baudet et al. (1968) for corn; by Munck et al. (1970) for barley; by Singh and Axtell (1973) for sorghum. On the other side, these studies do not enable a clear distinction between the physiological or the environmental effects. One of the ways towards such a discrimination consists in working with the same variety grown in very different controlled conditions of nutrients' application in order to get as large a range of seed protein content as possible. This has been done mainly on *Vicia faba* by Eppendorfer (1971), Mossé and Baudet (1977) and Baudet and Mossé (1980) and also on *Pisum sativum* by Eppendorfer (1974) and Holt and Sosulski (1979). By the changing application of sulphur, phosphorus and nitrogen levels, Eppendorfer (1971) has obtained *Vicia faba* seeds of the same variety the protein content of which ranged from 22.7 to 29.7 per cent. His results suggest that highly significant correlation coefficients are obtained between the amino acid content of seed dry matter and the total nitrogen content (i.e. protein content). The occurrence of linear relationships between the amount of each amino acid and protein content was pointed out by Mossé and Baudet (1977 and 1980) with about 70 different samples of a wider range of protein content (23 to 38 per cent). These authors have found that many varieties have identical behaviour concerning these relationships. This can easily be explained by the plausible similarity of genomes of many varieties in what concerns protein synthesis and its regulation. They have shown that for each proteic amino acid i, its amount A_i (in the per cent of seed dry matter) changes linearly as a function of protein content P (Fig. 3.2) and obeys the equation:

$$A_i = \alpha_i P + \beta_i$$

α_i and β_i being constant for one amino acid but differing for another. The coefficient α_i is obviously positive or null. It is impossible that any increase of protein content produces a decrease of any amino acid content of the seed. But coefficient β_i can have as well positive as null or negative values. The amino acid content α_i of seed proteins (in g per 16 g N, i.e. per 100 g of crude protein) is related to A_i by the equation:

$$a_i = 100 \times A_i/P$$

and the combination of the two preceding relations gives:

$$a_i = 100 (\alpha_i + \beta_i/P)$$

which means that a_i varies in function of P following an hyperbolic law. So

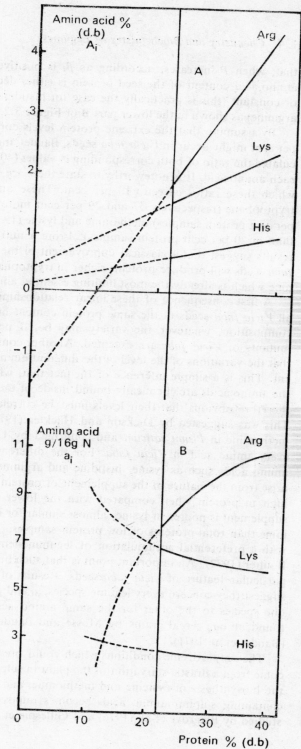

Fig. 3.2. The effect of *Vicia faba* seed protein content (N × 6.25g per 100 g of seed, on a dry basis) on their amino acid level.

A—basic amino acid content of seed (g per 100g of seed, d.b.); B—basic amino acid composition of seed proteins (g per 100g of protein) (*Source:* Mossé and Baudet 1977).

that, when P increases, according as β_i is positive, null or negative, the amino acid content of the seed protein is either decreasing (like for lysine), or constant (this is practically the case for histidine), or increasing (like for arginine) as shown in the lower part **B** of Figure 3.2.

By assuming that the extreme protein levels corresponding to 20 and 40 per cent might occur in *Vicia faba* seeds, Baudet and Mossé (1980) have calculated the ratio of both corresponding a_i values (40 per cent/20 per cent) for each amino acid. It is noteworthy to state there are only four amino acids for which these ratios exceed ± 10 per cent. These amino acids are arginine, tryptophane (respectively 57 and 29 per cent more in the 40 than in the 20 per cent protein samples) methionine and lysine (19 and 17 per cent less in 40 than in 20 per cent protein samples). From a nutritive point of view, these results suggest that a classical improvement of the protein content of *Vicia faba* seeds will produce proteins richer in tryptophane but poorer in methionine which is already the most limiting essential amino acid in broad beans.

A first consequence of these linear relationships consists in the fact that all *Vicia faba* seeds of the same protein content have the same amino acid composition, whatever the variety may be, if possible but yet unknown mutants of *Vicia faba* are excepted. Another consequence of these laws is that the variations of the level of the different amino acids are not independent. This is a simple inference of the fact that, with a very few exceptions, the amino acids are chemically bound inside of the polypeptide chains. It is therefore obvious that their levels must be correlated one with each other. This was suggested by Dickson and Hackler (1975) concerning lysine and methionine in *Pisum sativum* and confirmed by Lafiandra et al. (1979) for every amino acid in *Vicia faba*. For the differences observed between the amino acids such as lysine, histidine and arginine shown in Fig. 3.2, they arise from the nature of the supplement of protein accumulated in the seeds high in protein when compared with the lower ones. In *Vicia faba*, this supplement is poorer in lysine, almost similar for histidine but richer in arginine than total proteins of low protein samples. This is in good agreement with a preferential accumulation of legumin demonstrated by Wright and Boulter (1972). An important point is that such linear relationships are not a particular feature of *Vicia faba* seeds. Results of Holt and Sosulski (1979) suggest they concern every legume species, α_i and β_i coefficient differing from one species to the other for the same amino acid. In fact they were first found among cereal grains by Mossé and Baudet (1969), then in oilseeds (Baudet et al. 1971).

The only extreme conditions which could break these correlations could come from a drastic starvation of the plant in sulphur. Without this nutrient, the biosynthesis of cysteine and methionine, and also that of many proteins containing sulphur amino acids becomes impossible. This is just the case studied by Blagrove et al. (1976) and Gillespie et al. (1978) with *Lupinus an-*

gustifolius seeds. Because of the lack of sulphur, the lupine seeds accumulate proteins which are almost devoid of methionine and cysteine instead of others which are richer in these two amino acids. But until now, we do not know what is becoming of the amino acid/nitrogen relationships in such exceptional conditions.

Beside the studies which involve the complete amino acid analyses (with or without tryptophane), a few other ones are dealing with one or two particular amino acids generally determined by a rapid screening method and sometimes investigated on numerous samples. For instance, Sodek et al. (1975) have proposed a rapid determination of tryptophane in *Phaseolus vulgaris*, while Duncan et al. (1976) have propounded another rapid method for methionine in legume seeds. The most frequently studied are sulphur amino acid. Methionine and/or cysteine have been systematically studied in *Phaseolus vulgaris* by Adams (1975); Dickson and Hackler (1975); Alves-Moreira et al. (1976); Evans et al. (1978) who analysed 544 samples by a microbiological procedure, and Evans and Boulter (1980), *in Pisum sativum* by Jermyn et al. (1979) and also by Evans and Boulter (1980), *in Vicia faba* by Frölich et al. (1974). These latter authors and several others mentioned for *Phaseolus* and *Pisum* have also found a negative correlation between each sulphur amino acid and nitrogen content of the seeds.

A few non-protein amino acids have been determined. Pipecolic acid has been found in several species of *Phaseolus*, *Lablab* and *Vigna* by Casimir and Le Marchand (1966). S-methyl-L-cysteine content can reach 0.43 g/16 g N in *Phaseolus lunatus*, 0.56 in *Vigna unguiculata* and 0.87 in *Phaseolus vulgaris* (Evans and Boulter 1975). This shows that the use of the total sulphur content, either for the evaluation of the nutritive quality of grain legumes, as suggested by Porter et al. (1974) or as a screening criterion for sulphur amino acids (Bhatty et al. 1977), might be used with some caution. It is of course impossible to pass over canavanine in silence. In *Canavalia ensiformis* seeds, its amount reaches an exceptionally high level with 11 g/16 g N (Busson 1965). Besides the free non-protein amino acids, several dipeptides are progressively detected and identified in different legume seeds. Many of them are γ-glutamylpeptides: γ-glutamylmethionine in *Vigna mungo* (Otoul et al. 1975; Dardenne 1976). Such uncommon amino acids suggest the occurrence of particular biosynthesis pathways and constitute phylogenetic markers (Bell 1971; 1976) and very useful tools in chemotaxonomy.

3.2.4. CONCLUSION

In conclusion, the data of literature already allow to point out many different reasons and conditions in variation of the legume seed protein content. It changes from one single seed to another even inside of an apparently well defined sample. It can vary broadly among the different seeds of a single plant. It can also vary among the different plants of the same field. It changes from one cultivar or one line to another and it often differs from one species to another too. Nevertheless, because of the need to work with representative samples of some ten, hundred or sometimes thousand seeds (especially for small-sized seeds), the estimation of protein content is always a buffered average. The protein content within a species can vary as much as two- to threefold between the lowest and the highest values.

It is obvious that the protein content is related to the amino acid content from which it mainly results. For the same reason, i.e. the fact that amino acids are found inside of polypeptide chains, the levels of the different (proteic) amino acid are related between themselves. On another side, it is striking to state that individual amino acid contents of seeds are always linearly correlated to their protein content. Such linear relationships are particularly simple. But in spite of the very general feature of those relationships among many species, the question is to understand what the regulation mechanisms which arise into such a behaviour are. And this question still remains unanswered. One can only expect the explanation to stem from a better knowledge of the main storage protein subunits and of the molecular biology of their biosynthesis. We are now going to discuss these latter basic topics in the following sections.

3.3. The Main Storage Proteins: Properties and Subunit Structure and Composition

During the last decade, several authors have reviewed more or less specifically the different kinds of problems and results concerning the structure and composition of legume seed proteins (Boulter and Derbyshire 1971; Inglett 1972; Müntz et al. 1972; Miège 1975; Millerd 1975; Derbyshire et al. 1976; Wolf 1977; Thomson and Doll 1979).

The extraction and fractionation of these storage proteins are generally agreed as being hard and difficult. The different steps involved and conditions used pose some problems and need caution. The storing of seeds, removing of testas, defatting and milling, protein interaction with phytate or polyphenols, the use of reductants and of alkylating or chelating agents, salting out by ammonium sulphate or zonal isoelectric precipitation, exclusion and ion exchange chromatography, immuno-electrophoresis—all these steps or conditions have been exhaustively reviewed by Derbyshire et al. (1976), and it does seem necessary to give again a description of this subject. As noticed by Derbyshire et al. (1976), legumin and vicilin used to be distinguished by several general and frequent properties, some of which have been already pointed out by Danielsson (1949). Legumin sedimentation coefficient and molecular weight are respectively near 11 S and 300,000–450,000 daltons, while for vicilin they are nearly 7 S and 150,000–250,000 daltons.

Compared to vicilin, legumin is less soluble into salt solutions, it coagulates less easily at 95°C and it has larger amounts of nitrogen and sulphur.

Although the field of characterization of legume storage proteins covers a wide range of questions like size and sedimentation coefficient, dissociation-association systems, immunochemical properties, pepsic and trypsic hydrolysis, cryoprecipitation, etc. this section will be rather dealing with basic data about their subunits, their quaternary structure and their polymorphism.

3.3.1. *Arachis Hypogaea*

There have been many early studies on the storage proteins of *Arachis hypogaea* seed. This is mainly due to its importance as an oil seed in industrial nations. Moreover, the story of peanut globulins is one of the best examples of the still few legume seed proteins, the structure and composition of which have been almost completely determined, until the ultimate step of subunit isolation and characterization. For this reason and also because they were not or only briefly surveyed by Millerd (1975) and Derbyshire et al. (1976), they are particularly worth reviewing over. They have been indeed among the first of legume seed proteins for which location inside of protein bodies, genetic polymorphism, dissociation-association systems and an occurrence of a quaternary structure have been early and now unambiguously demonstrated. They were isolated a century ago for the first time by Ritthausen (1880) and fractionated in two groups, *arachin* and *conarachin* by Johns and Jones (1916) and Jones and Horn (1930). These two proteins are also called α-arachin and α-conarachin (Dechary et al. 1961 and Neucere and Ory 1970), suggesting that although major, they do not correspond to the whole storage globulins of peanut. When a salt extract (10 per cent NaCl buffered with 0.01 M sodium phosphate, pH 7.9) is precipitated by saturated ammonium sulphate solution, arachin is the fraction salted out by 40 per cent saturation, and conarachin is that one by 40 to 85 per cent saturation.

Arachin is largely predominant, making up 73 per cent, of the extractable proteins which correspond themselves to 93 per cent of the total defatted seed protein (Yamada et al. 1979a). It was shown to be localized within protein bodies (first called aleurone grains) by Altschul et al. (1961 and 1964); Dieckert et al. (1962) and Bagley et al. (1963). The existence of a dissociation-association system has been suggested as early as thirty years ago by Johnson and Shooter (1950). The existence of a quaternary structure with different kinds of subunits has been shown in a noteworthy study by Tombs (1965). This author has also claimed that arachin was the first example of polymorphism to be reported among plant proteins, with occurrence of genetic variants which have been detected by single seed electrophoretic analyses. Peanut globulins have been among the first legume seed proteins to be investigated by immunoelectrophoretic methods (Daussant et al. 1969a and b). Many authors have investigated their extraction, preparation and isolation, often using DEAE-Cellulose or DEAE-Sephadex chromatography (Dechary et al. 1961; Tombs 1965; Neucere 1969; Cherry et al. 1973; Shetty and Rao 1974; Yamada et al. 1979a).

Rather than to detail every step and result, it appears preferable to summarize the main recent conclusions of Yamada et al. (1979a and b) who have confirmed and completed the preceding studies by achieving the isolation and characterization of all arachin subunits. They have found that purified arachin can be separated in two components, A_I and A_{II}, through DEAE-Sephadex chromatography and that the ratio of these components depends on the variety. The two components have sedimentation coefficients of 9.1 S and 14.4 S and molecular weights of 170,000–180,000 and 340,000–350,000 respectively. By changing either ionic strength or pH, A_I undergoes a reversible interconversion between 9 S and 14 S and A_{II} only a partially reversible one. But A_I and A_{II} have identical amino acid compositions and exhibit the same subunit pattern on SDS-gel electrophoresis, with six main bands, when pretreated with 6–7 M urea in the presence of 0.2 M β-mercaptoethanol. The six kinds of subunits, S_1 to S_6, have been isolated by isoelectric focusing in sucrose density gradient and characterized (Table 3.3) by their isoelectric

TABLE 3.3. Physicochemical properties and amino acid composition (residues per 1000) of arachin subunits

M.W. (kilodalton) N-terminal amino acid isoelectric point	S_1	S_2	S_3	S_4	S_5	S_6
	35.5	37.5	40.5	19.5	19.5	19.5
	Val	Ile	Ile	Gly	Gly	Gly
	5.8	6.0	6.3	7.1	7.4	8.3
Gly	75.3	78.9	78.7	56.0	62.0	62.7
Ala	40.4	51.7	49.2	86.6	85.0	86.5
Val	48.8	34.5	33.3	78.2	63.6	71.3
Leu	45.2	52.0	49.3	80.0	85.7	85.5
Ile	31.7	30.6	32.8	45.0	44.7	47.3
Ser	66.7	58.3	58.9	76.1	75.5	77.3
Thr	32.8	23.3	26.0	25.2	26.3	26.7
Tyr	33.5	32.0	34.4	34.7	36.1	40.3
Phe	55.4	51.7	53.3	38.9	55.1	39.5
Trp	8.8	9.4	10.3	7.9	8.6	7.9
Pro	54.1	51.4	51.4	63.4	56.1	60.4
Met	2.4	0	1.9	9.3	0	0
Cys	12.9	7.6	7.4	10.1	0	0
Lys	12.1	8.5	10.7	28.4	31.3	32.4
His	15.9	14.7	17.2	30.3	21.3	19.9
Arg	102.7	123.2	116.9	55.2	63.4	59.9
Asx	115.3	121.8	123.8	140.9	148.3	161.0
Glx	245.8	250.4	244.8	133.8	137.1	121.2

Source: Yamada et al. (1979b)—Amino acid composition has been recalculated.

points, molecular weights, N-terminal amino acids, molar ratios and by their amino acid composition. The first three ones are acidic subunits. They have different molecular weights, ranging from 35,500 to 40,500, two N-terminal amino acids (valine and isoleucine) and amino acid compositions which are significantly different in spite of rough similarities. The three other ones are basic subunits. They have the same molecular weight (around half of that of acid subunits), the same N-terminal amino acid (glycine) and their amino acid compositions are rather similar, particularly concerning S_5 and S_6 which are both devoid of cysteine and methionine. Moreover, the three acidic subunits have higher content of glutamic acid or glutamine, but lower content of alanine, valine and leucine. By an estimation of the molar ratios of these subunits in the 9 S component (172,000 dalton), it was concluded that arachin consists of six different, noncovalently bound, subunits in equimoles. In other words, arachin is an heterohexamer which is itself a kind of "supermonomer" partially dimerized, forming an heterododecamer (a "superdimer" of twelve subunits in the 14 S component). Such a quaternary structure is a full confirmation of that already suggested by Tombs (1965) and Tombs and Lowe (1967). The latter have detected four different kinds of subunits (α to δ) which seem to be associated with several different possible molar ratios (for instance $\alpha_2\beta_2\gamma\delta$, $\alpha_4\gamma\delta$ and $\beta_4\gamma\delta$). They have found the three same N-terminal amino acids as these analysed by Yamada et al. (1979a and b), but no carbohydrate, contrary to Shetty and Rao (1974) and Basha and Cherry (1976) who have found traces of neutral and amino sugars. A last important point is the ringlike structure revealed by electron microscopy of the heterohexamer when investigated in *in vitro* salt-water systems (Tombs et al. 1974): one cannot refrain to compare this result with the structure of *Glycine max* globulins (glycinin) discussed beyond.

Conarachin, a much less abundant protein in the seed of *Arachis hypogaea*, is usually divided into two subgroups which precipitate with increasing amounts of saturated ammonium sulphate: 40 to 65 per cent of saturation for conarachin I and 65 to 85 per cent for conarachin II. A few data show that conarachin, in spite of the same electrophoretic mobility as arachin in polyacrylamide gel electrophoresis, has a lower sedimentation coefficient (7.8 S, Dechary et al. 1961, Shetty and Rao 1977) with a different amino acid composition. In an investigation of the content of peanut protein bodies, Neucere and Ory (1970) have shown that a part of conarachin is cytoplasmic and accumulated outside the protein bodies. But more elaborative conclusions could be drawn from isolation and characterization of its subunits.

One can wonder whether arachin is a single combination of the six subunits evidenced by Yamada et al. (1979b). In our opinion, the most likely assumption is the existence of several heterohexamers (multiple forms), more especially as very similar subunits like S_5 and S_6 might be interchangeable. Moreover, the electrophoretic patterns obtained by Yamada et al. (1979a) suggest a few minor components revealed by faint bands besides the main ones. But arachin and conarachin seem to be respectively legumin-like and vicilin-like proteins, as proposed by Derbyshire et al. (1976).

3.3.2. *Glycine max*

Glycine max (soybean) is the legume species of which the seed storage proteins have been the most intensively studied.

Catsimpoolas and Ekenstam (1969) have shown that four major immunochemically different globulins are present in *Glycine max* seeds: glycinin and three kinds of conglycinin (α, β and γ). They all exhibit different number and pattern of subunits by electrophoresis (Catsimpoolas 1970).

Glycinin, the 11 S component, has been classified as legumin-like by Derbyshire et al. (1976). It is extracted by water or by a phosphate buffer 0.4 NaCl, 0.01 M β-mercaptoethanol pH 7.6 and purified by chromatography on DEAE-Sephadex or on hydroxyapatite (Catsimpoolas et al. 1967; Wolf and Sly 1967; Koshiyama 1972; Kitamura et al. 1974; Badley et al. 1975; Fisher et al. 1976; Draper and Catsimpoolas 1977).

It was first demonstrated to be composed of twelve subunits (Catsimpoolas et al. 1967), corresponding to six different kinds of polypeptide chains (Catsimpoolas et al. 1971), with molecular weights of about 22,000 and 37,000. The subunits seem to be devoid of carbohydrate (Koshiyama and Fukushima 1976b). They have been isolated by isoelectric focusing and their amino acid composition analysed to enable them to divide them into two groups: three acidic subunits have a higher content of glutamic acid or glutamine and proline, but a lower content of alanine, valine, leucine, phenylalanine and tyrosine than the three basic subunits. It is possible that subunit molecular weights are overestimated when determined by SDS-polyacrylamide gel electrophoresis. This may be related to the high content of acidic (or amide) amino acids, as noticed by Ochiai-Yanagi et al. (1977) who have found 28,000 and 18,000 (instead of 37,000 and 22,000).

The quaternary structure of glycinin has been elucidated by Badley et al. (1975) and by Koshiyama and Fukushima (1976c) who have confirmed the results of Saio et al. (1970) and of Catsimpoolas et al. (1971). By means of electron microscopy, X-ray scattering measurements and by determining several physical parameters, Badley et al. (1975) have demonstrated that the twelve subunits are really packed in two identical hexagons placed one on the other in such a way that every acidic subunit is associated with three basic ones and vice versa: two in the same hexagon and the third in the other one. The structure involved is roughly similar to an oblate cylinder of 55Å radius and 75Å thickness with a sedimentation coefficient $S_{20w}^0 = 12.3$ and a molecular weight of 320,000. Any possible disulphide links between subunits have been suggested by Badley et al. (1975) and studied by Kitamura et al.

(1976), Draper and Catsimpoolas (1978) and Mori et al. (1979). It appears that some inter-chain disulphide bonds could exist between some of the subunits. However Catsimpoolas and Wang (1971) have found a microheterogeneity for glycinin and Kitamura et al. (1976) have reported results consistent with the hypothesis that there are at least four different basic subunits and possibly four different acidic ones.

The fact that the structure of glycinin is more complex than the previous suggested data has been recently illustrated by Moreira et al. (1979). They have indeed purified and isolated six different acidic polypeptides and four basic ones of which amino acid compositions (Table 3.4) and N-terminal sequences (15 to 27 successive residues) have been determined. As partially

TABLE 3.4. Amino acid composition of the 6 acid and 4 basic subunits of glycinin recalculated in residues per 1000 total residues

	A_1	$F_{2(1)}$	A_2	A_3	$F_{2(2)}$	A_4	B_1	B_2	B_3	B_4
Gly	90.0	80.3	86.3	77.6	87.9	65.6	62.5	59.0	76.6	89.1
Ala	41.8	46.8	52.3	28.7	55.6	18.2	87.8	81.1	70.9	62.0
Val	34.6	36.5	44.2	45.8	42.3	35.5	64.2	61.3	97.2	106.3
Leu	58.4	55.0	57.8	57.4	119.0	41.0	100.8	98.7	103.5	100.2
Ile	51.1	25.6	44.2	32.1	56.7	30.5	51.8	55.6	40.0	40.4
Ser	53.2	55.9	47.4	71.3	85.7	68.9	76.0	70.3	69.2	68.7
Thr	34.9	36.5	35.5	40.8	43.4	34.6	45.6	51.6	35.4	29.9
Tyr	21.2	25.6	19.1	14.7	22.2	12.9	15.8	14.2	33.2	46.5
Phe	35.4	52.6	35.5	31.6	10.0	22.6	48.4	51.6	34.3	31.6
Pro	69.7	73.5	61.5	89.2	90.1	80.0	59.1	61.3	58.3	50.4
Met	10.5	12.1	16.7	6.3	12.2	4.1	13.0	15.3	0	7.2
Cys	13.1	N.D.	12.4	9.5	N.D.	2.1	9.6	8.5	1.1	8.3
Lys	61.6	46.8	43.0	38.9	43.4	55.1	33.2	33.5	40.0	36.0
His	17.4	14.1	7.5	37.1	31.1	27.8	11.8	15.3	27.4	23.3
Arg	52.6	62.4	65.6	58.4	34.5	83.2	50.1	56.2	62.3	69.2
Asx	107.0	102.0	122.0	120.0	100.0	149.0	144.0	138.0	110.0	115.0
Glx	248.0	251.0	250.0	241.0	166.0	271.0	127.0	129.0	142.0	116.0

N.D.: not determined.
Source: Moreira et al. (1979).

found in earlier studies (Mitsuda et al. 1964; Catsimpoolas et al. 1967), acidic subunits have Phe, Leu, Ile and Arg at the NH_2 termini (Fig. 3.3) whereas all the basic subunits have Gly at the NH_2 termini (Fig. 3.4). N-terminal sequences show considerable homology between the members of individual families. These results suggest that members of each family arose from a common ancestor. Such a homology was already suggested by electrophoretic examination of the size of peptide fragments generated by cyanogen bromide digestion of purified acidic polypeptides by Kitamura and Shibasaki (1975 and 1977). Gilroy et al. (1979) have also reported the N-terminal

F₂(1) 1 Phe Ser *Phe* Arg 5 Glu Gln Pro Gln Gln 10 Asn Glu Cys Gln Ile 15 Gln

A₁ 1 Phe Ser Arg Glu 5 Gln Pro Gln Gln Asn 10 Glu Cys Gln Ile Gln 15 Lys Leu Asn Ala Leu 20 Lys Pro Asp Asn X 25 Ile

A₂ 1 Leu Arg Glu Gln 5 *Ala* Gln Gln Asn Glu 10 Cys Gln Ile Gln Lys 15 Leu Asn Ala Leu *Glu* 20 Pro Asp Asn Arg Ile

A₃ 1 *Ile* Thr Ser Ser 5 *Lys Phe* ... 10 Asn Glu Cys Gln *Leu* 15 Asn Asn Leu Asn Ala 20 Leu *Glu* Pro Asp *His* 25 Arg *Val*

F₂(2) 1 *Ile* Ser Ser *Lys* 5 *Leu* ... 10 Asn Glu Cys Gln *Leu* 15 Asn Asn Leu Asn Ala 20 Leu

A₄ 1 Arg Arg *Gly* Ser 5 Arg Ser *Gln Lys* Gln 10 Gln *Leu* Gln *Asp* (Ser) 15 *His* Lys Lys *Ile* (Arg) 20 *His Phe* Asn Glu *Gly* 25 Asp Gly

Fig. 3.3. The N-terminal amino acid sequences of 6 different glycinin acidic subunits aligned for optimal homology. Maximal homology is underlined with full line and partial homology with dotted line. Partial or non-homology is printed in italic type (*Source:* Moreira et al. 1979).

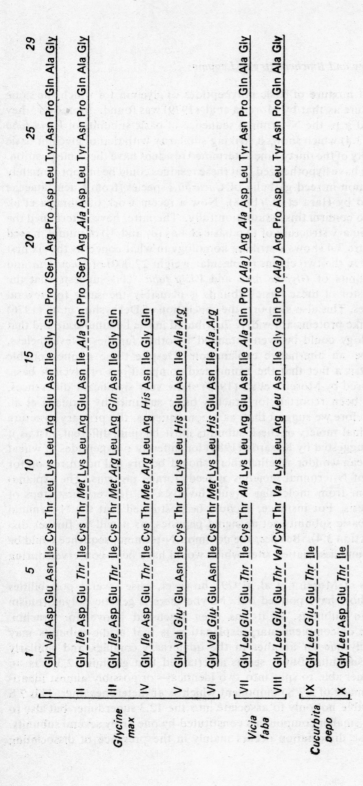

Fig. 3.4. The N-terminal amino acid sequences of 10 different basic subunits aligned for maximal homology. I to VI: glycinin basic subunits (Moreira et al. 1979); VII and VIII: *Vicia faba* legumin basic subunits (Gilroy et al. 1979); IX and X: *Cucurbita pepo* globulin basic subunits (Ohmiya et al. 1980). Maximal homology is underlined with full line and partial line and partial homology with dotted line. Partial or non-homology is printed in italic type.

sequences of a mixture of basic polypeptides of glycinin for which the same primary structure as that by Moreira et al. (1979) was found. Moreover, they have compared it to the N-terminal sequence of basic subunits of *Vicia faba* legumin (Fig. 3.4) which shows a striking similarity with that of glycinin basic subunits: twenty of the thirty-nine determined residues have the same position. Therefore they have hypothesized that these residues could be in approximately the same position in seed globulin of *Cucurbita* species (from Cucurbitaceae family) studied by Hara et al. (1978). Now a recent work of Ohmiya et al. (1980) seems to confirm this assumption fully. The latter have determined the N-terminal primary structure of two basic chains (δ_1 and δ_2) of pumpkin seed globulin. Figure 3.4 shows a striking homology in what concerns the six first residues between the two chains (molecular weight 22,000) of *Cucurbita* and the basic subunits of *Glycine max* and *Vicia faba*. This suggests that the common ancestor of these basic subunits is plausibly the same for several different species. This also supports the conclusion of Derbyshire et al. (1976) that legumin-like proteins are widely distributed in the Leguminosae and that protein homology could be even extended to other families. Nevertheless, there could be an alternative explanation. Despite these unquestionable homologies, it is a fact that the amino acid compositions of glycinin basic subunits analysed by Moreira et al. (1979) show very significant differences, as it has also been reported for arachin basic subunits by Yamada et al. (1979b). Therefore we suggest that, as a consequence, the primary structure of the C-terminal moiety of these subunits must be quite different, just as it was recently suggested by Kasarda (1980) for different polypeptides of wheat gliadins. One can wonder whether there should be any unknown reason for the similarity of N-terminal moieties of seed storage proteins. The explanation could stem from molecular events involved in the very first steps of their biosynthesis. For instance, it could be postulated that the N-terminal sequences of basic subunits act as signal peptides, as it will be further discussed (see Section 3.4). Besides, the common N-terminal sequence could be one of the subunit association sites which would have been conserved during evolution.

The results of Moreira et al. (1979), however, raise several possibilities that these authors have pointed out. Maybe there is genetic polymorphism among glycinin subunits, as it has been discussed above for arachin. An alternative or complementary explanation is that acidic subunits may substitute freely one for another in the quaternary complex and similarly for the basic subunits. But it seems ascertained that glycinin 12.3 S is an heterododecamer able to split into two identical—or possibly almost identical—moieties made of a 7 S component which is an heterohexamer. This 7 S component is able not only to associate into the 12.3 superdimer but also to dissociate into smaller components constituted by one or by several subunits. Such a stepwise dissociation occurs mainly in the presence of dissociating agents like guanidine chloride, urea, SDS and also at extreme pH or by heating. This plausibly explains the extraordinary variety of sizes found by many authors who have found as various sedimentation coefficients as 1.2, 1.5, 2, 2.2, 2.6, 3, 7, 7.5, 9, 10, 10.7, 11, 11.8, 12, 12.5 and 15 S (Wolf and Briggs 1959; Wolf and Sly 1967; Anderson et al. 1973; Hill and Breibenbach 1974; Okubo et al. 1976; Diep and Boulet 1977a; see also Chakraborty et al. 1979).

Conglycinins are less abundant in *Glycine max* seeds than glycinin and they have not been as intensively studied. Like glycinin, they appear as being located inside of proteins bodies (Catsimpoolas et al. 1968; Koshiyama 1972a). Only a few data are available concerning α- and γ- conglycinins. The first one appears as a low molecular weight component 2 S which could be devoid of quaternary structure: α-conglycinin should be made of one single polypeptide chain (Catsimpoolas and Ekenstam 1969; Catsimpoolas 1970). The two other ones are 7 S components. Concerning γ-conglycinin it has been fractionated by Catsimpoolas and Ekenstam (1969); Fisher et al. (1976) and isolated by Koshiyama and Fukushima (1976b). The last two have shown it is a glycoprotein (5.5 per cent of carbohydrate) with a sedimentation coefficient of 6.55 S and a molecular weight of 104,000. Although there are not yet any further experimental data, one can expect it has a quaternary structure with the occurrence of several subunits which could be identical or not.

The structure of β-conglycinin, the major of the three known conglycinins, has been partially elucidated by Thanh et al. (1975a and b), Thanh and Shibasaki (1976a and b, 1977, 1978a and b, and 1979) and also by Iibuchi and Imahori (1978a and b). Yamauchi et al. (1975 and 1976) and Yamauchi and Yamagishi (1979) have shown it is a glycoprotein. It consists of isotrimers of molecular weights ranging from 140,000 to 175,000. It is made of three main kinds of subunits called α, α' and β[1]. Among the ten theoretically possible multiple forms, six (B_1 to B_6) have been demonstrated to occur (Thanh and Shibasaki 1978a). They always contain either one α or one α' subunit associated with either two α or two β or one α plus one β subunits as follows: B_1, $\alpha'\beta_2$; B_2, $\alpha\beta_2$; B_3, $\alpha\alpha'\beta$; B_4, $\alpha_2\beta$; B_5, $\alpha_2\alpha'$; B_6, α_3. It can be seen that B_1 to B_5 are heterotrimers, B_3 being the only one to contain each kind of subunit, and B_6 is an homotrimer. Moreover, these six trimers are able to reversibly dimerize at low ionic strength or in the pH range 4.8–11.0 (Thanh and Shibasaki 1979). The resulting 9 S form is a superdimer (an hexamer) of two trimers facing each other. At extreme pH (2.0 and 12.0), dissociation into subunits (α, α' and β) is also reversible, so that the six molecular species B_1 to B_6 can be reconstituted by mixing the three subunits in urea solution and subsequently dialysing the solution against phosphate buffer.

[1]In the literature, the letters β and γ indicate both two kinds of 7 S conglycinins and two kinds of subunits of β-conglycinin, which unfortunately may cause some confusion.

In fact, the real quaternary structure is probably slightly more complex, for the possible number of subunit species is higher than three. Thanh and Shibasaki (1977) have isolated two other minor subunits (γ and δ) and evidenced the microheterogeneity of β subunits which is a mixture of four very similar components (β_1 to β_4).

All the main subunits α, α' and β are acidic. Their main characteristics are given in Table 3.5. The two subunits α and α' are very similar in amino acid composition with the same molecular weight (54,000), and are devoid of cysteine like β in which methionine is lacking. The three ones contain 4–5 per cent of carbohydrates of which several sequences have been determined by Yamauchi and Yamagishi (1979). They are Asn (GlcNAc)$_2$–(Man)$_n$ with 7, 8 or 9 for n values.

TABLE 3.5. **Physicochemical properties and amino acid composition (residues per 1000) of the three main β-conglycinin subunits**

	α	α'	β
M.W. (kilodalton)	54	54	40.5
N-terminal amino acid	Val Leu	Val Leu	Val
isoelectric point	4.9	5.2	5.6–6
Gly	44.2	48.7	44.1
Ala	45.7	43.1	53.3
Val	44.2	39.8	54.0
Leu	86.8	72.4	104.2
Ile	57.5	41.7	62.1
Ser	66.0	68.5	72.0
Thr	21.9	22.7	26.8
Tyr	23.1	22.3	25.8
Phe	52.1	46.0	63.9
Pro	70.7	43.9	50.9
Met	4.2	2.7	1.1
Cys	N.D.	0	0
Lys	62.2	65.5	53.4
His	11.7	36.1	20.2
Arg	87.0	62.2	70.9
Asx	117.6	120.2	129.8
Glx	204.9	264.2	167.3

Source: Thanh and Shibasaki (1977) and, for α' amino acid composition, Thanh and Shibasaki (1978a).

3.3.3. *Lupinus*

Physicochemical properties of *Lupinus* seed proteins have been first studied by Joubert during the fifties (see Blagrove and Gillespie 1975; Derbyshire et al. 1976). Three species were taken up: *Lupinus albus*, *L. angustifolius* and

L. luteus. It is only since the last five years that they have been more intensively investigated.

Several studies deal with the occurrence, extraction and fractionation of mainly vicilin-like and also legumin-like globulins (Klimenko and Ageeva 1977; Mironenko et al. 1978). The seed proteins of *Lupinus albus* have been extracted and partially fractionated by Cerletti et al. (1978) and by Sgarbieri and Galeazzi (1978). These authors have pointed out many subunit bands by SDS-polyacrylamide gel electrophoresis, with an estimation of corresponding molecular weights. Cerletti et al. (1978) have found at least three main proteins (α, β, and γ) just like Blagrove and Gillespie (1974 and 1975) which had previously shown to occur in *Lupinus angustifolius* seeds. As first suggested by Morawieska (1961), who named them α-, β- and γ-conglutins, *Lupinus angustifolius* seeds indeed contain mainly these three kinds of globulins which have been thoroughly studied by Blagrove and Gillespie (1975).

The first one, α-conglutin is an 11 S component consisting of four to five non-covalently linked subunits, having molecular weights within the approximate range 55,000–80,000. It can be fractionated into cold soluble and cold insoluble fractions. Both fractions split into subunits of 40,000, 53,000 and 63,000 after reduction, with extra subunits of molecular weight 36,000 and 50,000 for the cold soluble fraction. Now it seems too early to decide if α-conglutin is a legumin-like globulin or not.

The second globulin, β-conglutin, is the major one. On SDS-polyacrylamide gels, its electrophoretic pattern shows some fifteen different bands with molecular weights ranging from 20,000 to 60,000. Four of these bands correspond to major subunits which are not disulphide bonded.

The third globulin, γ-conglutin, is a minor constituent. Curiously it exists in the aqueous solution as a member of oligomeric species with sedimentation coefficients up to 10 S. But it gives only one sharp band by electrophoresis, as well as on SDS-polyacrylamide gel as on cellulose acetate in 6 M urea buffered at pH 7. Nevertheless this monomer can be either mainly associated into an homohexamer (Blagrove et al. 1980) or split by reduction into two small polypeptide chains of molecular weights 17,000 and 30,000. In other words, just like insulin, it is a dicatenar protein containing two disulphide bounded polypeptide chains, and it appears to be the simplest of the *Lupinus* seed globulins. Elleman (1977) has determined the primary structure of its smallest chain (154 amino acid residues). It is the first sequence of one polypeptide chain of a legume storage protein to have been determined.

Blagrove and Gillespie (1975) have also pointed out several other interesting points. They have shown that the three kinds of conglycinin seem to be located within protein bodies. They conclude it is the best evidence in favour of checking their storage protein status. This is a very plausible point of view into which we should want to introduce lights and shades. Concerning the amino acid composition, there are large differences between the three globulins. Conglutin γ has a much lower content of arginin (28 instead of 90–98 residues per 1000 for conglutins α or β) and glutamic acid or glutamin (83 instead of 208–249). But it has a much greater content of methionine (10 instead of nearly 1.5 for α and 0 for β) and of cysteine or half-cystine (29 instead of nearly 6). They all appear as being glycoproteins with a carbohydrate amount of 1.2 to 6.4 per cent. All three conglutins contain mannose, galactose and glucosamine and small quantities of fucose could be associated with γ-conglutin (Eaton-Mordas and Moore 1978). Moreover, Eaton-Mordas and Moore (1979) have studied the glycosylation of the main subunits: among four α- and six β-subunits and also the two α-moieties, all are glycosylated. The main glycopeptide released by proteolysis from α- and β-conglutins have molecular weight near 5,000 with similar galactose/mannose ratios.

From the point of view of polymorphism, conglutins α and β are characterized by a rather broad microheterogeneity, contrary to conglutin γ which is homogeneous. Moreover, they show a wide interspecific variability in proportion and type of subunits, while, among twelve different *Lupinus* species investigated by Gillespie and Blagrove (1975), γ-conglutin shows the same molecular size or electrophoretic mobility in the genus.

3.3.4. *Phaseolus vulgaris*

Storage proteins of *Phaseolus vulgaris* have not been systematically studied. Its seed protein bodies are not easy to isolate, as it will be seen further, because their preparation requires solvents devoid of water. The cotyledons are poor in legumin-like proteins, but they are rich in phytohemagglutinin (or lectin). These peculiarities have given birth to foggy data with a proliferation of names introduced for the possible main storage proteins involved.

Unfortunately nobody has still completed isolation and characterization of the subunits evidenced by several authors, which would be the only way to clarify the question. Nevertheless, several authors like Barkers et al. (1976); Bollini and Chrispeels (1978 and 1979); Murray and Crump (1979) have contributed to bring order to this confusion. The main conclusions already suggested by Derbyshire et al. (1976) seem to be rather confirmed, as noticed by Bollini and Chrispeels (1978) who have very clearly explained several discrepancies and summarized the present state of our knowledge.

Pusztai (1966) and Pusztai and Watt (1970) have successively isolated two kinds of globulins having carbohydrates: Glycoproteins I and II. Glycoproteins II was later suggested to be a vicilin-like protein by Derbyshire et al. (1976). Two proteins very similar to glycoprotein II were prepared, the first one by Racusen and Foote (1971) and the other one, named *Globulin G₁*, by McLeester et al. (1973). Like Danielsson (1949), the latter have first assigned it a sedimentation coefficient of 12 S, but they have shown later it is close to

7 S at pH 7, and associates to an 18 S component at pH 4.5 (Sun et al. 1974). Contrary to Pusztai and Watt (1970) who concluded that *glycoprotein II* was made of a single subunit of 40,000 daltons, McLeester et al. (1973) have shown that G_1 globulin is made of three different polypeptides of 43,000, 47,000 and 50,000. They have called it G_1 *phaseolin*.

The same number of subunits, with practically the same molecular weights were then evidenced by Barker et al. (1976) for the *major storage proteins* of *Phaseolus vulgaris* which is a vicilin, by Bollini and Chrispeels (1978) and by Murray and Crump (1979). All these authors were first inclined towards an heterotrimeric structure for the first step of association between subunits. Nevertheless Sun et al. (1974), and Stockman et al. (1976) suggested a tetrameric structure which seems to have been confirmed by Bollini and Chrispeels (1979).

Finally, as proposed by Bollini and Chrispeels (1978), the glycoprotein II of Pusztai and Watt (1970), the G_1 globulins of Hall and his associates (McLeester et al. 1973; Sun et al. 1974; Romero et al. 1975; Sun and Hall 1975, Stockman et al. 1976; Hall et al. 1978; Ma and Bliss 1978; Mutschler et al. 1980), the major storage protein, the major globulin fraction (Ishino and Ortega 1975), and the euphaseolin of Kloz and Klozova (1974) and Murray and Crump (1979), all these differently named proteins correspond to the same 7 S vicilin of *Phaseolus vulgaris*. It is a glycoprotein of which glycosylation has been described by Davies and Delmar (1979) as discussed later, and it seems to be devoid of interchain disulphide bonds (Barker et al. 1976).

Another glycoprotein which seems to play a storage function is a phytohemagglutinin accounting for about 10 per cent of seed protein content (Bollini and Chrispeels 1978). It corresponds to the fraction named G_2 globulin by McLeester et al. (1976) and is an isolectin, made of two different subunits (34,000 and 36,000) able to associate into the five possible homo- or heterotetramers. It is the second most abundant protein of *P. vulgaris* seed. It is present in protein bodies and could be tightly associated with their membrane (Pusztai et al. 1979). Legumin, an 11 S protein fraction, insoluble at pH 4.7, appears to be a minor protein component, made of at least three different subunits of 37,000, 34,000 and 20,000 (Derbyshire and Boulter 1976). Like several other legumins, it appears to contain intersubunit disulphide bonds.

3.3.5. *Pisum sativum*

Since remote times, *Pisum sativum* storage proteins (Pate and Flinn 1977, for a brief review) have been known to be composed of two major globulins, *legumin* and *vicilin*, first described by Osborne and Campbell (1896), then characterized following their sizes (12 S and 7 S) by Danielsson (1949) who has already shown that they are not synthesized at equal rates and they differ in the sulphur content, as later confirmed by Boulter and Derbyshire (1971) and Millerd et al. (1979). Two other important points have been also pointed out during the sixties. They are present in protein bodies (Varner and Schidlovsky 1963). They undergo reversible dissociation and the ten subunits evidenced by SDS-polyacrylamide gel electrophoresis (four in vicilin, six in legumin) all show at least two distinct N-terminal amino acids (Grant and Lawrence 1964). If legumin subunits are assumed to be dicatenar proteins made up of two disulphide linked polypeptides, the results of Grant and Lawrence (1964) are in agreement with the conclusions of Vaintraub and Gofman (1961) who had already suggested that legumin contains twelve polypeptide chains of average molecular weight of 33,000. Heterogeneity of vicilin subunits has also been demonstrated by Ghetie and Buzila (1968).

This heterogeneity was much emphasized during the last three years. Thomson and Doll (1979) and Thomson et al. (1979) have partially summarized evidence that both vicilin and legumin of *Pisum sativum* comprise a series of related heterooligomers. Among different *Pisum* lines Thomson et al. (1978) and Przybylska et al. (1975) have distinguished electrophoretically on cellulose acetate and polyacrylamide gels, many strong and faint bands. These globulins show a genetically controlled variation (Thomson and Schroeder 1978). Their accumulation has been studied by Millerd et al. (1978) and Thomson et al. (1979) who pointed out different genetic dependent patterns by immunoelectrophoresis analysis.

Legumin was purified by immunoaffinity chromatography (Casey 1979a) and by isoelectrofocusing (Gatehouse et al. 1980). These latter have suggested a glycinin-like quaternary structure, involving oligomerisation of several different acidic and basic subunits (Casey 1979b; Croy et al. 1979; Krishna et al. 1979). In spite of these numerous elaborate studies, there is not any unquestionable demonstration of such a quaternary structure, for nobody has still completely isolated and characterized the involved subunits. Until now, several of these subunits have been detected, their N-terminal amino acids identified, and their isoelectric points and molecular weight determined by different electrophoresis techniques, after partial fractionations. But everybody agrees with the very plausible structure summarized by Gatehouse et al. (1980). The legumin of *Pisum sativum* is an heterooligomer of circa 350,000 daltons, made up of six pairs of subunits (40,000 and 20,000 daltons) probably linked by one or more disulphide bridges. Both types of subunits are heterogeneous in charge and size. It is a fact that SDS-polyacrylamide gel with 1 per cent mercaptoethanol reveals about two dozen bands among which 40,000 and 20,000 bands are only a little stronger than the other ones (Croy et al. 1979). Recently Croy et al. (1980a and b), investigating *in vitro* proteosynthesis of legumin (see Section 3.4), have concluded that the disulphide bound subunits (40,000 and 20,000) are synthesized as a unique 60,000-dalton polypeptide chain. As we inferred above from the results of Vaintraub and Gofman (1961) and Grant and Lawrence (1964), it would mean that each legumin pair of (apparent) subunits consists of one dicatenar protein (like insulin) with two chains covalently linked by disulphide bridges. In our opinion, although split by reducing agents in electrophoretic studies, the basic and acidic molecules are not native subunits but only consist of the moieties of dicatenar proteins. Strictly speaking, *Pisum sativum* legumin would be composed of heterohexamer multiple forms.

Controversy surrounds the glycoprotein nature of legumin. Basha and Beevers (1976) have reported that both legumin and vicilin are glycosylated and Browder and Beevers (1978) have shown that the glycopeptide linked to legumin is GlcNAc-Asn. The same group (Beevers and Mense 1977; Nagahashi et al. 1980) has confirmed these results in a study of *in vitro* biosynthesis of glycoproteins in *Pisum sativum* cotyledons, as discussed later (see Section 3.4). Davey and Dudman (1979) have confirmed that legumin, as well as four vicilin fractions all contain covalently bound carbohydrates: glucose, mannose and glucosamine in legumin, and mainly mannose with lesser amounts of glucose, galactose and glucosamine in the four vicilin fractions. On the other hand, Casey (1979a) has found a negligible neutral-sugar content and Gatehouse et al. (1980) have shown that a radioactive-labelling technique failed to detect any carbohydrate bound to legumin subunits. It can be added that Basha and Beevers (1976) have noticed that not all the component subunits of the storage proteins are glycosylated to the same extent. It is likely that there may be genetic differences in the amount of carbohydrate among the different cultivars, which could partially explain literature discrepancies, but the question still remains open.

Besides the two main globulins, *Pisum sativum* seeds also contain albumins which have been investigated by Grant et al. (1976) and Murray (1979). These albumins also consist of many subunits (ranging from 18,000 to 90,000 daltons) of which certain major polypeptides function as reserve proteins during germination. Murray (1979) seizes this opportunity for proposing a seed storage protein definition, which will be discussed later (see Section 3.6).

3.3.6. *Vicia faba*

Beyond identical cross reactivity between *Pisum sativum* and *Vicia faba* globulins (Dudman and Millerd 1975), there are several analogies between the stories of their reserve proteins, although the latter have not been as much investigated as the former. The preparation and purification of *Vicia faba* globulins and mainly of legumin which predominates in the seed, have been studied by Danielson (1949); Vaintraub et al. (1962); Kloz et al. (1966); Jackson et al. (1969); Bailey and Boulter (1970); Graham and Gunning (1970); Millerd et al. (1971); Wright and Boulter (1973); Scholz et al. (1974); and Mori and Utsumi (1979). Salt or isoelectric precipitation after neutral or acid extraction, sucrose density centrifugation and ion-exchange chromatography have been used.

Vaintraub et al. (1962) have proposed that *Vicia faba* legumin contains twelve polypeptide chains of an average molecular weight of 40,000. Bailey and Boulter (1970) have then suggested a ten-chain model, made up of three kinds of subunits (56,000, 42,000 and 23,000 in molecular weight with a molar ratio of 1/3/6). But Wright and Boulter (1974) inclined towards a glycinin-like structure. They have indeed identified two kinds of acidic (α) chains of 37,000 daltons and three kinds of basic (β) chains with molecular weights of about 20,000, 21,000 and 24,000, with equimolar amounts of disulphide-bound α and β chains. In agreement with Vaintraub et al. (1962) they have postulated an $\alpha_6\beta_6$ model. We prefer to write it $(\alpha - \beta)_6$ in consequence of the fact that α and β chains are moieties—rather than subunits—of dicatenar subunits. It can be noticed that this occurrence of subunits in which one acidic chain and one basic chain are held together by disulphide bridges explains the equimolar amounts of these two kinds of chains. The 11 S legumin (approximately 350,000 daltons) is therefore a heterohexamer. The very low amount of carbohydrate (0.1 per cent of neutral sugars and less than 0.1 per cent of amino-sugars) found by Bailey and Boulter (1970) indicates that many, if not all, of the legumin subunits are not glycoproteins. Mori and Utsumi (1979) have confirmed this model of quaternary structure. They have also detected three sizes of basic chains with molecular weights of 19,000, 20,500 and 23,000 very close to the ones found by Wright and Boulter (1974). But, as it could be expected, they have found a number of acidic chains higher than that indicated by the latter: four instead of two, with molecular weights of 36,000, 36,000, 49,000 and 50,000. This fully supports the existence of multiple forms of legumin heterohexamers. Moreover, by the use of sucrose density centrifugation, they have demonstrated that *Vicia faba* legumin, characterized by a sedimentation coefficient of 12.3 S and molecular weight of 319,000 is able either to associate into a superdimer (17.2 S and 599,000 daltons) which is an heterododecamer, or to split into heterotrimers of 6.7 S.

The amino acid composition of this legumin has been determined by Bailey and Boulter (1970), Wright and Boulter (1974) and Mori and Utsumi (1979). They are in good agreement, although they do not coincide, probably as a result of differences between the cultivars used. All the polypeptide chains detected have been revealed and their molecular weights measured by using SDS-polyacrylamide gel electrophoresis with mercaptoethanol. But in the absence of this reducing agent, only three bands are detected with molecular weights of about 62,000, 60,000 and 48,000. These three bands are presumed by Mori and Utsumi (1979) to correspond to the intermediary subunits. We think they can be considered as the real subunits, each being made up of a basic chain linked to an acidic one.

The similarity between the three basic chains already suggested by their very close sizes is strongly confirmed by the great homology of their N-terminal amino acid sequences which have been determined by Gilroy et al. (1979), as already shown in Fig. 3.4. And the heterogeneity of legumin 12 S is also supported by the investigation of Utsumi and Mori (1980). Inside of three different cultivars and by the use of gradient gel electrophoresis, the latter have found the existence of four similar groups of molecular species, with molecular weights of 320,000, 350,000, 380,000 and 400,000. From these results, they have tentatively proposed seven different possible molecular forms of legumin heterohexamers. Moreover, by isoelectrofocusing, they have observed four (instead of three) main bands as basic chains and the kinds of main band of basic chains are different among the three cultivars.

The vicilin of *Vicia faba* seed gives four main bands on SDS-polyacrylamide gels and has four different N-terminal amino acids (Bailey and Boulter 1970). Wright and Boulter (1972) have shown that, contrary to legumin, vicilin reveals subunit patterns which change during seed development. This affords another evidence that it cannot be considered as a single protein as it has also been confirmed by Scholz et al. (1974). Schlesier et al. (1978) explain the series of bands obtained on polyacrylamide gel electrophoresis as association products and conclude in favour of the existence of monomeric, dimeric and trimeric molecules. It seems too early to answer the question whether these latter models can be compared with the quaternary structure of β-conglycinin suggested by Thanh and Shibasaki (1978a).

In *Vicia narbonensis* which is often considered to be of the same species as *Vicia faba*, Schlesier and Scholz (1974) have isolated a crystalline globulin fraction of a molecular weight below 50,000 daltons (probably a 2 S component). After purification by zonal precipitation on Sephadex G 25 and exclusion chromatography on Sephadex G 150, the protein precipitates during dialysis against distilled water at 5°C in the form of rhomboidal crystals.

Like for *Pisum sativum*, it can be seen that nobody has still isolated and purified any of the subunits of *Vicia faba* globulins. Only Gilroy et al. (1979) have worked with a mixture of very similar basic chains of legumin. Nevertheless the model of quaternary structure with which several authors agree is already supported by different converging results. It seems to be frequently encountered among different legume species, but it also appears as able to exhibit a very broad heterogeneity.

3.3.7. OTHER LEGUME SPECIES

Little recent information is available in the literature on seed storage proteins of other leguminosae genera or species.

Cajanus cajan seed contains mainly globulins which have been first studied by Krishna et al. (1977). Two major globulin fractions give three peaks by sucrose density centrifugation, corresponding to 9.3, 10.6 and 12.6 S. A third fraction is a 3.9 S component. Polyacrylamide gel electrophoresis exhibits nine bands with unreduced globulins and seven with reduced globulins. This suggests that the two major globulin fractions are made up of the same subunits with molecular weights ranging from 20,000 to 50,000. The third fraction is composed of three smaller subunits (with molecular weights from 13,000 to 20,000).

Cicer arietinum seed globulins seem to behave like those of *Pisum sativum* or of *Vicia faba,* with which they have been compared by Jackson et al. (1969). They consist of legumin and vicilin which have been characterized by their amino acid composition, N-terminal amino acids (Ser, Asp and Thr for vicilin and Gly, Leu and Thr for legumin), polyacrylamide gel electrophoresis pattern with 8 M urea, and also peptide maps and tryptic digests. The evolution of the globulin fractions during germination has been studied by Kumar and Venkataraman (1978).

Lens culinaris (also called *Lens esculenta*) storage proteins of six different genotypes have been extracted by Bhatty et al. (1976), but there is still no indication of their subunit composition.

Phaseolus angularis seed contains a major 7 S globulin of about 150,000 daltons, which has been purified by exclusion chromatography followed by DEAE cellulose chromatography (Sakakibara et al. 1979). It is devoid of cysteine or cystine. With SDS-polyacrylamide gel electrophoresis, one main and two minor components are detected, with molecular weights of respectively 55,000, 28,000 and 25,000. It is a glycoprotein containing 5 per cent of neutral sugar and 0.5 per cent of amino sugar. A circular dichroism study of its secondary structure indicates that it is a β-form rich protein.

Psophocarpus tetragonolobus storage proteins have been studied by Gillespie and Blagrove (1978) who have succeeded into a description of their subunit composition. The seed appears as practically devoid of legumin and contains three major storage proteins with sedimentation coefficients of 2 S, 6 S and 8 S respectively named *psophocarpin* B, C and A. Psophocarpin A is a single protein which probably consists into a homotetramer of 160,000 daltons. The monomer appears as a dicatenar protein of 40,000 daltons made up of two disulphide bonded polypeptide chains of 16,000 and 24,000. Compared to B and C fractions, psophocarpin A is the richest in sulphur amino acids and its amino acid composition and structure is relatively similar to those of conglutin γ of *Lupinus angustifolius*, already discussed above. Psophocarpin B gives one major band corresponding to a molecular weight of 20,000 beside several other minor bands of higher molecular weight. It has no quaternary structure. Psophocarpin C shows a series of components which are not covalently bound and which range from 15,000 to 80,000. Only one minor component (of about 17,000 daltons) is split by reducing agents and has therefore plausibly disulphide bound chains. A comparison of storage proteins of eighty pure lines of *Psophocarpus tetragonolobus* (Blagrove and Gillespie 1978) confirms the occurrence of some twenty distinct bands in their SDS-polyacrylamide gel electrophoretic pattern. There are no significant differences between the lines except for psophocarpin A which is almost absent in several lines.

Vigna mungo (classified in the past as *Phaseolus mungo*) has a seed containing mainly globulins. Some six different component sizes have been detected by SDS-polyacrylamide gel electrophoresis with molecular weights of 21,000, 25,000, 34,000, 55,000, 140,000 and 200,000 (Padhye ánd Salunkhe 1979). The major component, corresponding to the 55,000 daltons band, represents 65 per cent of total globulins. Isoelectrofocusing in 6 M urea with 2 per cent mercaptoethanol and 5 per cent sucrose shows that these globulins contain two basic and a series of acidic polypeptide chains. Padhye and Salunkhe (1979) have found only traces of cysteine in globulins and they have concluded that they were devoid of this amino acid. This could not to be in agreement with the occurrence of basic and acidic chains. Indeed if *Vigna mungo* globulins are glycinin-like proteins, as suggested by the latter, they are plausibly disulphide bound, unless they are arachin-like globulins.

Vigna radiata (classified in the past as *Phaseolus aureus*) seed proteins are both legumin and vicilin. Vicilin is the most abundant fraction (Ericson 1975). It has a sedimentation coefficient of 8 S and contains four different subunits of molecular weights 63,000, 50,000, 29,500 and 24,000. The latter three subunits are very similar in their amino acid composition. Two of them (50,000 and 29,500 daltons) contain one carbohydrate group which is linked by a glucosamine residue to the amide nitrogen of an asparagine residue. The carbohydrate unit contains two residues of glucosamine and ten of mannose. The legumin fraction has been studied by Derbyshire and Boulter (1976). It is an 11 S component dissociated (by SDS with mercaptoethanol) into subunits or polypeptide chains with molecular weights of 37,000, 34,000 and 20,000.

Vigna unguiculata (sometimes named *Vigna sinensis*) has also seed storage proteins consisting of both 7 S and 11 S globulins, which have been investigated by Carasco et al. (1978) and Sefa-Dedeh and Stanley (1979a and b). It is a 7 S protein which largely predominates. By SDS-polyacrylamide gel electrophoresis after treatment with mercaptoethanol, about twenty different bands range from 10,000 to 100,000 daltons. The components corresponding to the three main bands (52,000, 54,000 and 56,000 daltons) have been isolated by ion exchange chromatography in 8 M urea by Carasco et al. (1978). Their amino acid compositions are significantly different and two of them have different N-terminal amino acid and circa 2 per cent of bound neutral sugar. On another side, the prominent electrophoretic components of thirteen different species of *Phaseolus* genera have been compared with the three species of *Vigna* we have discussed (*V. mungo, V. radiata* and *V. unguiculata*) by Derbyshire et al. (1976a).

3.3.8. Conclusion: Maximal Packing Hypothesis and Polymorphism

The main classical distinctions between legumin and vicilin have been reminded at the beginning of the present section. They principally concern physico-chemical properties of these two kinds of globulins.

Both kinds of globulins can now also be characterized by their quaternary structure and subunit composition. From this point of view, if it is a little early to tabulate the available data as a synopsis, it becomes nevertheless possible to draw some common features inside of each kind of legume seed globulins, and to postulate general models.

A legumin generally consists of multiple forms of heterohexamers made of the least partially different subunits. These subunits seem to be only non covalently bound, as in classical quaternary structure (Sund and Weber 1966). They have molecular weights of around 60,000 and are almost always devoid of bound carbohydrate groups. With reducing agents, they split into two different polypeptide chains: one acidic chain with a molecular weight of nearly 40,000, and one basic chain with a molecular weight of nearly 20,000. The N-terminal sequence of a basic polypeptide is probably very constant through many genera of Leguminosae and even of other plant families, with only discrete point mutations (the N-terminal amino acid seems to be always glycine). Each subunit may result from a single pro-protein which is cut (like proinsulin into insulin) into two distinct polypeptide chains remaining bound by disulphide bridges. Moreover, legumin hexamers can not only stepwise dissociate into subunits with often stable intermediary trimers but also associate into dodecamers or possibly superoligomers of larger degrees of polymerization. Such a model is in agreement with the presently available data that we have reported above on the legumin of *Arachis hypogaea* (α-arachin), *Glycine max* (glycinin), *Pisum sativum*, *Vicia faba*, *Vigna mungo*, probably *Cicer arietinum* and possibly *Lupinus angustifolius* (α-conglutin).

Due to the small number of legume species in which vicilin has been investigated, it is perhaps reckless to draw any model concerning this protein. Contrary to legumin, vicilin contains always a bound carbohydrate but its subunits are often devoid of cysteine or cystine (except for β-conglutin of the *Lupinus* seed), which precludes any disulphide bond. The only other common feature could be its aptitude to associate into heterotrimers, as shown with the vicilin of *Glycine max* (β-conglycin), *Phaseolus vulgaris* (trimers or tetramers) and *Vicia faba*. Furthermore, β-conglycin is able to associate to a higher degree in heterohexamers.

On another side, the γ-conglutin of *Lupinus angustifolius* which seems to escape to classification into legumin and vicilin, appears to be a homohexamer.

It is striking to state the high frequency of subunit association into either trimers or hexamers. Now one cannot refrain to notice that these two kinds of quaternary structure are exactly those which allow the most compact packing: such a characteristic is plausibly an important one for storing the protein inside of protein bodies. We think it is not by chance that reserve proteins are able to associate into these kinds of quaternary structures. Two other arguments are in favour of this explanation. First is the sequence homology of basic polypeptide chains of legumin from *Glycine max*, *Vicia faba* and of globulins of *Cucurbita pepo* shown by Gilroy et al. (1979) and Ohmiya et al. (1980). It is indeed known that the quaternary structure is mainly a consequence of primary structure. Secondly the fact that globulin subunits of plant species as distant as *Glycine max* and *Sesamum indicum* can associate between themselves and undergo *in vitro* molecular hybridization, as recently shown by Mori et al. (1979). Moreover, it is interesting to mention the particular case of urease. It could act as an enzyme during seed development and as a storage protein during germination. It is indeed present in many legume seeds, and amounts to 1 per cent of total seed protein of *Canavalia ensiformis* (Bailey and Boulter 1971). It consists of one single kind of subunits (about 80,000 in molecular weight) which associate into two main homooligomers: cyclic homotrimers (protein A_1 of 240,000 daltons) and homohexamers (protein α of 480,000 daltons) as it has been shown by Fishbein et al. (1977). The latter have pointed out it is also able to associate into homopolymers in which α molecules are aligned and should be bound by disulphide bridges. This is an exceptional case in which disulphide bonds participate to the stabilization of quaternary structure (Sund and Weber 1966).

In other words, we arrive at the conclusion that legume storage proteins exhibit a particular fitness to compacting.

Another conclusion to draw is their heterogeneity and their polymorphism. The different reasons of plant protein heterogeneity have already been analysed and reviewed elsewhere (Mossé 1973). The occurrence of genetic variants which often explains the secondary minor bands of globulin electrophoretic patterns is a first cause of heterogeneity, as well as the multiplicity of a few kinds of subunits. Other causes are post-translational, among which one can mention: the occurrence of several degrees of oligomerization (which results into a series of sedimentation coefficients); the possible interchange of some of the subunits, as shown for β-conglycin which arises into multiple forms; the contingent variation in size or nature of bound carbohydrate groups (of which the exact function remains still unknown). All these reasons combine into producing a great heterogeneity both of size and charge.

3.4. Accumulation, Location and Biosynthesis

The study of the seed proteosynthesis involves the solution of three main questions: (i) Where do the storage proteins accumulate? The answer to this question corresponds to locate storage proteins within specialized organelles (protein bodies); (ii) When are these proteins synthesized? The answer to this question is not only a global accumulation pattern as a function of time but also the detailed understanding of the time-proteosynthesis processes relationships; (iii) How are seed proteins synthesized? This question is a molecular biology problem which implies not only to acknowledge the global proteosynthesis regulations but also to have precise informations about nucleic acid transcription and maturation and about their translation and the post-translational modifications of nascent polypeptide chains.

3.4.1. ACCUMULATION OF STORAGE PROTEINS DURING SEED DEVELOPMENT

The protein accumulation within legume seeds has been reviewed by Dure (1975); Millerd (1975) and Chrispeels et al. (1979). It is well established that the embryogenesis is remarkably constant from one legume species to another: the deposition of storage proteins takes place with other storage product synthesis during the phase of growth (cell expansion) which follows the rapid cell division phase of the embryo in the course of which essentially no storage protein is synthesized. During the desiccation phase no or few proteins are biosynthesized. The data from the literature are nevertheless difficult to compare because of the genotypical and phenotypical variations from one study to another. Nevertheless the synthesis of storage proteins usually begins on the ninth to twelfth day after flowering. Most of the accumulation of storage proteins takes place between the fifteenth and thirtieth day after anthesis.

It is interesting to notice that the two globulin fractions of legume globulins, i.e. legumin and vicilin, are not synthesized at the same time, like it has been shown for the cereal proteins (Bishop 1930; Landry and Moureaux 1976). In the species accumulating both legumin and vicilin like globulins, the beginning of the synthesis of vicilin always precedes that of legumin. The rate and the extent of legumin biosynthesis is far higher than the vicilin one, so this globulin fraction is the most important one at the end of the seed development. This important temporal difference has been shown as well in *Pisum sativum* (Millerd et al. 1978); *Vicia faba* (Graham and Gunning 1970); *Glycine max* (Hill and Breidenbach 1974; Ochiai-Yanagi et al. 1978) and *Arachis hypogaea* (Basha et al. 1976). From a general point of view, the proteins begin to be synthesized following the increasing order of their sedimentation coefficient.

As previously mentioned, the reserve protein deposition is affected by phenotypical factors: Randall et al. (1979) have shown that the pea storage protein synthesis may be dependent on extreme environmental conditions (the deficiency in some mineral elements results in variations of the protein fraction ratio) and Thompson et al. (1977) have evidenced that the nature of the nitrogen source provided to *in vitro* cultured soybean cotyledons influences greatly the amount of synthesized proteius. Recent data by Mutschler et al. (1980) have put forward the genotype influence in vicilin accumulation rate in *Phaseolus* cotyledons.

3.4.2. LOCATION OF STORAGE PROTEINS WITHIN PROTEIN BODIES

ULTRASTRUCTURE

Seed storage proteins accumulate within subcellular organelles called protein bodies—general term which must be preferred to many other terms such as aleurone grains, aleurone vacuoles, protein vacuoles or protein grains which are—sometimes confusing. These organelles have recently been reviewed (Pernollet 1978) and only later studies are to be discussed here. It is worth reminding that legume seed protein bodies are membrane bound spherical organelles, a few microns in diameter, filled with proteins and phytates, generally devoid of internal inclusions (globoid or crystalloid) except for *Arachis hypogaea* (Sharma and Dieckert 1975) and *Lupinus luteus* (Sobolev et al. 1977). In this case, several protein body types have been evidenced and their relative amount shown to depend on the distance from the axis (Mlodzianowski 1978). Barker et al. (1976) and Pusztai et al. (1978) have shown that *Phaseolus vulgaris* protein bodies are 2 to 3 μm in diameter without inclusion. Figure 3.5 represents scanning electron microscopic views of protein bodies of *Pisum sativum* cotyledons. A recent paper by Lott and Buttrose (1978) reviews the occurrence of globoids in protein bodies of legume seed cotyledons. These authors have shown that the cotyledon protein bodies in some legumes have large and frequent globoids (*Arachis, Clianthus, Cassia*), whereas others have only small and rare inclusions (*Aracia, Glycine, Phaseolus, Pisum, Vicia*). The membrane of these protein bodies is generally smooth, but in some cases (Barker et al. 1976) some granules, comparable to ribosomes attached to maize protein bodies (Burr and Burr 1976) are noticeable. At last, *Vigna radiata* protein bodies have been shown by Thomson (1979) to be homogeneous and their membrane smooth.

The membrane of *Vicia faba* protein bodies has been recently isolated and shown to be a typical phospholipid bilayer, about 100 Å in thickness, in which about 30 different proteins are embedded (Weber et al. 1979).

PROTEIN BODY ISOLATION

Many authors have isolated protein bodies by different techniques which have been reviewed by Mascherpa (1975). These are mainly differential sedimentation and density gradient ultracentrifugation methods using either aqueous or non-aqueous media (in order to preserve the protein body structure of some species). The methods used are summarized in Table 3.6 where the apparent density of protein bodies is shown to be almost constant for the main protein body fraction whichever the species may be. This is an indication that the structure of legume protein bodies is rather homogeneous. As a matter of fact, the constancy of the density is an index that the ratio between the membrane lipids and the proteins is constant. Large scale protein body isolation is now being performed by the use of a zonal rotor (Begbie 1979), even on an industrial level for *Vicia faba* protein body isolation (Olsen 1980). Nevertheless it is important to notice that the purest protein bodies are necessarily obtained by means of density gradient centrifugation.

An important point to elucidate is the heterogeneity of protein bodies as well as their integrity. From Table 3.6 it appears that some authors have isolated several kinds of protein bodies depending on their density: Is it an artifact or does it represent a real *in vivo* heterogeneity? The case of the pea seed

TABLE 3.6. Isolation techniques and apparent density of legume
seed cotyledon protein bodies

Species	Authors	Method and medium[a]	Apparent density of protein bodies $(g.cm^{-3})$
Pisum sativum	Varner and Schidlovsky (1973)	DS ; A	—
	Alekseeva and Kovarskaya (1978)	DS ; NA	—
	Thomson et al. (1978)	DS ; A	—
Vicia faba	Morris et al. (1970)	GU ; A	1.8
	Koroleva et al. (1973)	DS ; NA	—
	Weber et al. (1978)	GU ; A	1.36
	Van der Wilden et al. (1980)	GU ; A	—
Phaseolus sp.	Ericson and Chrispeels (1973)	GU ; A	1.27
	Harris and Chrispeels (1975)	DS ; A	—
	Miège and Mascherpa (1976)	GU ; A	1.27 and heavier
	Barker et al. (1976)	GU ; A	1.30
	Pusztai et al. (1977)	DS ; A	—
	Pusztai et al. (1978)	DS, GU ; A	1.32
	Bollini and Chrispeels (1978)	DS ; GA	—
	Begbie (1979)	GU ; NA	1.37
Lablab purpureus	Miège and Mascherpa (1976)	GU ; A	1.2 and 1.4
Glycine max	Saio and Watanabe (1966)	DS ; NA	—
	Tombs (1967)	GU ; A	1.3 and >1.32
	Wolf and Baker (1972)	GU ; A	—
Arachis hypogaea	Dieckert et al. (1962)	DS ; NA	—
Lupinus luteus	Sobolev et al. (1977)	GU ; NA	1.27 and 1.50

[a]DS = differential sedimentation, GU = gradient ultracentrifugation; A = aqueous medium, NA = non-aqueous medium.

protein bodies is the simplest: no heterogeneity has been put forward even by aqueous media isolation. On the contrary, lupine protein bodies have been shown to be structurally heterogeneous and Sobolev et al. (1977) have proved that there is an actual polymorphism among cotyledon protein bodies depending on their location with regard to the axis. Results by Tombs (1967) are not so convincing and a real heterogeneity is not likely to occur in *Glycine max*. *Phaseolus* protein bodies are certainly very flimsy as it has been shown by Miège and Mascherpa (1976). Nevertheless a protein body polymorphism depending on their location in the cotyledon has also been put forward by Barker et al. (1976) on a qualitative protein content basis. More recently Begbie (1979) has proved that it is possible to isolate and purify protein bodies from *Phaseolus vulgaris* cotyledons by density gradient centrifugation in non-aqueous media (anhydrous glycerol) without any occurrence of heterogeneity.

A point which is still ambiguous is the important specific difference of protein body stability in the aqueous media. If *Pisum sativum* protein bodies are very stable in water, *Phaseolus* ones rapidly lose their integrity if they are not extracted with non-aqueous solvents (Pusztai et al. 1978; Barker et al. 1976; Begbie 1979). This is also the case for *Glycine max* (Wolf and Baker 1972) and *Arachis hypogaea* (Mikola et al. 1975). The weakness of *Phaseolus* protein bodies can be tentatively correlated with the lower membrane content (evidenced by the greater density) which could be interpreted in terms of disruption or discontinuity of this membrane as it has been shown for *Pisum* and cereal protein bodies by Miflin et al. (1980).

PROTEIN CONTENT OF PROTEIN BODIES

From a quantitative point of view, the protein content of legume protein bodies is *ca.* 80 ± 10 per cent. The results obtained by many authors are given in Table 3.7. Some of them are lower due to protein body disruption during isolation as it has been said above.

On a qualitative point of view, most authors have used chromatographic, electrophoretic and immunological methods to identify the polypeptide chains which are stored within protein bodies. These proteins are principally globulins and phytohemagglutinins as storage proteins, and some enzymes and enzyme inhibitors. We shall first deal with globulins and phytohemagglutinins which are the most important proteins by weight.

In *Pisum sativum*, vicilin and legumin are the only proteins which are mainly present within protein bodies and none or very few of these storage proteins are located outside (Varner and Schidlovsky 1963; Alekseeva and Kovarskaya 1978; Thomson et al. 1978). No phytohemagglutinin can be located within pea protein bodies (Millerd et al. 1978; Rougé 1977). In *Vicia faba*, Graham and Gunning (1970) have clearly shown that most protein bodies contain both legumin and vicilin but they could only detect vicilin in some storage organel-

**TABLE 3.7. Protein content of legume cotyledon protein bodies
(percentage of dry weight)**

Species	Authors	Percentage of protein
Pisum sativum	Thomson et al. 1978	90
Vicia faba	Morris et al. 1970	88
	Weber et al. 1978	80
Phaseolus (sp.)	Pusztai et al. 1977	57
	Pusztai et al. 1978	82
	Begbie 1979	78
Glycine max	Saio and Watanabe 1966	71
	Tombs 1967	95 and 75
Arachis hypogaea	Dieckert et al. 1962	67
	Sharma and Dieckert 1975	69
Lupinus luteus	Sobolev et al. 1977	60 and 66

les and in others, neither vicilin nor legumin. This immunofluorescence study has been confirmed by biochemical analyses performed by Morris et al. (1970). The phytohemagglutinins are found restricted to protein bodies in *Vicia faba*; they represent 1 per cent of protein bodies extracted by Weber et al. (1978) who have used serological techniques. In the case of *Phaseolus*, phytohemagglutinins (isolectins) are nearly all located within protein bodies (Barker et al. 1976; Pusztai et al. 1977 and 1978; Bollini and Chrispeels 1978; Begbie 1979), reaching 10 per cent of the proteins which seem largely to be constituted by much vicilin and few legumin (Ericson and Chrispeels 1973). In *Glycine max*, Catsimpoolas et al. (1968) have shown that protein bodies are filled with larger amounts of legumin, some vicilin and five other components. Tombs (1967) had previously evidenced that all glycinins (legumins) are stored within protein bodies. These results have been confirmed by Alekseeva and Kovarskaya (1978). Studies on *Lupinus* are less advanced but Sobolev et al. (1977) have shown that globulins are the main reserve proteins within protein bodies. We end this description with *Arachis hypogaea* protein bodies which have been the subject of earliest studies (Bagley et al. 1963; Dieckert et al. 1962; Sharma and Dieckert 1975; Dieckert and Dieckert 1976 for review). The presence of globulins (arachin and conarachin) within these organelles has been emphasised by Altschul et al. (1964) who have proposed a new classification for seed storage proteins, calling "aleurins", the proteins filling the protein bodies.

It is now more and more usual to define storage proteins as those which are stored within protein bodies. If it is globally correct, one must not forget

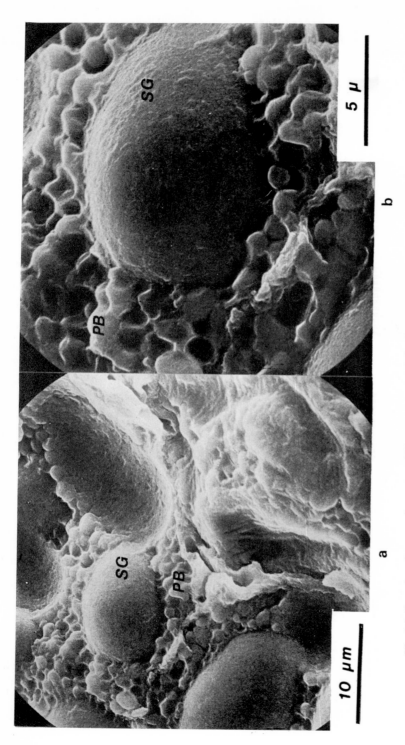

Fig. 3.5. Scanning electron microscopy of *Pisum sativum* cotyledon fracture. SG=starch granules; PB=protein bodies.
a: G=2 400, b: G=5 400.

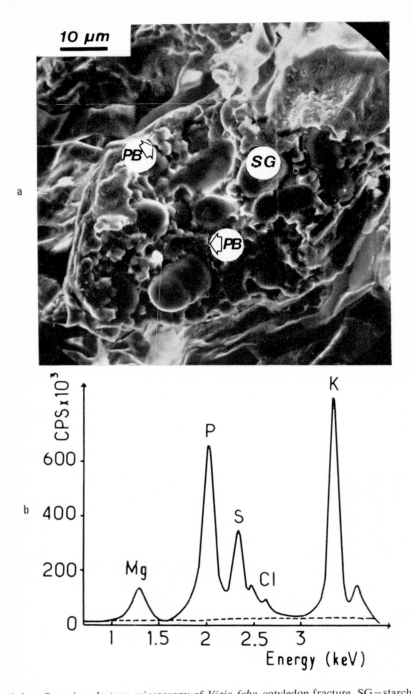

Fig. 3.6. a: Scanning electron microscopy of *Vicia faba* cotyledon fracture. SG=starch granules; PB=protein body. G=1 000.
b: Energy dispersive X-ray analysis. —— protein body spectrum; - - - starch granule spectrum; the elements corresponding to the peaks are named using their usual symbol.

that a few proteins which are not storage proteins, such as enzymes, are present within protein bodies, and that a non-negligible part of nitrogenous reserves is found outside these organelles. Nevertheless it is now very likely that all storage globulins (legumins and vicilins) are principally contained by protein bodies in most studied legume seeds. As a matter of fact the only storage (and non lectin) globulin which has been found in a cytoplasmic compartment is conarachin (see Section 3.3). Contrarily, lectins are located within or without these organelles depending on the species; their role as storage proteins is certainly of importance.

The occurrence of enzymes (amylases, proteases, phosphatases) within protein bodies has been reviewed elsewhere (Ashton 1976; Pernollet 1978) and we will only insist on proteolytic enzymes and enzyme inhibitors. Proteases nave been shown to be located within protein bodies of various legume seeds: in *Pisum* about half the proteolytic activity is associated with protein bodies (Hobday et al. 1973; Matile 1968), in *Vicia* Koroleva et al. (1973) and Shutov et al. (1978) have also shown that proteases were partly stored within protein bodies. In mung bean (*Vigna radiata*), Baumgartner and Chrispeels (1977) have evidenced that a globulin specific protease, called vicilin peptido-hydro-lase, accumulates into protein bodies at the beginning of germination. Correlatively, Pusztai et al. (1977) have concluded that *Phaseolus vulgaris* protein bodies possess very little autolytic activity. Recent works on *Vigna radiata* by Harris and Chrispeels (1975) and by Van der Wilden et al. (1980) have put to evidence that, as it had first been postulated by Matile (1968), protein bodies are autophagic organelles, but they have shown that globulin endoproteases come from the cytoplasm in the course of germination. Sobolev et al. (1977) have also evidenced a leucine aminopeptidase in *Lupinus* protein bodies. The biological significance of the presence of enzymes within protein bodies will be discussed in the part dealing with seed storage protein degradation during germination (see Section 3.5).

It is yet worthy to end by the presence of enzyme inhibitors within protein bodies: in *Glycine max*, Tombs (1967) and Knypl and Radziwonowska-Jozwiak (1979) have shown that trypsin inhibitors are found outside these organelles, while in *Vigna radiata*, Chrispeels and Baumgartner (1978) have located trypsin inhibitor in the cytoplasmic compartment. In *Pisum sativum*, Hobday et al. (1973) have evidenced that 25 per cent of enzyme inhibitors are associated with protein bodies, but they have supposed they are outside the membrane. Contrarily, Pusztai et al. (1977) have found these inhibitors within *Phaseolus vulgaris* protein bodies. It is likely that, as for lectins, the enzyme inhibitors are variably distributed among species.

Other compounds such as phytin and mineral elements are present within protein bodies, but they are usually not gathered in a globoid, except for *Arachis* and *Lupinus* (Pernollet 1978). They can be evidenced by the use of X-ray dispersive analysis combined to scanning electron microscopy (Lott and Buttrose 1978 for review). Figure 3.6 illustrates the use of element analysis to identify protein bodies when using scanning electron microscopy. The sulphur content is a good index for protein location (Pernollet and Mossé 1980).

Nucleic acids have also been shown to be more or less associated with protein bodies. Their presence is only clearly evidenced during seed development and, as it will be explained by the protein body ontogeny, nucleic acids are linked but external to protein bodies which are not to be compared to plastids.

3.4.3. Storage Protein Biosynthesis

The question is to understand the processes involved in the synthesis of storage proteins and in their transport into protein bodies where they are deposited. The many evidences that protein bodies are not plastids have already been reviewed (Pernollet 1978). It is now clear that proteins are synthesized on ribosomes located outside protein bodies and undergo a carrying mechanism by which they accumulate into these organelles. As it has already been pointed out by several authors, seed storage proteins are therefore comparable to secretory proteins (Chrispeels 1976; Millerd 1975). The protein transport process across membranes is now known to be frequently associated with post-translational modifications of the polypeptide chains. It has first been hypothesized by Blobel and Dobberstein (1975), and then evidenced by several authors (for review, see Leader 1979), that secretory proteins are synthesized with an extra N-terminal sequence of 15–25 amino acids (equivalent to a 1,000–2,000 daltons extra molecular weight). This extra sequence, called signal peptide or leader sequence, is mainly constituted of hydrophobic residues (Austen 1979) and allows the initiation of the phospholipid membrane crossing. Once this extra part of the polypeptide chain has crossed the membrane, it is cut out by a specific enzyme (signal peptidase). Other modifications such as glycosylation may accompany the membrane crossing.

In the case of cereal grains, it has been proved that some prolamins, the endosperm specific alcohol-soluble storage proteins, undergo post-translational modifications, with signal peptide hydrolysis (Burr et al. 1978; Larkins and Hurkman 1978; Melcher 1979). In the cereal endosperm it has also been shown that protein bodies originate from endoplasmic reticulum expansion (for review see Pernollet 1978; Mossé and Landry 1980). On the contrary, the findings in legume seed storage protein are more intricate and a controversy still exists both about the origin of the protein body and about the proteosynthesis mechanism. So we will try to draw some hypothetical models from the data taken either from the cytological studies or from the biochemical analyses.

Microscopical and biochemical observations during seed maturation

It seems that protein bodies have more than one origin and it is clear that these organelles come as well from the splitting of the cell vacuoles in which proteins accumulate at the beginning of the expansion phase of the seed development, as from new vesicles formed from the endoplasmic reticulum.

First of all, it is important to emphasize again that legumin-like proteins are not synthesized at the same time than vicilin-like globulins: at the beginning of the expansion phase and after albumin synthesis vicilin polypeptide chains accumulate before legumin proteosynthesis starts (Beevers and Poulson 1972). It is the case of all studied legume seeds which contain both kinds of globulins, i.e. *Pisum sativum* (Millerd et al. 1978), *Vicia faba* (Graham and Gunning 1970; Neumann and Weber 1978), *Glycine max* (Ochiai-Yanagi et al. 1978) and *Arachis hypogaea* (Basha et al. 1976).

Another result to point out is that the endoplasmic reticulum proliferation coincides with globulin synthesis: it has been shown in *Pisum sativum* (Bain and Mercer 1966; Craig et al. 1979), in *Vicia faba* (Briarty et al. 1969; Payne and Boulter 1969; Bailey et al. 1970; Neumann and Weber 1978); in *Phaseolus vulgaris* (Opik 1968; Sun et al. 1978; Bollini and Chrispeels 1979) and in *Vigna unguiculata* (Harris and Boulter 1976). Moreover membrane bound polyribosomes, and therefore rough endoplasmic reticulum, play an important role in globulin synthesis (Payne and Boulter 1969; Payne et al. 1971; Bailey et al. 1970; Craig et al. 1979; Harris 1979). In spite of disagreements between authors, it seems that *Phaseolus vulgaris* would not follow the same rule: Sun et al. (1978) have shown that the free/bound ribosome ratio is 18:1, while Bollini and Chrispeels (1979) have found 1:2; the latter have not evidenced any functional difference between free and bound ribosomes using *in vitro* tests. This question is not clear-cut since Bollini and Chrispeels (1979), as well as Harris and Boulter (1976) in the case of *Vigna unguiculata*, have confirmed the important role of the rough endoplasmic reticulum in proteosynthesis.

Contrary to cereal endosperm protein bodies, legume protein storage organelles would exhibit a double origin. At the beginning of protein synthesis, some protein bodies originate by the subdivision of the cell vacuoles once they have been filled with storage proteins. Other later formed organelles take their origin in the endoplasmic reticulum. As early as in 1966, Bain and Mercer have shown in *Pisum sativum* by the use of electron microscopy, that at the beginning of their synthesis, storage proteins accumulate within the cell vacuoles which divide into "vacuole protein bodies" and that protein bodies originating from the endoplasmic reticulum are formed later. The recent results by Craig et al. (1979) are in agreement with this description, bringing the important data that rough endoplasmic reticulum and membrane bound polysomes are involved in the synthesis of the second kind protein bodies. At last Miflin et al. (1980) have pointed out that pea cotyledon storage proteins appear to be synthesized on the endoplasmic reticulum, but the protein bodies in which they are deposited are bounded by a membrane which lack NADH-cytochrome c reductase, an endoplasmic reticulum marker enzyme. In

Lupinus albus, a legume seed which stores both legumin and vicilin, Davey and Van Staden (1978) have reached the same conclusions as in the case of pea. In *Vicia faba*, the role of vacuole division which occurs during protein body formation was first evidenced by Briarty et al. (1969); the role of the rough endoplasmic reticulum in the formation of a second class of protein bodies has been pointed out by Neumann and Weber (1978). It is worth noticing that Graham and Gunning (1970) have clearly shown that some protein bodies of the mature seed contain only vicilin, while the major part is filled both with legumin and vicilin. Contrary to these results, a single origin has been shown for *Phaseolus vulgaris* by Öpik (1968) and for *Vigna unguiculata* (Harris and Boulter 1976) i.e. accumulation in vacuoles which later divide into protein bodies. It is important to notice that these two species synthesize mainly large amounts of vicilin and very few legumin polypeptide chains.

The role of the Golgi apparatus in protein synthesis and transport has been emphasized in the *Vigna unguiculata* cotyledon development by Harris and Boulter (1976). In other species, which accumulate both legumin and vicilin, the importance of these organelles is only transient. Active dictyosomes are only seen at the beginning of globulin synthesis and are contemporaneous with vicilin production. It has been shown by Bain and Mercer (1966) for *Pisum sativum*, by Davey and Van Staden (1978) for *Lupinus albus*, and by Neumann and Weber (1978) for *Vicia faba*. Therefore the Golgi apparatus seems to play a role in vicilin-like proteins biosynthesis and transport towards vacuole but not in legumin-like polypeptide chain accumulation into *de novo* synthesized protein bodies.

The transport of storage protein and mainly of legumin into protein bodies was intriguing until the last few years. As a matter of fact, using radioactive labelling and autoradiography of *Vicia faba* cotyledon slices, Bailey et al. (1970) have shown that proteosynthesis occurs in the rough endoplasmic reticulum and that proteins are translocated towards protein bodies within 25 minutes. If in *Pisum sativum* (Craig et al. 1979) and in *Lupinus albus* (Davey and Van Staden 1978) connection and continuity between endoplasmic reticulum and protein bodies precursors (vesicles) have been shown, the study by Harris (1979) in *Vicia faba* cotyledons constitutes a real progress in storage protein synthesis understanding. This author has used a thick section electron microscope technique which provides stereoscopical views of the developing cotyledon. He has shown that cisternal endoplasmic reticulum vesicles are artefactual structures of thin section techniques which represent the interconnection of cisternal and tubular endoplasmic reticulum. At the beginning of the proteosynthesis, the endoplasmic reticulum is smooth and tubular; then ribosomes bind parts of this tubular reticulum where cisternae will later appear, connected together by the smooth tubular endoplasmic reticulum. During dehydration the rough endoplasmic reticulum becomes smooth. So storage proteins accumulate within cisternae which become protein bodies linked together by a network formed by the tubular endoplasmic reticulum. This latter structure would be responsible for protein transport. Harris hypothesizes that different classes of storage protein would undergo different transport processes and postulates a role for the Golgi apparatus in protein transport towards vacuoles.

A last point to emphasize is that, contrary to cereal endosperm protein bodies (Burr and Burr 1976; Larkins et al. 1979), legume seed organelles are devoid of membrane bound ribosomes (Beachy et al. 1978). Only Yokoyama et al. (1972a and b) in *Glycine max* mature seed protein bodies, have evidenced functional ribosomes associated with these organelles. We postulate that the globulins which are synthesized by scattered ribosomes could likely diffuse through the tubular endoplasmic reticulum to the cisternae. On the contrary, cereal storage proteins, which are insoluble, would accumulate at the place where they are synthesized. This would explain the presence of ribosomes bound to protein bodies. Finally, as it has been reviewed by Millerd (1975), ribosomes vanish at the end of seed maturation. It is therefore likely that, as it is hypothesized by Harris (1979), a smooth endoplasmic reticulum would persist during the dehydration of the seed, since it appears very abundantly at the early stage of imbibition.

To summarize all the data which have been compiled, we can tentatively, but reasonably, draw a hypothesis: Vicilin-like proteins, synthesized on rough endoplasmic reticulum, would be transported via the Golgi apparatus towards the vacuoles which later divide into protein bodies. As the legumin-like protein synthesis begins, another kind of protein bodies, originating from rough endoplasmic reticulum cisternae, are formed. They are mainly filled with legumin, but since vicilin-like proteins are still being produced, these proteins would also accumulate besides legumin in the second kind of protein bodies, perhaps due to the decrease of Golgi apparatus activity. This hypothesis is particularly consistent with the heterogeneity of protein bodies, with the differences between legumin and vicilin accumulating legume species and vicilin only storing species and mainly with temporal relationships between events occurring in the developing cotyledons. Experiments to test this hypothesis will come either from microscopical studies dealing with precise developing steps and with the comparison between legume species or from *in vitro* proteosynthesis studies at a molecular level.

In vitro PROTEOSYNTHESIS BIOCHEMICAL STUDIES

In vitro proteosynthesis studies consist in the isolation of messenger RNAs coding for the concerned proteins, and then in translating these mRNAs into proteins by the use of an acellular translation system. Isolation of mRNAs is either performed directly or is subsequent to a previous ribosomes isolation. Classically the acellular translation systems are isolated from wheat germ or from rat reticulocyte lysate. The translation can be performed within a

Xenopus laevis oocyte, but, as far as we know, if it has been used for maize prolamins (Park et al. 1980), it has never been performed for legume seed storage proteins.

Much information can be obtained through these techniques. Firstly, the mRNAs and the ribosomes involved in the studied protein synthesis can be characterized. Secondly, the *in vitro* translation products can be compared to the native (alternatively called authentic) proteins. Thirdly, the mRNAs are allowed to be copied into copy or complementary DNAs (cDNAs) using an enzyme called reserve-transcriptase. After insertion within a plasmid the cDNAs may be cloned in micro-organisms, usually in a bacterium, which results in a very efficient purification, and then studied by the use of molecular biology methods. It is therefore possible to investigate the molcular events involved in the proteosynthesis and to understand the regulations of its mechanism.

Many authors have shown by these methods that the ribosomes involved in globulin synthesis are associated into polysomes. It has been first hypothesized by Beevers and Poulson (1972) who have observed by electron microscopy that the proteosynthesis occurs when polysomes are present in *Pisum sativum* cotyledon and it has later been shown in developing seeds of *Pisum sativum* (Higgins and Spencer 1977; Evans et al. 1979; Croy et al. 1980a and b), of *Vicia faba* (Muntz et al. 1978; Püchel et al. 1979), of *Phaseolus vulgaris* (Hall et al. 1978; Hall et al. 1980; Sun et al. 1979) and of *Glycine max* (Beachy et al. 1978; Mori et al. 1979). The number of ribosomes per polysomes seems to be characteristic of the storage protein classes: Beachy et al. (1978) have shown in *Glycine max* that legumin synthesis is performed on little polysomes while a higher degree of association (eight or more ribosomes per polysome) is necessary for vicilin synthesis. Most authors ascertain that only membrane bound polysomes are able to synthesize globulins particularly in the case of *Vicia faba* (Püchel et al. 1979) and of *Phaseolus vulgaris* (Bollini and Chrispeels 1979). Nevertheless Beachy et al. (1978) claim that in *Glycine max* no polysome association with a membrane is necessary for globulin synthesis, which is quite intriguing. Although no definite proof has already been obtained that legumin and vicilin polypeptide chains are synthesized by different systems, we think that it is likely that subsequent experiments will permit to distinguish these systems, which are certainly not very different from one another.

When comparing nascent products to native proteins, four kinds of differences may be expected if the studied proteins are secretory ones: the modification of the length of the polypeptide chains by the cleavage of the signal sequence of pre-proteins (the precursor is longer and its life-time short), glycosylation of the holoproteins which results in an increase in molecular weight, enzymatic cleavage of proproteins (a process which may be longer than the leader sequence cleavage) and association of subunits giving rise to a quaternary structure with likely further involvements for the polypeptide chain folding. The legume seed storage proteins are not unambiguously easily compared to the *in vitro* products when they are glycoproteins.

On a molecular weight basis comparisons may lead to erroneous data if the peptide signal cleavage decrease is balanced by the glycosylation increase. On the other hand glycosylation may alter the electrophoretic mobility, the main technique used in molecular weight determination. In term of tertiary structure, the use of immunochemical techniques cannot give clear-cut and comparable results: for instance, if the involved antigenic site is only located on sugar residues, a lack of recognition of the nascent protein will occur before the glycosylation process, though the nascent polypeptide chain is identical to the authentic one. Reciprocally, if the cleavage of a part of the polypeptide sequence does not alter the rest of the molecule and if the antigenic site is located on this region, an unmodified protein would not then be differentiated from an altered one. So the data which are mainly obtained through molecular weight determination and immuno-chemistry need further physico-chemical characterization to be clearer: these are peptide mapping and sequencing. Since these techniques are not frequently utilized, consequently the following results are not clear-cut, whatever the considered species may be.

In *Pisum sativum* Millerd et al. (1978) have suggested that precursors may exist and that they are not recognized by the antisera used in their system. Higgins and Spencer (1977) have found that *in vitro* products synthesized either from polysomes or poly-A containing RNAs are different from native proteins through molecular weight or immunochemistry, but similar through peptide mapping. They have not excluded the existence of precursors, but they have thought early termination of translation may be responsible for differences. Higgins and Spencer (1980) have proved two kinds of precursors: for 30,000 molecular weight vicilin and 20,000 molecular weight legumin they have shown shorter precursors which are slowly transformed into native proteins in several days. On the contrary, for 50,000 and 75,000 molecular weight vicilin subunits, they have evidenced larger nascent products, difference which is reduced by addition of membanes to their *in vitro* translation system; these findings are in agreement with a peptide signal cleavage associated with membrane crossing. On the contrary, Evans et al. (1979) have shown that part of the translation products are identical to native legumin and vicilin polypeptide chains. They accounted for early termination in order to explain the differences encountered with the native products. The last results by Croy et al. (1980a and 1980b) exclude the existence of a signal peptide either for legumin or vicilin polypeptide chains but show the occurrence of post-translational modifications: the 20,000 and 40,000 molecular weight legumin subunits are synthesized as a unique 60,000 molecular weight, long-lived precursor (proprotein) which is slowly split into the native subunits. This would explain the *in vivo* occurrence of 60,000 molecular weight subunits. As a matter of fact, the *in vitro* legumin precursor is very similar to the cleaved disulphide bonded 60,000 molecular weight subunit of native legumin. For vicilin polypeptide chains the situation is different: no difference (thus no signal peptide) is evidenced for 70,000 and 50,000 molecular weight proteins. The authors have showed a 47,000 molecular weight chain which is absent from the mature seed. As it is synthesized at the beginning of the *in vivo* proteosynthesis, it could be hydrolysed into the low molecular weight (33,000) chains which reciprocally are not encountered among the *in vitro* translation products.

In *Phaseolus vulgaris* post-translational modifications are currently observed. Hall et al. (1980) using poly-A RNAs as template, have shown that only the β and γ chains of vicilin are synthesized *in vitro*. They are slightly smaller than native protein and are immunologically identical to them. The differences are interpreted by these authors to be due to a lack of glycosylation. Sun et al. (1979) have evidenced a difference of the products based on the template they used: if, like Hall et al. (1980), they obtained only the β and γ vicilin chains by translating total poly-A RNA, they succeeded in producing the three vicilin subunits by the use of polysome isolated RNAs. The β and γ chains synthesized *in vitro* are slightly smaller than native ones. The authors have compared the *in vitro* products to β and γ polypeptides obtained *in vivo* by incubating cotyledon slices in the presence of a glycosylation inhibitor and have shown that the *in vitro* product difference is due to a lack of glycosylation. Bollini and Chrispeels (1979) have put to evidence that all the vicilin polypeptide chains obtained *in vitro* by the use of membrane bound polysomes are comparable in number but not in size to native proteins. As the *in vitro* products which are recognized by antisera against native proteins, are smaller, the authors have postulated a lack of glycosylation and discarded the existence of a signal peptide. On the contrary, they have shown that one chain of nascent phytohemagglutinin is longer than the native one. In this case the presence of a leader sequence could be responsible for the difference.

In *Glycine max* Ochiai-Yanagi et al. (1978) have suggested that legumin and vicilin post-translational modifications occur *in vivo* through the protein accumulation period. Mori et al. (1978), using poly-A RNA from mature cotyledons, have put to evidence that the *in vitro* translational products are only partially recognized by antisera against native proteins and contain polypeptide chains other than storage proteins. An interesting result arrived at by these authors is that the subunit polymerization occurs spontaneously in the wheat germ translation system. Mori et al. (1979) have used polysomes isolated from developing cotyledons and shown no difference between the *in vitro* and native products, even for the subunit polymerization. Beachy et al. (1980) have evidenced that the *in vitro* products are nearly identical to native proteins, but they are less in number than authentic globulins. Barton et al. (1979) have used poly-A RNA as template to synthesize the α and α' subunits of vicilin. By the use of peptide mapping, they have not evidenced any important difference between the polypeptide chains of the *in vitro* products and the native proteins.

From these intricate data one can partially draw some conclusions: if differences are obviously depending on species, on templates and on characterization methods, it is clear that many post-translational modifications do occur in legume seed storage protein synthesis. No signal peptide has ever been evidenced, but only N-terminal sequencing will ascertain this point. Modifications due to glycosylation are very frequent and it is likely, but not definite, that this event is delayed from the polypeptide chain synthesis. The last point is that, as the slow cleavage of precursors gives rise to native products, the complexity of the studies is increased. At last, it results from these investigations that the physicochemical and biological characterizations of the reserve proteins are still a necessary step toward the proteosynthesis studies.

The glycosylation processes of legume seed storage proteins have been studied during the last years. In *Pisum sativum*, Nagahashi and Beevers (1978) have shown that these processes are located on the rough endoplasmic reticulum which is the site where glycosyl transferases are located. These enzymes are involved in the formation of lipid-linked sugar intermediates utilized in glycoprotein biosynthesis. Beevers and Mense (1977) had shown that, like in mammalians, a lipid intermediate (mannolipid) is necessary to the glycosylation processes. This compound has been purified and studied by Delmer et al. (1978). More recently, Davies and Delmer (1979) have performed *in vitro* vicilin and phytohemagglutinin glycosylation. They conclude that the *in vitro* glycosylation is probably a transfert of N-acetyl glucosamine alone rather than in combination with mannose as preformed oligosaccharide. These studies enforce one point: at least part of the post-translational modifications are located on rough endoplasmic reticulum like proteosynthesis, but it is not possible to ascertain that such glycosylations are rapidly processed just after polypeptide chain termination or if the nascent proteins are translocated before being glycosylated.

NUCLEIC ACIDS AND PROTEOSYNTHESIS REGULATIONS

The regulations of protein synthesis do not only consist in the translation process in itself but also in nucleic acid metabolisms. We will principally deal with events which occur when the seed proteosynthesis starts and stops. Millerd (1975) has reviewed the events associated with the protein synthesis beginning: a large increase in nucleic acids precedes the protein synthesis. Scharpé and Van Parijs (1973) have shown in *Pisum sativum* that, just before the expansion phase of the developing cotyledon which corresponds to reserve synthesis, the storage cells reach a high polyploidy level (up to five endoreduplications). A gene dosage effect has been evoked as an explanation of the

for ovalbumin (an animal storage protein), the NH_2-terminal sequence of which is identical whatever it is synthesized, *in vitro* or *in vivo* (Gagnon et al. 1978); in this case the leader sequence has been shown to be internal and not located at the NH_2-terminal end (Lingappa et al. 1979). In *Pisum sativum*, it has been seen that a 60,000 dalton legumin subunit is a two-chain (or dicatenar) protein which seems to be synthesized as a unique chain (Croy et al. 1980a). We wonder whether a possible short connection polypeptide (like in proinsulin) could act as an internal signal sequence and then should be cleaved out, leaving two disulphide bonded chains. But, whatever they may be, post-translational modifications do occur in legumin and vicilin polypeptide chains: the nascent chains, comparatively to pro-proteins as prohormones or proenzymes, undergo cleavages of some peptide bonds giving rise to the authentic storage proteins. This process is a slow maturation which takes a few days to be completed. Glycosylation of these chains is another modification that they undergo. The final step is the entry into protein bodies. Except for glycosylation which occurs in the rough endoplasmic reticulum, the location of the other post-translational modifications is still unknown. We are still almost completely ignorant of the order these processes occur and about their biological significance.

It results from these observations that many controls are likely to occur. The study of the proteosynthesis regulation steps will certainly bring new possibilities for legume seed storage protein improvement: some control steps are certainly too basic to be altered either by phenotypical or by genetic means, but it is likely that some of them may be modified. Few results have still been obtained about regulation but these by Randall et al. (1979) on mineral nutrition influence on *Pisum sativum* proteins and other ones on genetical variability (e.g. Mutschler et al. 1980 and see Section 3.2) allow to ascertain that real progress can be hoped from the study of the proteosynthesis regulations at the molecular level.

Another important point to emphasize is that different pathways are used for vicilin and legumin synthesis and accumulation which complicate the experimental assays and allow some confusions in the early data. New experiments which could take the vicilin legumin difference into account will be more efficient for protein body and proteosynthesis studies. This fact emphasizes the necessity of species comparison in legume seed storage protein studies: the case of *Phaseolus* which is almost devoid of legumin is intriguing and certainly rich in informations.

As it has been pointed out by Croy et al. (1980a), the heterogeneity of legume seed storage proteins would partially find its origin in post-translational modifications. It will be interesting to compare the mRNA coding for these proteins in order to determine how many, and how much different, they are, as it has just been done for maize zein and mRNA heterogeneity by Park et al. (1980).

The formation of legume cotyledon protein bodies is still not clear and, as it is likely depending on the kind of proteins (legumin or vicilin) which are stored within them, therefore new experiments are not easy. Nevertheless the role of the different subcellular organelles will certainly be elucidated by electron microscopy and biochemistry in the near future. The transport of proteins is not clear-cut and probably also depends on the protein classes: What is the relative role of tubular endoplasmic reticulum and Golgi apparatus in this process is still a topical question.

The last point is about proteins which are not located within protein bodies. They are now often neglected as storage proteins and most authors only consider protein body globulins to be storage proteins. This simplification is necessary to understand the main events, but one must not forget that the seed protein nutritional quality is depending on the sum of all the polypeptide chains contained within the seed.

3.5. Storage Protein and Protein Body Evolution during Germination

Seed germination involves the imbibition of water, the reactivation of metabolism and the resetting of physiological processes which allow the embryo to develop. We only want to gather the events in which storage proteins or protein bodies are implicated. As recent reviews have dealt with storage protein mobilization during germination (Ashton 1976; Chrispeels et al. 1979) we will merely sum up the essential points about these processes.

3.5.1. PROTEIN BODY EVOLUTION DURING GERMINATION

Numerous authors have studied the fate of protein bodies during germination. During the first few days of embryo growth, the protein bodies swell and approximately by the fourth day the autolysis of storage proteins begins while their external membrane remains unaltered. Protein bodies, which are then more or less transformed in little vacuoles, fuse together to give the cell central vacuoles.

The persistency of the protein body membrane during storage protein degradation is consistent with the comparison of these organelles to lysosomes which have been first proposed by Matile (1968). A recent work by Weber et al. (1979) has evidenced that the *Vicia faba* protein body membrane is not altered during the first six days of seed germination and that the main membrane proteins do not undergo any detectable change. Harris et al. (1975) have shown that the initiation of protein mobilization appears to be associated with the fusion of cytoplasmic vesicles with the protein bodies in *Vigna unguiculata*. In *Vigna radiata*, Harris and Chrispeels (1975) have studied the autolytic capacity of isolated protein bodies; since their surrounding membranes have been shown to remain intact while the protein reserves disappear, they have concluded that new enzymatic activities, not present in the protein bodies isolated from dry seeds, must be activated or synthesized outside these organelles. Using the same legume species, Van der Wilden et al. (1980) have emphasized the lytic role of protein bodies: they have evidenced that these organelles constitute the principal lytic compartment in *V. radiata* cotyledon cells and proposed that they play a role in cellular autophagy; as a matter of fact, on the third day of seedling growth, protein bodies have been shown by these authors to contain small vesicles with a cytoplasmic content in which proteolytic activities are found.

3.5.2. STORAGE PROTEIN DEGRADATION

During germination storage, proteins are hydrolyzed into small peptides and amino acids which are used by the embryo to give rise to newly synthesized proteins. In legume seeds, proteolysis usually does not occur before the third day, and is maximum after five to six days of seed germination. This process is completed two weeks later. Konopska (1978 and 1979) has shown that *Pisum sativum* vicilin is hydrolyzed more rapidly than legumin at the beginning of germination but contrary happens later. These findings are not in agreement with previous ones (Basha and Beevers 1975) which have indicated that the legumin degradation rate is always higher than the vicilin one. Basha and Beevers (1976) have suggested that the cleavage of the glycosyl peptide bonds is preceded by the degradation of peptide linkages in *Pisum sativum* and that the reserve globulins undergo modification prior to their hydrolysis.

In *Phaseolus vulgaris*, Bollini and Chrispeels (1978) have evidenced in the cotyledons of young seedlings substantial quantities of polypeptides smaller than vicilin subunits and have suggested that these polypeptides represent partial breakdown of the vicilin prior to its complete catabolism. These results are in agreement with those by Basha and Cherry (1978) who have studied *Arachis hypogaea* storage protein degradation during germination. The breakdown of phytohemagglutinins is concomitant with globulin degradation in many legume seeds (Chen et al. 1977). The legume seed protein degradation seems therefore quite similar to the evolution of *Avena sativa* storage proteins (Kim et al. 1979). It is noteworthy that in this cereal species, the protein reserves are located, at least partly, within the protein bodies in mature endosperm (Pernollet and Mossé 1980).

All these results are not detailed at the molecular level and the *in vitro* studies using specific proteolytic enzymes will bring new information on the catabolism of storage proteins. The understanding of the existence of intermediates and of transport processes will be a consequence of these investigations which imply a better knowledge on storage protein structure.

3.5.3. PROTEOLYTIC ENZYMES

In the seed many proteolytic enzymes are involved in the mobilization of storage proteins: they may be classified as endopeptidases or exopeptidases (for review, see Ashton 1976). Very few proteases have been isolated and characterized neither from a structural nor from a functional point of view. The main limitations in these investigations lie in the lack of endopeptidase specific assays. The most important results which are presently available have been obtained and reviewed by Chrispeels et al. (1979).

The location of proteolytic enzymes within protein bodies has previously been mentioned. Their role and the way they are activated is not yet understood. On the contrary it has been evidenced clearly by Baumgartner et al. (1978) that a vicilin specific enzyme, which has previously been isolated and characterized by Baumgartner and Chrispeels (1977), is synthesized in the cytoplasm and later transported into protein bodies where it hydrolyzes the vicilin subunits.

Since a few detailed investigations have still been done on storage protein breakdown, we are compelled to follow the reasonable hypothesis proposed by Chrispeels et al. (1979): the protein bodies contain both storage proteins and proteases but lack endopeptidases necessary for internal protein degradation. These necessary enzymes would be synthesized in the cytoplasm and transported into the protein bodies where they would initiate the storage protein catabolism. The degradation would then be carried on by the exopeptidases already located within the protein bodies. The resulting amino acids are further metabolized into amides which are responsible for the nitrogen transport to the growing axis (Kern and Chrispeels 1978 and see also Chapter 4).

3.5.4. REGULATION OF STORAGE PROTEIN DEGRADATION

Numerous controls may be applied in the regulation of the storage protein hydrolysis and it is likely that all proteolytic enzymes are not controlled in the same way.

The first control likely to occur is a hormonal control of *de novo* enzyme synthesis which has been evidenced in *Pisum sativum*: both the shoot and the root affect hydrolysis and transport of nitrogen to the axis (Guardiola and Sutcliffe 1971). Another control may lie in end product inhibition: it has first been suggested by Oota et al. (1956). Yomo and Varner (1973) have suggested that the accumulation of amino acids would control endopeptidase activity.

Enzyme inhibitors have widely been studied in order to understand their role in the protein breakdown regulation (see Chrispeels et al. 1979, for review). Direct evidence supporting such a function has never been obtained. Recently Chrispeels and Baumgartner (1978) have shown that *Vigna radiata* trypsin inhibitor inhibits trypsin but does not inhibit the major cotyledon endoprotease of this species. Trypsin inhibitor regulation role is the more unlikely since it is located outside of the protein bodies where the proteolysis takes place.

Other regulations may be involved in storage protein catabolism such as the existence of zymogens or substrate specificity, but no proof of these likely roles has still been obtained.

Compartmentalization has now been proved to play an important role in the control mechanism of protein breakdown (Baumgartner et al. 1978). This emphasizes the importance of the persistency of the protein body membrane and of the compartmentalization role played by these organelles. It is likely that the membrane is involved both in the regulation of the entrance of newly synthesized cytoplasmic proteases (Van der Wilden et al. 1980) and in the regulation of final nitrogenous product exportation from the protein bodies. In addition, this compartmentalization may allow other regulation steps to occur, such as end product enzyme inhibition.

3.5.5. Conclusion

If it is well known that storage proteins are hydrolyzed in order to provide amino acids to the young seedling, the mechanisms of this degradation are yet poorly understood. Proteolytic enzyme isolation is not frequent and endopeptidase assays are lacking. The knowledge of protein breakdown regulation will need many more investigations and a great deal of hypotheses are to be tested.

Three main points are nevertheless noteworthy. The first one is that the degradation mechanism is stepwise. In this, endoproteases play a major initiation role. The second point is the very important part taken by the protein bodies, the unaltered membrane of which allows to consider them as lytic compartments. The stability of this membrane is likely responsible for protein degradation control. This fact is an argument to distinguish legume storage protein breakdown from cereal endosperm protein reserves mobilization since the integrity of protein bodies is not preserved during the dehydration of some cereal caryopses (Pernollet and Mossé 1980). The last point is the intriguing biological role of the trypsin inhibitors which do not appear to be involved in the regulation of legume seed endoproteases.

3.6. General Conclusion

Several conclusions have already been drawn at the end of preceding sections of the present chapter. It could have seemed more logical to begin with giving first a definition of storage proteins of legume seeds. This has already been attempted in recent reviews devoted to them (Millerd 1975; Derbyshire et al.

1976 a; Thomson and Doll 1979). But it is now time to discuss this definition to which there are no perfectly sharp frontiers, by reminding the several features which characterize them.

From a physiological point of view, storage proteins have a primordial trophic role, which is to supply amino acids and nitrogen compounds to the young seedling for the short but critical period during which it is still heterotrophic. In this acceptation, every protein which is really degraded during germination is storage protein, like noticed for pea albumins by Murray (1979). But this trophic role may not be exclusive. Some of them have a lectin function, as it was already shown in *Phaseolus vulgaris* (Bollini and Chrispeels 1978) and in *Pisum sativum* (Horisberger and Vonlanthen 1980). And it is known that in drastic nitrogen deficiencies, living organisms utilize progressively structural or functional proteins for trophic purposes in order to supply amino acids for the proper turnover of the most vital proteins. Some others may have still undiscovered functions.

Cytologically speaking, a large amount of storage proteins are accumulated and sequestered in protein bodies, such as starch in starch granules or oil in spherosomes: these reserve organelles are themselves located in reserve tissues. But some of these storage proteins like conarachin, a vicilin-like protein of *Arachis hypogaea* are accumulated outside of protein body, as evidenced by Neucere and Ory (1970). And the reverse is true also: protein bodies do contain some enzymes. Moreover, it must be remembered that storage proteins can also be deposited in roots (Newcomb 1967) and tubers and the question remains as to what their nature and the aspects of their accumulation, location and biosynthesis is.

At the molecular level, a third feature that we have emphasized concerns their quaternary structure. They are frequently composed of subunits associated into trimers or hexamers the shape and structure of which are particularly well fitted to a dense packing inside of protein bodies.

From a biochemical point of view, they are represented by a rather small number of molecular species but they account for a large proportion of seed proteins.

Physiologically and biochemically, they behave like secretory proteins, with nevertheless some intriguing differences: no signal peptide has yet been evidenced. The formation of polysomes which is involved obeys an unknown mechanism. The steps of synthesis and deposition of legumin and vicilin differ one from another.

If their quaternary structure appears as resulting in very few models of association and packing, it also results in a great multiplicity of these proteins due to the numerous multiple forms involved, which is a complementary source of heterogeneity, besides genetic variants.

Lastly, they appear as both stable and variable. Stable at the life scale, in the sense that their phenotypical expression almost does not depend on the environmental conditions. Very variable on the genetic level at the evolutionary scale, due to the fact that the non-lethal point mutations may reach a much higher level than in functional proteins. For this reason, they appear as noteworthy genetic markers.

References*

Abdel-Fattah, A.F., Zaki, D.A., Edrees, M. and Abbassy, M.M. 1974. Investigations on some constituents of *Lupinus termis* seeds. *Qual. Plant. Pl. Fds. Hum. Nutr*. **23**: 359–368.

Adams, M.W. 1975. On the quest for quality in the field bean. In: *Nutritional Improvement of Food Legumes by Breeding*. M. Milner, ed. pp. 143–149. Wiley-Interscience.

Alekseeva, M.V. and Kovarskaya, N.V. 1978. Proteins of aleurone grains from the axial part and cotyledons of pea and soybean seeds: A comparative study. *Sov. Plant Physiol*. **25**: 365–370.

Altschul, A.M. 1978. Plant protein food models in industry. In: *Plant Proteins*. G. Norton, ed. pp. 267–282. Butterworths.

Altschul, A.M., Neucere, N.J., Woodham, A.A. and Dechary, J.M. 1964. A new classification of seed proteins: application to the aleurins of *Arachis hypogaea*. *Nature* **203**: 501–504.

Altschul, A.M., Snowden, J.E., Manchon, D.D. and Dechary, J.M. 1961. Intracellular distribution of seed proteins. *Arch. Biochem. Biophys*. **95**: 402–404.

Alves Moreira, M., Brune, W. and Martins Batista, C. 1976. Evaluation of methionine contents of beans (*Phaseolus vulgaris* L.). *Turrialba* **26**: 225–231.

Anderson, R.L., Wolf, W.J. and Glover, D. 1973. Extraction of soybean meal proteins with salt solutions at pH 4.5. *J. Agr. Food Chem*. **21**: 251–254.

Anonymous, 1978. Interlaboratory comparison of amino acid analysis. In: *Seed Protein Improvement by Nuclear Techniques* (*Proc. two res. coord. meet. Baden 1977*). pp. 393–402. IAEA Vienna.

Ashton, F.M. 1976. Mobilization of storage proteins of seeds. *Ann. Rev. Plant Physiol*. **27**: 95–117.

Austen, B.M. 1979. Predicted secondary structures of amino terminal extension sequences of secreted proteins. *FEBS Lett*. **103**: 308–313.

Badley, R.A., Atkinson, D., Hauser, H., Oldani, D., Green, J.P. and Stubbs, J.M. 1975. The structure, physical and chemical properties of the soybean protein glycinin. *Biochim. Biophys. Acta* **412**: 214–228.

*Bibliography was drawn up in June 1980.

Bagley, B.W., Cherry, J.H., Rollins, M.L. and Altschul, A.M. 1963. A study of protein bodies during germination of peanut (*Arachis hypogaea*) seed. *J. Bot.* **50**: 523–532.

Bailey, C.J. and Boulter, D. 1970. The structure of legumin, a storage protein of broad bean (*Vicia faba*) seed. *Eur. J. Biochem.* **17**: 460–466.

Bailey, C.J. and Boulter, D. 1971. Urease, a typical seed protein of the Leguminosae. In: *Chemotaxonomy of Leguminosae.* J.B. Harborne, D. Boulter and B.L. Turner, eds. pp. 485–502. Academic Press, London and New York.

Bailey, C.J., Cobb, A. and Boulter, D. 1970. A cotyledon slice system for the electron autoradiographic study of the synthesis and intracellular transport of the seed storage protein of *Vicia faba*. *Planta* **95**: 103–118.

Bain, J.M. and Mercer, F.V. 1966. Subcellular organization of the developing cotyledons of *Pisum sativum* L. *Austr. J. Biol. Sci.* **19**: 49–67.

Bajaj, S., Mickelsen, O., Baker, L.R. and Markarian, D. 1971. The quality of protein in various lines of peas. *Br. J. Nutr.* **25**: 207–212.

Ballester, D., Yanez, E., Garcia, R., Erazo, S., Lopez, F., Haardt, E., Cornejo, S., Lopez, A., Pokniak, J. and Chichester, C.O. 1980. Chemical composition, nutritive value, and toxicological evaluation of 2 species of sweet Lupine (*Lupinus albus* and *Lupinus luteus*). *J. Agric. Food Chem.* **28**: 402–405.

Barker, R.D.J., Derbyshire, E., Yarwood, A. and Boulter D. 1976. Purification and characterization of the major storage proteins of *Phaseolus vulgaris* seeds, and their intracellular and cotyledonary distribution. *Phytochemistry* **15**: 751–757.

Barton, K.A., Beachy, R.N., Thompson, J.F. and Madison, J.T. 1979. Identification of polypeptides synthesized *in vitro* from soybean messenger RNA. *Supp. N°5 Plant Physiol: Annual Meeting: American Society of Plant Physiology*. Ohio State University, Columbus, Ohio.

Basha, S.M.M. and Beevers, L. 1975. The development of proteolytic activity and protein degradation during the germination of *Pisum sativum* L. *Planta* **124**: 77–87.

Basha, S.M.M. and Beevers, L. 1976. Glycoprotein metabolism in the cotyledons of *Pisum sativum* during development and germination. *Plant Physiol.* **57**: 93–97.

Basha, S.M.M. and Cherry, J.P. 1976. Composition, solubility and gel electrophoretic properties of protein isolated from Florunner (*Arachis hypogaea*) peanut seeds. *J. Agric. Food Chem.* **24**: 359–365.

Basha, S.M.M. and Cherry, J.P. 1978. Proteolytic enzyme activity and storage protein degradation in cotyledons of germinating peanut (*Arachis hypogaea* L.) seeds. *J. Agric. Food Chem.* **26**: 229–234.

Basha, S.M.M., Cherry, J.P. and Young, C.T. 1976. Changes in free amino acids, carbohydrates and proteins of maturing seeds from various peanut

(*Arachis hypogaea* L.) cultivars. *Cereal Chem.* **53**: 586–597.

Baudet, J., Cauderon, A., Fauconneau, G., Mossé, J. and Pion, R. 1968. Sur un troisième gène m utant (amylose-extender) qui accroît la teneur en lysine du grain de Maïs et sur son effet cumulatif avec le gène opaque 2. *C.R. Acad. Sci.* **266D**: 2260–2263.

Baudet, J., Cousin, R. and Mossé, J. 1977. Topographical study of weight and protein content of the pea seed (*Pisum sativum*) following its localization on the plant. In: *Protein Quality from Leguminous Crops* (Commission of the European Communities, *Coordination of agricultural research*) *EUR. 5686 EN*, 316–323.

Baudet, J., Leclercq, P. and Mossé, J. 1971. Sur la richesse en lysine des graines de Tournesol en fonction de leur teneur en protéines. *C.R. Acad. Sci.* **273D**: 1112–1113.

Baudet, J. and Mossé, J. 1980. Amino acid composition of different cultivars of broad beans (*Vicia faba*): comparison with other legume seeds. In: *World Crops: Production, Utilization and Description*, Vol. 3, *Vicia faba: Feeding Value, Processing and Viruses*. D.A. Bond, ed. pp. 67–82. Martinus Nijhoff Pub. The Hague, Boston, London for the CEE.

Baumgartner, B. and Chrispeels, M.J. 1977. Purification and characterization of vicilin peptidohydrolase, the major endopeptidase in the cotyledons of mung-bean seedlings. *Eur. J. Biochem.* **77**: 223–233.

Baumgartner, B., Tokuyasu, K.T. and Chrispeels, M.J. 1978. Localization of vicilin peptidohydrolase in the cotyledons of mung-bean seedlings by immunofluorescence microscopy. *J. Cell. Biol.* **79**: 10–19.

Beachy, R.N., Barton, K.A., Madison, J.T., Thompson, J.F. and Jarvis, N. 1980. Messenger RNAs that code for soybean seed proteins. In: *Genome Organization and Expression in Plants, NATO Advanced Studies Institute Series A*. C.J. Leaver ed. Vol. 29, pp. 273–282. Plenum Press, New York.

Beachy, R.N., Thompson, J.F. and Madison J.T. 1978. Isolation of polyribosomes and messenger RNA active in *in vitro* synthesis of soybean seed proteins. *Plant Physiol.* **61**: 139–144.

Beevers, L. and Mense, R.M. 1977. Glycoprotein biosynthesis in cotyledons of *Pisum sativum*. *Plant Physiol.* **60**: 703–708.

Beevers, L. and Poulson, R. 1972. Protein synthesis in cotyledons of *Pisum sativum* L. I. Changes in cell-free amino acid incorporation capacity during seed development and maturation. *Plant Physiol.* **49**: 476–481.

Begbie, R. 1979. A non-aqueous method for the subcellular fractionation of cotyledons from dormant seeds of *Phaseolus vulgaris* L. *Planta* **147**: 103–110.

Bell, E.A. 1971. Comparative biochemistry of non-protein amino acids. In: *Chemotaxonomy of Leguminosae*. J.B. Harborne, D. Boulter and B.L. Turner, eds. pp. 179–206. Academic Press, London and New York.

Bell, E.A. 1976. Uncommon amino acid in plants. *FEBS Lett.* **64**: 29–35.

Bhatty, R.S. 1974. Chemical composition of some *Faba* bean cultivars. *Can. J. Plant Sci.* **54**: 413–421.

Bhatty, R.S., Finlayson, A.J. and Mackenzie, S.L. 1977. Total sulphur as a screening criterion for sulphur amino acids in legumes. *Can. J. Plant Sci.* **57**: 177–183.

Bhatty, R.S., Slinkard, A.E. and Sosulski, F.W. 1976. Chemical composition and protein characteristics of lentils. *Can. J. Plant Sci.* **56**: 787–794.

Bishop, L.R. 1930. The composition and determination of the barley proteins. III. Fourth report on barley proteins. The proteins of barley during development and storage and in the mature grain. *J. Inst. Brewing* **36**: 336–349.

Blagrove, R.J. and Gillespie, J.M. 1974. Lupin storage proteins. *Proc. Aust. Biochem. Soc.* **7**: 3.

Blagrove, R.J. and Gillespie, J.M. 1975. Isolation, purification and characterization of the seed globulins of *Lupinus angustifolius*. *Aust. J. Plant Physiol.* **2**: 13–27.

Blagrove, R.J. and Gillespie, J.M. 1978. Variability of the seed globulins of winged bean, *Psophocarpus tetragonolobus* (L.) DC. *Aust. J. Plant Physiol.* **5**: 371–375.

Blagrove, R.J., Gillespie, J.M., Lilley, G.G. and Woods, E.F. 1980. Physicochemical studies of conglutin γ, a storage globulin from seeds of *Lupinus angustifolius*. *Aust. J. Plant Physiol.* **7**: 1–13.

Blagrove, R.J., Gillespie, J.M. and Randall, P.J. 1976. Effect of sulphur supply on the seed globulin composition of *Lupinus angustifolius*. *Aust. J. Plant Physiol.* **3**: 173–184.

Bliss, F.A. 1975. Cowpea in Nigeria. In: *Nutritional Improvement of Food Legumes by Breeding*. M. Milner, ed. pp. 151–158. John Wiley and Sons.

Blixt, S. 1979. Natural and induced variability for seed protein in temperate legumes. In: *Seed Protein Improvement of Cereals and Grain Legumes* (*Proc. Symp. Neuherberg, 1978*), IAEA, Vienna, Vol. II, pp. 3–21.

Blobel, G. and Dobberstein, B. 1975. Transfer of protein across membrane. I. Presence of proteolytically and unprocessed nascent immunoglobulin light chains on membrane bound ribosomes of murine myeloma. *J. Cell. Biol.* **67**: 835–851.

Bollini, R. and Chrispeels, M.J. 1978. Characterization and subcellular localization of vicilin and phytohemagglutinin, the 2 major reserve proteins of *Phaseolus vulgaris* L. *Planta* **142**: 291–298.

Bollini, R. and Chrispeels, M.J. 1979. The rough endoplasmic reticulum is the site of reserve protein synthesis in developing *Phaseolus vulgaris* cotyledons. *Planta* **146**: 487–501.

Boulter, D. and Derbyshire, E. 1971. Taxonomic aspects of the structure of legume proteins. In: *Chemotaxonomy of Leguminosae*. J.B. Harborne, D. Boulter and B.L. Turner, eds. pp. 285–308. Academic Press, London and New York.

Boulter, D. and Derbyshire, E. 1978. The general properties, classification and distribution of plant proteins. In: *Plant Proteins*. G. Norton, ed. pp. 3–24. Butterworths.

Boulter, D., Evans, I.M., Thompson, A. and Yarwood, A. 1975. The amino acid composition of *Vigna unguiculata* (cowpea) meal in relation to nutrition. In: *Nutritional Improvement of Food Legumes by Breeding*. M. Milner, ed. pp. 205–216. John Wiley and Sons.

Briarty, L.G. 1978. The mechanisms of protein body deposition in legumes and cereals. In: *Plant Proteins*. G. Norton, ed. pp. 81–106. Butterworths.

Briarty, L.G., Coult, D.A. and Boulter, D. 1969. Protein bodies of developing seeds of *Vicia faba. J. Exp. Bot.* **20**: 358–372.

Browder, S.K. and Beevers, L. 1978. Characterization of the glycopeptide bond in legumin from *Pisum savitum* L. *FEBS Lett.* **89**: 145–148.

Burr, B. and Burr, F.A. 1976. Zein synthesis in maize endosperm by polyribosomes attached to protein bodies. *Proc. Nat. Acad. Sci. USA* **73**: 515–519.

Burr, B., Burr, F.A., Rubenstein, I. and Simon, M.N. 1978. Purification and translation of zein messenger RNA from maize endosperm protein bodies. *Proc. Nat. Acad. Sci. USA* **75**: 696–700.

Busson, F. 1965. Plantes alimentaires de l'Ouest africain, étude botanique, biologique et chimique. *Min. Coop., Rech. Sci. and Techn. et Armées*.

Byers, M., Kirkman, M.A. and Miflin, B.J. 1978. Factors affecting the quality and yield of seed protein. In: *Plant Proteins*. G. Norton, ed. pp. 227–243. Butterworths.

Carasco, J.F., Croy, R., Derbyshire, E. and Boulter, D. 1978. The isolation and characterization of the major polypeptides of the seed globulin of cowpea (*Vigna unguiculata* L. Walp) and their sequential synthesis in the developing seed. *J. Exp. Bot.* **29**: 309–323.

Casey, R. 1979a. Immunoaffinity chromatography as a means of purifying legumin from *Pisum* (pea) seeds. *Biochem. J.* **177**: 509–520.

Casey, R. 1979b. Genetic variability in the structure of α subunit of legumin from *Pisum*: A two dimensional electrophoresis study. *Heredity* **43**: 265–272.

Casimir, J. and Le Marchand, G. 1966. Présence d'acide pipécolique dans les graines de *Vigna. Bull. Jard. Bot. Etat. (Bruxelles)*, 36–53.

Catsimpoolas, N. 1970. Note on dissimilar subunits present in dissociated soybean globulins. *Cereal Chem.* **47**: 70–71.

Catsimpoolas, N., Campbell, T.G. and Meyer, E.W. 1968. Immunochemical study of changes in reserve proteins of germinating soybean seeds. *Plant Physiol.* **43**: 799–805.

Catsimpoolas, N. and Ekenstam, C. 1969. Isolation of α, β, γ-conglycinins. *Arch. Biochem. Biophys.* **129**: 490–497.

Catsimpoolas, N., Keeney, J.A., Meyer, E.W. and Szubaj, B.F. 1971. Molecular weight and amino acid composition of glycinin subunits. *J. Sci. Fd.*

Agric. **22**: 448–450.

Catsimpoolas, N., Rogers, D.A., Circle, J. and Meyer, E.W. 1967. Purification and structural studies of the 11 S component of soybean protein. *Cereal Chem.* **44**: 631–637.

Catsimpoolas, N. and Wang, J. 1971. Analytical scanning isoelectric focusing. 5. Separation of glycinin subunits in urea-dithiothreitol media. *Anal. Biochem.* **44**: 436–444.

Cerletti, P., Fumagalli, A. and Venturin, D. 1978. Protein composition of seeds of *Lupinus albus.* *J. Food Sci.* **43**: 1409–1411.

Chakraborty, P., Sosulski, F. and Bose, A. 1979. Ultracentrifugation of salt soluble proteins in 10 legume species. *J. Sci. Food Agric.* **30**: 766–771.

Chen, L.H., Thacker, R.R. and Pan, S.H. 1977. Effect of germination on hemagglutinating activity of pea and bean seeds. *J. Food Sci.* **42**: 1666–1668.

Cherry, J.P. 1977. Potential sources of peanut seed proteins and oil in the genus *Arachis.* *J. Agric. Food Chem.* **25**: 186–193.

Cherry, J.P., Dechary, J.M. and Ory, R.L. 1973. Gel electrophoretic analysis of peanut proteins and enzymes. I. Characterization of DEAE-cellulose separated fractions. *J. Agric. Food Chem.* **21**: 652–655.

Chrispeels, M.J. 1976. Biosynthesis, intracellular transport and secretion of extracellular macromolecules. *Ann. Rev. Plant Physiol.* **27**: 19–38.

Chrispeels, M.J. and Baumgartner, B. 1978. Trypsin inhibitor in mung bean cotyledons. *Plant Physiol.* **61**: 617–623.

Chrispeels, M.J., Bollini, R. and Harris, N. 1979. Biosynthesis, accumulation and catabolism of reserve proteins in legume seeds. *Ber. Deutsch Bot. Ges.* **92**: 535–551.

Coffman, C.W. and Garcia, V.V. 1977. Functional properties and amino acid content of protein isolate from mung bean flour. *J. Food Technol.* **12**: 473–484.

Craig, S., Goodchild, D.J. and Hardham, A.R. 1979. Structural aspects of protein accumulation in developing pea cotyledons. I. Qualitative and quantitative changes in parenchyma cell vacuoles. *Aust. J. Plant Physiol.* **6**: 81–98.

Croy, R.R.D., Derbyshire, E., Krishna, T.G. and Boulter, D. 1979. Legumin of *Pisum sativum* and *Vicia faba. New Phytol.* **83**: 29–35.

Croy, R.R.D., Gatehouse, J.A., Evans, I.M. and Boulter, D. 1980a. Characterization of the storage protein subunits synthesized *in vitro* by polyribosomes and RNA from developing pea (*Pisum sativum* L.). *I. Legumin. Planta* **148**: 49–56.

Croy, R.R.D., Gatehouse, J.A., Evans, I.M. and Boulter, D. 1980b. Characterization of the storage protein subunits synthesized *in vitro* by polyribosomes and RNA from developing pea (*Pisum sativum* L.). *II. Vicilin. Planta* **148**: 57–63.

Danielsson, C.E. 1949. Seed globulins of the Gramineae and Leguminosae. *Biochem. J.* **44**: 387–400.

Dardenne, G.A. 1976. Recherche, isolement et structure de nouveaux acides aminés libres dans les végétaux. *Mémoire Agr. Ens. Sup. (Fac. Sci. Gembloux).*

Daussant, J., Neucere, N.J. and Yatsu, L.Y. 1969a. Immunochemical studies on *Arachis hypogaea* proteins with particular reference to the reserve proteins. I. Characterization, distribution and properties of α-arachin and α-conarachin. *Plant Physiol.* **44**: 471–479.

Daussant, J., Neucere, N.J. and Conkerton, E.J. 1969b. Immunochemical studies on *Arachis hypogaea* proteins with particular reference to the reserve proteins. II. Protein modification during germination. *Plant Physiol.* **44**: 480–484.

Davey, J.E. and Van Staden, J. 1978. Ultrastructural aspects of reserve protein deposition during cotyledonary cell development in *Lupinus albus*. *Z. Pflanzenphysiol* **89**: 259–271.

Davey, R.A. and Dudman, W.F. 1979. The carbohydrate of storage glycoproteins from seeds of *Pisum sativum*: characterization and distribution of component polypeptides. *Aust. J. Plant Physiol.* **6**: 435–447.

Davies, D.R. and Brewster, V. 1975. Studies of seed development in *Pisum sativum*. II. rRNA contents in reciprocal crosses. *Planta* **124**: 303–309.

Davies, H.M. and Delmar, D.P. 1979. Seed reserve protein glycosylation in an *in vitro* preparation from developing cotyledons of *Phaseolus vulgaris*. *Planta* **146**: 513–520.

Dechary, J.M., Talluto, K.F., Evans, W.J., Carney, W.B. and Altschul, A.M. 1961. α-conarachin. *Nature* **190**: 1125–1126.

Delmer, D.P., Kulow, C. and Ericson, M.C. 1978. Glycoprotein synthesis in plants. II. Structure of the mannolipid intermediate. *Plant Physiol.* **61**: 25–29.

Derbyshire, E. and Boulter, D. 1976. Isolation of legumin-like protein from *Phaseolus aureus* and *Phaseolus vulgaris*. *Phytochemistry* **15**: 411–414.

Derbyshire, E., Wright, D.J. and Boulter, D. 1976a. Legumin and vicilin, storage proteins of legume seeds. *Phytochemistry* **15**: 3–24.

Derbyshire, E., Yarwood, J.N., Neat, E. and Boulter, D. 1976b. Seed proteins of *Phaseolus* and *Vigna*. *New Phytol.* **76**: 283–288.

Dickson, M.H. and Hackler, L.R. 1975. Protein quantity and quality in high-yielding beans. In: *Nutritional Improvement of Food Legumes by Breeding*. M. Milner, ed. pp. 185–192. John Wiley and Sons.

Dieckert, J.W. and Dieckert, M.C. 1976. The chemistry and cell biology of the vacuolar proteins of seeds. *J. Food Sci.* **41**: 475–482.

Dieckert, J.W., Snowden, J.E., Moore, A.T., Heinzelman, D.C. and Altschul, A.M. 1962. Composition of some subcellular fractions from seeds of *Arachis hypogaea*. *J. Food Sci.* **27**: 321–325.

Diep, O. and Boulet, M. 1977a. Study of variations in sedimentation coefficient and molecular dimensions of 11 S soybean globulin as a function of ionic strength at acid and alkaline pH and in the presence of n-butanol. *Can. Inst. Food Sci. Technol. J.* **10**: 169–175.

Diep, O. and Boulet, M. 1977b. Study of variations in the sedimentation coefficient and molecular dimensions of 7 S soybean globulin as a function of ionic strength at acid and alkaline pH and in the presence of n-butanol. *Can. Inst. Food Sci. Technol. J.* **10**: 176–189.

Draper, M. and Catsimpoolas, N. 1977. Isolation of the acidic and basic subunits of glycinin. *Phytochemistry* **16**: 25–27.

Draper, M. and Catsimpoolas, N. 1978. Disulphide and sulfhydryl groups in glycinin. *Cereal Chem.* **55**: 16–23.

Dudman, W.F. and Millerd, A. 1975. Immunochemical behaviour of legumin and vicilin from *Vicia faba*: Survey of related proteins in Leguminosae subfamily Faboideae. *Biochem. Syst. Ecol.* **3**: 25–33.

Duncan, A., McIntosh, A. and Ellinger, G.M. 1976. Determination of methionine in the seeds of legumes. *Proc. Nutr. Soc.* **80**: 148A–149A.

Duranti, M. and Cerletti, P. 1979. Amino acid composition of seed proteins of *Lupinus albus*. *J. Agric. Food Chem.* **27**: 977–978.

Dure, L.S. 1975. Seed formation. *Ann. Rev. Plant Physiol.* **26**: 259–278.

Earle, F.R. and Jones, Q. 1962. Analyses of seed samples from 113 plant families. *Econ. Bot.* **16**: 221–250.

Eaton-Mordas, C.A. and Moore, K.G. 1978. Seed glycoproteins of *Lupinus angustifolius*. *Phytochemistry* **17**: 619–621.

Eaton-Mordas, C.A. and Moore, K.G. 1979. Subunit glycosylation of *Lupinus angustifolius* seed globulins. *Phytochemistry* **18**: 1775–1779.

Elleman, T.C. 1977. Amino acid sequence of the smaller subunit of conglutin γ, a storage globulin of *Lupinus angustifolius*. *Aust. J. Biol. Sci.* **30**: 33–45.

Eppendorfer, W.H. 1971. Effects of S, N and P on amino acid composition of field beans (*Vicia faba*) and responses of the biological value of the seed protein to S-amino acid content. *J. Sci. Food. Agric.* **22**: 501–505.

Eppendorfer, W.H. and Bille, S.W. 1974. Amino acid composition as a function of total-N in pea seeds grown on two soils with P and K additions. *Plant Soil* **41**: 33–39.

Ericson, M.C. 1975. Isolation and characterization of a glucosamine-containing storage glycoprotein from mung beans (*Phaseolus aureus* Roxb). *Diss. Abstr.* **35**: 4767-B.

Ericson, M.C. and Chrispeels, M.J. 1973. Isolation and characterization of glucosamine-containing storage glycoproteins from the cotyledons of *Phaseolus aureus*. *Plant Physiol.* **52**: 98–104.

Ericson, M.C. and Chrispeels, M.J. 1976. The carbohydrate moiety of mung bean vicilin. *Aust. J. Plant Physiol.* **3**: 763–769.

Ericson, M.C. and Delmer, D.P. 1977. Glycoprotein synthesis in plants. I. Role of lipid intermediate. *Plant Physiol.* **59**: 341–347.

Evans, I.M. and Boulter, D. 1974. Amino acid composition of seed meals of yam bean (*Sphenostylis stenocarpa*) and Lima bean (*Phaseolus lunatus*). *J. Sci. Food Agric.* **25**: 919–922.

Evans, I.M. and Boulter, D. 1975. S-methyl-L-cysteine content of various legume meals. *Qual. Plant.* **24**: 257–261.

Evans, I.M. and Boulter, D. 1980. Crude protein and sulphur amino acid contents of some commercial varieties of peas and beans. *J. Sci. Food Agric.* **31**: 238–242.

Evans, I.M., Croy, R.R.D., Hutchinson, P., Boulter, D., Payne, P.I. and Gordon, M.E. 1979. Cell free synthesis of some storage protein subunits by polyribosomes and RNA isolated from developing seeds of pea (*Pisum sativum* L.). *Planta* **144**: 455–462.

Evans, R.J., Bauer, D.H., Adams, M.W. and Saettler, A.W. 1978. Methionine and cystine contents of bean (*Phaseolus*) seeds. *J. Agric. Food Chem.* **26**: 1234–1237.

F.A.O. 1964. Les protéines. Noeud du probleme alimentaire mondial. *Cahier*, No. 5, Rome.

F.A.O. 1970. Amino acid content of foods and biological data on proteins. *Nutritional studies*, No. 24, Rome.

Fishbein, W.N., Engler, W.F., Griffin, J.L., Scurzi, W. and Bahr, G. 1977. Electron microscopy of negatively stained Jack bean urease at three levels of quaternary structure, and comparison with hydrodynamic studies. *Eur. J. Biochem.* **73**: 185–189.

Fisher, C.E., Leach, I.B. and Wilding, P. 1976. Improved separation of the major water-soluble proteins of soy meal by a single-step chromatographic procedure. *J. Sci. Food Agric.* **27**: 1039–1043.

Frolich, W.G., Pollmer, W.G. and Christ, W. 1974. Variation of the contents of protein and of methionine and cystine in *Vicia faba* L.*Z. Pflanzenzüchtg.* **72**: 160–165.

Gagnon, J., Palmiter, R.D. and Walsh, K.A. 1978. Comparison of the NH_2-terminal sequence of ovalbumin as synthesized *in vitro* and *in vivo*. *J. Biol. Chem.* **253**: 7464–7468.

Gatehouse, J.A., Croy, R.R.D. and Boulter, D. 1980. Isoelectric-focusing properties and carbohydrate content of pea (*Pisum sativum*) legumin. *Biochem. J.* **185**: 497–503.

Georgi, B., Niemann, E.G., Brock, R.D. and Axmann, H. 1979. Comparison of analytical techniques for seed protein and amino acid analysis. In: *Seed Protein Improvement of Cereals and Grain Legumes* (*Proc. Symp. Neuherberg*, 1978), IAEA, Vienna, Vol. II, pp. 311–341.

Ghetie, V. and Buzila, L. 1968. Hétérogènéité de la viciline des graines de pois (*Pisum sativum*). *Rev. roum. Biochim.* **5**: 271–279.

Gillespie, J.M. and Blagrove, R.J. 1975. Variability in the proportion and type of subunits in lupin storage globulins. *Aust. J. Plant Physiol.* **2**: 29–39.

Gillespie, J.M. and Blagrove, R.J. 1978. Isolation and composition of the seed globulins of winged bean (*Psophocarpus tetragonolobus* L.). DC. *Aust. J. Plant Physiol.* **5**: 357–369.

Gillespie, J.M., Blagrove, R.J. and Randall, P.J. 1978. Effect of sulphur supply on the seed globulin composition of various species of lupin. *Aust. J. Plant Physiol.* **5**: 641–650.

Gilroy, J., Wright, D.J. and Boulter, D. 1979. Homology of basic subunits of legumin from *Glycine max* and *Vicia faba*. *Phytochemistry* **18**: 315–316.

Graham, T.A. and Gunning, B.E.S. 1970. Localization of legumin and vicilin in bean cotyledon cells using fluorescent antibodies. *Nature* **228**: 81–82.

Grant, D.R. and Lawrence, J.M. 1964. Effects of sodium dodecyl sulfate and other dissociating reagents on the globulins of peas. *Arch. Biochem. Biophys.* **108**: 552–561.

Grant, D.R., Sumner, A.K. and Johnson, J. 1976. An investigation of pea seed albumins. *Can. Inst. Food Sci. and Technol. J.* **9**: 84–91.

Griffiths, D.W. and Lawes, D.A. 1978. Variation in the crude protein content of field beans (*Vicia faba* L.) in relation to the possible improvement of the protein content of the crop. *Euphytica* **27**: 487–495.

Guardiola, J.L. and Sutcliffe, J.F. 1971. Control of protein hydrolysis in the cotyledons of germinating pea (*Pisum sativum* L.) seeds. *Ann. Bot.* **35**: 791–807.

Hall, T.C., Ma, Y., Buchbinder, B.U., Pyne, J.W., Sun, S.M. and Bliss. F.A. 1978. Messenger RNA for Gl protein of french bean seeds: cell free translation and product characterization. *Proc. Natl. Acad. Sci. U.S.A.* **75**: 3196–3200.

Hall, T.C., Sun, S.M., Buchbinder, B.U., Pyne, J.W., Bliss, F.A. and Kemp, J.D. 1980. Bean seed globulin messenger RNA-translation characterization and its use as a probe toward genetic engineering of crop plants. In: *Genome Organization and Expression in Plants.* NATO Advanced Study Institute Series A. C.J. Leaver, ed. Vol. 29. pp. 259–272, Plenum Press, New York.

Hara, I., Ohmiya, M. and Matsubara, H. 1978. Pumpkin (*Curcurbita* sp.) seed globulin. III. Comparison of subunit structures among seed globulins of various *Cucurbita* species and characterization of peptide components. *Plant Cell Physiol.* **19**: 237–243.

Harborne, J.B., Boulter, D. and Turner, B.L. 1971. *Chemotaxonomy of Leguminosae.* Academic Press, London and New York.

Harris, N. 1979. Endoplasmic reticulum in developing seeds of *Vicia faba*. *Planta* **146**: 63–69.

Harris, N. and Boulter, D. 1976. Protein body formation in cotyledons of developing cowpea (*Vigna unguiculata*) seeds. *Ann. Bot.* **40**: 739–744.

Harris, N. and Chrispeels, M.J. 1975. Histochemical and biochemical observations on storage protein metabolism and protein body autolysis in cotyledons of germinating mung beans. *Plant Physiol.* **56**: 292–299.

Harris, N., Chrispeels, M.J. and Boulter, D. 1975. Biochemical and histochemical studies on protease activity and reserve protein metabolism in the cotyledons of germinating cowpeas (*Vigna unguiculata*). *J. Exp. Bot.* **26**: 544–554.

Heywood, V.H. 1971. The Leguminosae. A systematic purview. In: *Chemotaxonomy of Leguminosae.* J.B. Harborne, D. Boulter and B.L. Turner, eds. pp. 1–30. Academic Press, London and New York.

Higgins, T.J.V. and Spencer, D. 1977. Cell free synthesis of pea seed storage proteins. In: *Acides Nucléiques et Synthèse des Protéines chez les végetaux* Ed. by CNRS, Paris, pp. 327–333.

Higgins, T.J.V. and Spencer, D. 1980. Biosynthesis of pea seed proteins. Evidence for precursor forms from *in vivo* and *in vitro* studies. In: *Genome Organization and Expression in Plants.* NATO Advanced Studies Institute Series A. C.J. Leaver, ed. Vol. 29. pp. 245-259. Plenum Press, New York.

Hill, J.E. and Breidenbach, R.W. 1974a. Proteins of soybean seeds. I. Isolation and characterization of the major components. *Plant Physiol.* **53**: 742–746.

Hill, J.E. and Breidenbach, R.W. 1974b. Proteins of soybean seeds. II. Accumulation of the major protein components during seed development and maturation. *Plant Physiol.* **53**: 747–751.

Hobday, S.M., Thurman, D.A. and Barber, D.J. 1973. Proteolytic and trypsic inhibitory activities in extracts of germinating *Pisum sativum* seeds. *Phytochemistry* **12**: 1041–1046.

Holt, N.W. and Sosulski, F.W. 1979. Amino acid composition and protein quality of field peas. *Can. J. Plant Sci.* **59**: 653–660.

Horisberger, M. and Vonlanthen, M. 1980. Ultrastructural localization of soybean agglutinin on thin sections of *Glycine max* (soybean). Var. Altona by the gold method. *Histochemistry* **65**: 181–186.

Hove, E.L. 1974. Composition and protein quality of sweet lupin seed. *J. Sci. Food Agric.* **25**: 851–859.

Iibuchi, C. and Imahori, K. 1978a. Interconversion between monomer and dimer of the 7S globulin of soybean seed. *Agric. Biol. Chem.* **42**: 25–30.

Iibuchi, C. and Imahori, K. 1978b. Heterogeneity and its relation to the subunit structure of the soybean 7S globulin. *Agric. Biol. Chem.* **42**: 31–36.

Inglett, G.E. 1972. *Symposium: Seed Proteins.* AVI Publishing Co. 320.

Ishino, K. and Ortega, D.M.L. 1975. Fractionation and characterization of major reserve proteins from seeds of *Phaseolus vulgaris*. *J. Agric. Food Chem.* **23**: 529–533.

Jackson, P., Boulter, D. and Thurman, D.A. 1969. A comparison of some properties of vicilin and legumin isolated from seeds of *Pisum sativum*, *Vicia faba* and *Cicer arietinum*. *New Phytol.* **68**: 25–33.

Jaffé, W.G. 1980. Hemagglutinins (lectins). In: *Toxic Constituents of Plant Foodstuffs.* I.E. Liener, ed. pp. 73–98. Academic Press,

Jermyn, W.A., MacKenzie, S.L. and Slinkard, A.E. 1979. Methods of reporting methionine content in peas. *Can. J. Plant Sci.* **59**: 231–232.

Johns, C.O. and Jones, D.B. 1916. The globulins arachin and conarachin. *J. Biol. Chem.* **28**: 77–87.

Johnson, P. and Shooter, E.M. 1950. The globulins of the groundnut. I. Investigation of arachin as a dissociation system. *Biochim. Biophys. Acta.* **5**: 361–375.

Jones, D.B. and Horn, M.J. 1930. The properties of arachin and conarachin and the proportionate occurrence in the peanut. *J. Agric. Res.* **40**: 673–682.

Kaldy, M.S. 1978. Amino acid composition and protein quality of 2 faba bean cultivars. *Can. Inst. Food Sci. Technol. J.* **11**: 97–98.

Kaldy, M.S. and Kasting, R. 1974. Amino acid composition and protein quality of 8 faba bean cultivars. *Can. J. Plant Sci.* **54**: 869–871.

Kasarda, D.D. 1980. Structure and properties of α-gliadins. *Ann. Technol. Agric.* **29**: 151–173.

Kelly, J.D. and Bliss, F.A. 1975. Quality factors affecting the nutritive value of bean seed protein. *Crop Sci.* **15**: 757–760.

Kern, R. and Chrispeels, M.J. 1978. Influence of the axis on the enzymes of protein metabolism in the cotyledons of mung bean seedlings. *Plant Physiol.* **62**: 815–819.

Kim, S.I., Pernollet, J.C. and Mossé, J. 1979. Evolution des protéines de l'albumen et de l'ultrastructure du caryopse d'*Avena sativa* au cours de la germination. *Physiol. Veg.* **17**: 231–245.

Kitamura, K., Okubo, K. and Shibasaki, K. 1974. The purification of soybean 11 S globulin with ConA-Sepharose 4B and Sepharose 6B. *Agric. Biol. Chem.* **38**: 1083–1085.

Kitamura, K. and Shibasaki, K. 1975. Homology between the acidic subunits of soybean 11 S globulin. *Agric. Biol. Chem.* **39**: 1509–1510.

Kitamura, K. and Shibasaki, K. 1977. Homology among the acidic subunits of soybean 11 S globulin. *Agric. Biol. Chem.* **41**: 351–357.

Kitamura, K., Takagi, T. and Shibasaki, K. 1976. Subunit structure of soybean 11 S globulin. *Agric. Biol. Chem.* **40**: 1837–1844.

Klimenko, V.G. and Ageeva, L.I. 1977. Isolation of vicilin-like proteins from lupin seeds and assay of their amino acid composition. *Prikl. Biokhim. Mikrobiol.* **13**: 730–737.

Kloz, J. and Klozova, E. 1974. The protein Euphaseolin in Phaseolinae. A chemotaxonomical study. *Biol. Plant.* **16**: 290–300.

Kloz, J., Turkova, V. and Klozova, E. 1966. Proteins found during maturation and germination of seeds of *Phaseolus vulgaris* L. *Biol. Plant.* **8**: 164–173.

Klozova, E. and Kloz, J. 1972. Distribution of the protein "Phaseolin" in some representatives of *Viciaceae*. *Biol. Plant.* **14**: 379–384.

Knypl, J.S. and Radziwonowska-Jozwiak, A. 1979. Electrophoregram patterns of proteins of embryo axes, cotyledons and protein bodies of soybean seeds. *Biochem. Physiol. Pflanzen.* **174**: 641–645.

Konopska, L. 1978. Proteins of aleurone grains in germinating seeds of *Pisum sativum* L. *Biochem. Physiol. Pflanzen.* **173**: 322–326.

Konopska, L. 1979. Changes in protein composition of aleurone grains of *Pisum sativum* L. during germination. *Biochem. Physiol. Pflanzen.* **174**: 275–282.

Koroleva, T.N., Alekseeva, M.V., Shutov, A.D. and Vaintraub, I.A. 1974. Localization of proteolytic enzymes in aleurone grains of vetch seeds. *Sov. Plant Physiol.* **20**: 650–653.

Koshiyama, I. 1972. A comparison of soybean globulins and the protein bodies in the protein composition. *Agric. Biol. Chem.* **36**: 62–67.

Koshiyama, I. and Fukushima, D. 1976a. Purification and some properties of γ-conglycinin in soybean seeds. *Phytochemistry* **15**: 161–164.

Koshiyama, I. and Fukushima, D. 1976b. A note on carbohydrates in the 11 S globulin of soybean seeds. *Cereal Chem.* **53**: 768–769.

Koshiyama, I. and Fukushima, D. 1976c. Physico-chemical studies on the 11 S globulin in soybean seeds: size and shape determination of the molecule. *Int. J. Peptide Protein Res.* **8**: 283–289.

Kourilsky, P. and Chambon, P. 1978. The ovalbumin gene: an amazing gene in 8 pieces. *Trends Biochem. Sci.* **3**: 244.

Krishna, T.G., Croy, R.R.D. and Boulter, D. 1979. Heterogeneity in subunit composition of the legumin of *Pisum sativum*. *Phytochemistry* **18**: 1879–1880.

Krishna, T.G., Mitra, R.K. and Bhatia, C.R. 1977. Seed globulins of *Cajanus cajan*. *Qual. Plant.* **27**: 313–325.

Kumar, K.G. and Venkataraman, L.V. 1975. Changes in reserve proteins of cowpea, chick pea and green gram during germination: physicochemical studies. *J. Food Sci. Technol.* **12**: 292–294.

Kumar, K.G. and Venkataraman, L.V. 1978. Chickpea seed proteins: modification during germination. *Phytochemistry* **17**: 605–609.

Lafiandra, L.D., Polignano, G.B. and Golaprico, G. 1979. Protein content and amino acid composition of seed in varieties of *Vicia faba* L. *Z. Pflanzenzüch.* **83**: 308–311.

Landry, J. and Moureaux, T. 1976. Quantitative estimation of accumulation of protein fractions in unripe and ripe maize grain. *Qual. Plant Mater. Veg.* **25**: 343–360.

Larkins, B.A. and Hurkman, W.J. 1978. Synthesis and deposition of zein in protein bodies of maize endosperm. *Plant Physiol.* **62**: 256–263.

Larkins, B.A., Pearlmutter, N.L. and Hurkman, W.J. 1979. Mechanism of zein synthesis and deposition in protein bodies of maize endosperm. In: *Plant Seed Development, Preservation and Germination.* I. Rubenstein, R.L. Phillips, C.E. Green and B.G. Gengenbach, eds. pp. 49–66. Academic Press, New York.

Leader, D.P. 1979. Protein biosynthesis on membrane bound ribosomes. *Trends Biochem. Sciences* **4**: 205–208.

Leaver, C.J. 1980. *Genome Organization and Expression in Plants.* NATO Advanced Study Institute Series A. Vol. 29. 607 pp. Plenum Press, New York.

Lester, B.R., Morris, R.O. and Cherry, J.H. 1979. Purification of leucine tRNA isoaccepting species from soybean cotyledons. II. RPC-2 purification, ribosome binding, and cytokinin content. *Plant Physiol.* **63**: 87–92.

Liener, I.E. 1976. Phytohemagglutinins (lectins). *Ann. Rev. Plant Physiol.* **27**: 291–319.

Liener, I.E. and Kakade, M.L. 1980. Protease inhibitors. In: *Toxic Constituents of Plant Foodstuffs.* I.E. Liener, ed. pp. 7–57. Academic Press.

Lillford, P.J. 1978. Physical properties of seed globulins with reference to meat analogue production. In: *Plant Proteins.* G. Norton, ed. pp. 289–298. Butterworths.

Lingappa, V.R., Lingappa, J.R. and Blobel, G. 1979. Chicken ovalbumin contains an internal signal sequence. *Nature* **281**: 117–121.

Lis, H. and Sharon, N. 1973. The biochemistry of plant lectins (phytohemagglutinins). *Ann. Rev. Biochem.* **42**: 541–574.

Lott, J.N.A. and Buttrose, M.S. 1978. Globoids in protein bodies of legume seed cotyledons. *Aust. J. Plant Physiol.* **5**: 89–111.

Ma, Y. and Bliss, F.A. 1978. Seed proteins of common bean. *Crop Sci.* **18**: 431–437.

McCurdy, S.M., Scheier, G.E. and Jacobson, M. 1978. Evaluation of protein quality of 5 varieties of lentils using *Tetrahymena pyriformis. J. Food Sci.* **43**: 694–697.

McLeester, R.C., Hall, T.C., Sun, S.M. and Bliss, F.A. 1973. Comparison of globulin proteins from *Phaseolus vulgaris* with those from *Vicia faba. Phytochemistry* **2**: 85–93.

Maneepun, S., Luh, B.S., and Rucker, R.B. 1974. Amino acid composition and biological quality of lima bean protein. *J. Food Sci.* **39**: 171–174.

Mascherpa, J.M. 1975. Contribution du fractionnement cellulaire à l'étude des corps protéiques, sphérosomes, microbodies et microsomes des graines. In: *Les Protéines des Graines.* J. Miège, ed. pp. 125–157. Georg Geneva.

Matile, P. 1968. Aleurone vacuoles as lysosomes. *Z. Pflanzenphysiol.* **58**: 365–368.

Mears, J.A. and Mabry, T.J. 1971. Alkaloids in the Leguminosae. In: *Chemotaxonomy of Leguminosae*. J.B. Harbone, D. Boulter and B.L. Turner eds. pp. 73–178. Academic Press, London and New York.

Melcher, U. 1979. *In vitro* synthesis of a precursor to the methionine rich polypeptide of the zein fraction of corn. *Plant Physiol*. **63**: 354–358.

Mertz, E.T., Bates, L.S. and Nelson, O.E. 1964. Mutant gene that changes protein composition and increases lysine content of maize endosperm. *Science* **145**: 279–280.

Miège, J. 1975. *Les Protéines des Graines*. Georg Geneva.

Miège, M.N., and Mascherpa, J.M. 1976. Isolation and analysis of protein bodies from cotyledons of *Lablab purpureus* and *Phaseolus vulgaris* (Leguminosae). *Physiol. Plant* **37**: 229–238.

Miflin, B.J., Matthews, J.A., Burgess, S.R., Faulks, A.J. and Shewry, P.R. 1980. The synthesis of barley storage proteins. In: *Genome Organization and Expression in Plants*. C.J. Leaver, ed. pp. 233–244. NATO Advanced Study Institute, Series A. Vol. 29, Plenum Press, New York.

Mikola, J., Yatsu, L.Y., Jacks, T.J., Hebert, J.J. 1975. Disruption of certain aleurone grains by various homogenization agents. *Plant Cell Physiol*. **16**: 933–937.

Millerd, A. 1975. Biochemistry of legume seed proteins. *Ann. Rev. Plant Physiol*. **26**: 53–72.

Millerd, A., Thomson, J.A. and Randall, P.J. 1979. Heterogeneity of sulphur content in the storage proteins of pea cotyledons. *Planta* **146**: 463–466.

Millerd, A., Thomson, J.A. and Schroeder, H.E. 1978. Cotyledonary storage proteins in *Pisum sativum*. III. Patterns of accumulation during development. *Aust. J. Plant Physiol*. **5**: 519–534.

Millerd, A. and Whitfeld, P.R. 1973. Deoxyribonucleic acid and ribonucleic acid synthesis during the cell expansion phase of cotyledon development in *Vicia faba* L. *Plant Physiol*. **51**: 1005–1010.

Milner, M. 1975. *Nutritional Improvement of Food Legumes by Breeding*. Proceedings of a symposium, Rome, 1972. 400 pp. Wiley-Interscience, New York.

Mironenko, A.V., Troitskaya, T.M., Shurkhai, S.F. and Domash, V.I. 1978. Vicilin- and legumin-like proteins of yellow lupin seeds. *Prikl. Biokhim. Mikrobiol*. **14**: 752–758.

Mitsuda, H., Kusano, T. and Hasegawa, K. 1964. Purification of the 11 S component of soybean proteins. *Agric. Biol. Chem*. **29**: 7–12.

Mlodzianowski, F. 1978. The fine structure of protein bodies in lupine cotyledons during the course of seed germination. *Z. Pflanzenphysiol*. **86**: 1–13.

Morawiecka, B. 1961. Changes in proteins of yellow lupin (*Lupinus luteus* L.) during germination. *Acta Biochim. Pol*. **8**: 313–319.

Moreira, M.A., Hermodson, M.A., Larkins, B.A. and Nielsen, N.C. 1979.

Partial characterization of the acidic and basic polypeptides of glycinin. *J. Biol. Chem.* **254**: 9921–9926.

Mori, T., Takagi, S. and Utsumi, S. 1979. Synthesis of glycinin in a wheat germ cell-free system. *Biochem. Biophys. Res. Commun.* **87**: 43–49.

Mori, T. and Utsumi, S. 1979. Purification and properties of storage proteins of broad bean. *Agric. Biol. Chem.* **43**: 577–583.

Mori, T., Utsumi, S. and Inaba, H. 1979. Interaction involving disulfide bridges between subunits of soybean seed globulin and between subunits of soybean and sesame seed globulins. *Agric. Biol. Chem.* **43**: 2317–2322.

Mori, T., Wakabayashi, Y. and Takagi, S. 1978. Occurrence of mRNA for storage protein in dry soybean seeds. *J. Biochem.* **84**: 1103–1111.

Morris, G.F.I., Thurman, D.A. and Boulter, D. 1970. The extraction and chemical composition of aleurone grains (protein bodies) isolated from seeds of *Vicia faba*. *Phytochemistry* **9**: 1707–1714.

Mossé, J. 1973. Hétérogénéité et polymorphisme des protéines et isoenzymes végétales: Aspects moléculaires et évolutifs. *Ann. Physiol. Vég.* **11**: 361–384.

Mossé, J. and Baudet, J. 1969. Etude intervariétale de la qualité protéique des orges: taux d'azote, composition en acides aminés et richesse en lysine. *Ann. Physiol. Vég.* **11**: 51–66.

Mossé, J. and Baudet, J. 1977. Relationship between amino acid composition and nitrogen contents of broad bean seed. In: *Protein Quality from Leguminous Crops.* Commission of the European Communities, Coordination of agricultural research, EUR 5686 EN, 48–56.

Mossé, J. and Fauconneau, G. 1973. Les aliments de demain. *La Recherche* **4**: 635–643.

Mossé, J. and Landry, J. 1980. Recent research on major maize proteins: zeins and glutelins. In: *Cereals for Food and Beverages. Recent progress in cereal chemistry.* (Proceedings of International Conference on cereals for food and beverages, Copenhagen, 1979). G.E. Inglett and L. Munck, eds. Academic Press.

Munck, L., Karlsson, K.E., Hagberg, A. and Eggum, B.O. 1970. Gene for improved nutritional value in barley seed protein. *Science* **168**: 985–987.

Müntz, K., Horstmann, C. and Scholz, G. 1972. Proteine und Protein Biosynthese in Samen von *Vicia faba* L. *Kulturpflanzen.* **20**: 277–326.

Müntz, K., Parthier, B., Aurich, O., Bassüner, R., Manteuffel, R., Püchel, N., Schmidt, P. and Scholz, G. 1978. Cell specialisation processes during biosynthesis and storage of proteins in plant seeds. In: *Regulation of Developmental Processes in Plants.* H.R. Gross, ed. pp. 70–97. Fisher Verlag, Iena.

Murray, D.R. 1979. A storage role for albumins in pea cotyledons. *Plant Cell Environ.* **2**: 221–226.

Murray, D.R. and Crump, J.A. 1979. Euphaseolin, the predominant reserve globulin of *Phaseolus vulgaris* cotyledons. *Z. Pflanzenphysiol.* **94**: 339–350.

Mutschler, M.A., Bliss, F.A. and Hall, T.C. 1980. Variation in the accumulation of seed storage protein among genotypes of *Phaseolus vulgaris* L. *Plant Physiol.* **65**: 627–630.

Nagahashi, J. and Beevers, L. 1978. Subcellular localization of glycosyl transferases involved in glycoprotein biosynthesis in the cotyledons of *Pisum sativum* L. *Plant Physiol.* **61**: 451–459.

Nagahashi, J., Browder, S.K. and Beevers, L. 1980. Glycosylation of pea cotyledon membranes. *Plant Physiol.* **65**: 648–657.

NAS. 1976. Genetic improvement of seed proteins (Proceedings of a workshop). National Acad. Sci., Washington DC.

Nelson, O.E., Mertz, E.T. and Bates, L.S. 1965. Second mutant gene affecting the amino acid pattern of maize endosperm proteins. *Science* **150**: 1469–1470.

Neucere, N.J. 1969. Isolation of α-arachin, the major peanut globulin. *Anal. Biochem.* **27**: 15–24.

Neucere, N.J. and Ory, R.L. 1970. Physicochemical studies of the proteins of the peanut cotyledon and embryonic axis. *Plant Physiol.* **45**: 616–619.

Neumann, D. and Weber, E. 1978. Formation of protein bodies in ripening seeds of *Vicia faba*. *Biochem. Physiol. Pflanzen.* **173**: 167–180.

Newcomb, E.H. 1967. Fine structure of protein-storing plastids in bean root tips. *J. Cell Biol.* **33**: 143–163.

Norton, G. 1978. *Plant Proteins*. Butterworths.

Ochiai-Yanagi, S., Fukazawa, C. and Harada, K. 1978. Formation of storage protein components during soybean seed development. *Agric. Biol. Chem.* **42**: 697–702.

Ochiai-Yanagi, S., Takagi, T., Kitamura, K., Tajima, M. and Watanabe, T. 1977. Reevaluation of the subunit molecular weights of soybean 11 S globulin. *Agric. Biol. Chem.* **41**: 647–653.

Ohmiya, M., Hara, I. and Matsubara, H. 1980. Pumpkin (*Cucurbita* sp.) seed globulin. IV. Terminal sequences of the acidic and basic peptide chains and identification of a pyroglutamyl peptide chain. *Plant Cell Physiol.* **21**: 157–167.

Okubo, K., Myers, D.V. and Iacobucci, G.A. 1976. Binding of phytic acid to glycinin. *Cereal Chem.* **53**: 513–524.

Olsen, S.H. 1980. A survey on recent developments in industrial processing of *Vicia faba* proteins. In: *Vicia faba*: feeding value, processing and viruses. D.A. Bond, ed. pp. 233–255. Published for EEC. M. Nijhopp.

Oota, J., Fujii, R. and Sunobe, Y. 1956. Studies on the connexion between sucrose formation and respiration in germinating bean cotyledons. *Physiol. Plant* **9**: 38–50.

Opik, H. 1968. Development of cotyledon cell structure in ripening *Phaseolus vulgaris* seeds. *J. Exp. Bot.* **19**: 64–76.

Orr, E. 1978. Sources of plant proteins—world supply and demand. In: *Plant Proteins*. G. Norton, ed. pp. 155–170. Butterworths.

Osborne, T.B. and Campbell, G.F. 1896. Legumin and other proteids of the pea and the vetch. *J. Am. Chem. Soc.* **18**: 583–609.

Otoul, E. 1973. Variabilité du spectre des acides aminés chez *Vigna unguiculata* (L.) WALP. *Bull. Rech. Agron. Gembloux.* **8**: 124–134.

Otoul, E. 1976. Spectres des acides aminés chez *Phaseolus lunatus* L. chez quelques espèces apparentées et chez l'amphidiploïde *P. lunatus × P. polystachyus* (L.) B.S. et P. *Bull. Rech. Agron. Gembloux.* **11**: 207–220.

Otoul, E., Marechal, R., Dardenne, G. and Desmedt, F. 1975. Des dipeptides soufrés différencient nettement *Vigna radiata* de *Vigna mungo. Phytochemistry* **14**: 173–179.

Padhye, V.W. and Salunkhe, D.K. 1979. Biochemical studies on black gram (*Phaseolus mungo* L.) seeds: Amino acid composition and subunit constitution of fractions of the proteins. *J. Food Sci.* **44**: 606–610.

P.A.G. 1975. Statement 22. In: *Nutritional Improvement of Food Legumes by Breeding.* M. Milner, ed. pp. 349–369. John Wiley and Sons, New York.

Palmer, R., MacIntosh, A. and Pusztai, A. 1973. The nutritional evaluation of kidney beans (*Phaseolus vulgaris*). The effect on the nutritional value of seed germination and changes in trypsin inhibitor content. *J. Sci. Food Agric.* **24**: 937–944.

Park, W.D., Lewis, E.D. and Rubenstein, I. 1980. Heterogeneity of zein messenger RNA and protein in maize. *Plant Physiol.* **65**: 98–106.

Pate, J.S. and Flinn, A.M. 1977. Fruit and seed development. In: *The Physiology of the Garden Pea.* J.F. Sutcliffe and J.S. Pate, eds. pp. 431–468. Academic Press, London and New York.

Paterson, B.M. and Rosenberg, M. 1979. Efficient translation of prokaryotic mRNAs in a eukaryotic cell-free system requires addition of a cap structure. *Nature* **279**: 692–701.

Payne, E.S., Brownrigg, A., Yarwood, A., and Boulter, D. 1971. Changing protein synthetic machinery during development of seeds of *Vicia faba. Phytochemistry* **10**: 2299–2303.

Payne, P.I. and Boulter, D. 1969. Free and membrane bound ribosomes of the cotyledons of *Vicia faba* (L.) I. Seed development. *Planta* **84**: 263–271.

Pernollet, J.C. 1978. Protein bodies of seeds: Ultrastructure, biochemistry, biosynthesis and degradation. *Phytochemistry* **17**: 1473–1480.

Pernollet, J.C. and Mossé, J. 1980. Caractérisation des corpuscules protéiques de l'albumen des caryopses de céréales par microanalyse électronique à balayage. *C.R. Acad. Sci. Paris* **290D**: 267–270.

Pillay, D.T.N. 1977. Protein synthesis in aging soybean cotyledons loss in translational capacity. *Biochem. Biophys. Res Commun.* **79**: 796–804.

Pirie, N.W. 1975. In: *Food Protein Sources.* Cambridge University Press.

Porter, W.M., Maner, J.H., Axtell, J.D. and Keim, W.F. 1974. Evaluation of the nutritive quality of grain legumes by an analysis for total sulfur. *Crop Sci.* **14**: 652–654.

Przybylska, J., Hurich, J. and Zimniak-Przybylska, Z. 1979. Comparative study of seed proteins in the genus *Pisum*. IV. Electrophoretic patterns of legumin and vicilin components. *Genet. Pol.* **20**: 517–528.

Püchel, M., Muntz, K., Parthier, B., Aurich, O., Bassuner, R., Manteuffel, R. and Schmidt, P. 1979. RNA metabolism and membrane bound polysomes in relation to globulin biosynthesis in cotyledons of developing field beans (*Vicia faba* L.). *Eur. J. Biochem.* **96**: 321–329.

Pusztai, A. 1966. The isolation of two proteins, glycoprotein I and a trypsin inhibitor from the seeds of kidney bean (*Phaseolus vulgaris*). *Biochem. J.* **101**: 379–384.

Pusztai, A., Clarke, E.M.W., King, T.P. and Stewart, J.C. 1979. Nutritional evaluation of kidney beans (*Phaseolus vulgaris*): Chemical composition, lectin content and nutritional value of selected cultivars. *J. Sci. Food Agric.* **30**: 843–848.

Pusztai, A., Croy, R.R.D., Grant, G. and Watt, W.B. 1977. Compartmentalization in the cotyledonary cells of *Phaseolus vulgaris* L. seeds: A differential sedimentation study. *New Phytol.* **79**: 61–71.

Pusztai, A., Croy, R.R.D., Stewart, J.S. and Watt, W.B. 1979. Protein body membranes of *Phaseolus vulgaris* L. cotyledons: Isolation and preliminary characterization of constituent proteins. *New Phytol.* **83**: 371–378.

Pusztai, A., Stewart, J.C. and Watt, W.B. 1978. A novel method for the preparation of protein bodies by filtration in high (over 70% w/v) sucrose-containing media. *Plant Sci. Lett.* **12**: 9–15.

Pusztai, A. and Watt, W.B. 1970. The isolation and characterization of a major antigenic and non-haemagglutinating glycoprotein from *Phaseolus vulgaris*. *Biochim. Biophys. Acta* **207**: 413–431.

Racusen, D. and Foote, M. 1971. The major glycoprotein in germinating bean seeds. *Can. J. Bot.* **49**: 2107–2111.

Randall, P.J., Thomson, J.A. and Schroeder, H.E. 1979. Cotyledonary storage proteins in *Pisum sativum*. IV. Effects of sulfur, phosphorus, potassium, and magnesium deficiencies. *Aust. J. Plant Physiol.* **6**: 11–24.

Richter, E. 1976. Merkmalsdifferenzen zwischen Pal- und Markerbsen I. Gehalt reifer Samen an primaren Inhaltsstoffen. *Gartnenbauwissenschaft.* **41**: 72–78.

Ritthausen, H. 1880. Über die Eiweisskörper verschiedenen Ölsamen. *Pflugers Arch. Ges. Physiol.* **21**: 81–104.

Romero, J., Sun, S.M., MacLeester, R.C., Bliss, F.A. and Hall, T.C. 1975. Heritable variation in a polypeptide subunit of the major storage protein of the bean. *Phaseolus vulgaris* L. *Plant Physiol.* **56**: 776–779.

Rougé, P. 1977. *Biologie des hémaggutinines chez le Pois.* Thèse de Doctorat d'Etat. n°746, Université Paul Sabatier, Toulouse.

Rubenstein, I., Phillips, R.I., Green, C. and Gengenbach, B.G. 1979. The plant seed: Development, preservation and germination. Academic Press.

Saio, K., Matsuo, T. and Watanabe, T. 1970. Preliminary electron microscopic investigation on soybean 11 S protein. *Agric. Biol. Chem.* **34**: 1851–1854.

Saio, K. and Watanabe, T. 1966. Preliminary investigation on protein bodies of soybean seeds. *Agric. Biol. Chem.* **30**: 1133–1138.

Sakakibara, M., Aoki, T. and Noguchi, H. 1979. Isolation and characterization of 7 S protein-I of *Phaseolus angularis* (Adzuki bean). *Agric. Biol. Chem.* **43**: 1951–1957.

Sauvaire, Y., Baccou, J.C. and Besançon, P. 1976. Nutritional value of the proteins of a leguminous seed: fenugreek (*Trigonella foenum-graecum* L.). *Nutrition Reports International* **14**: 527–537.

Scharpé, A. and Parijs, R.V. 1973. The formation of polyploid cells in ripening cotyledons of *Pisum sativum* L. in relation to ribosome and protein synthesis. *J. Exp. Bot.* **24**: 216–222.

Schlesier, B., Manteuffel, R. and Scholz, G. 1978. Studies on seed globulins from legumes. VI. Association of vicilin from *Vicia faba* L. *Biochem. Physiol. Pflanzen.* **172**: 285–290.

Scholz, G., Richter, J. and Manteuffel, R. 1974. Studies on seed globulins from legumes. I. Separation and purification of legumin and vicilin from *Vicia faba* L. by zone precipitation. *Biochem. Physiol. Pflanzen.* **166**: 163–172.

Schwenke, K.D. 1975. Pflänzliche Samenglobuline als dissozierende und assozierende Protein-systeme. *Nahrung.* **19**: 69–82.

Sefa-Dedeh, S. and Stanley, D. 1979a. Cowpea proteins. 1. Use of response surface methodology in predicting cowpea (*Vigna unguiculata*) protein extractability. *J. Agric. Food Chem.* **27**: 1238–1243.

Sefa-Dedeh, S. and Stanley, D. 1979b. Cowpea proteins 2. Characterization of water extractable proteins. *J. Agric. Food Chem.* **27**: 1244–1247.

Sgarbieri, V.C. and Galeazzi, M.A.M. 1978. Some physicochemical and nutritional properties of a sweet lupin (*Lupinus albus* var. multolupa) protein. *J. Agric. Food Chem.* **26**: 1438–1442.

Sharma, C.B. and Dieckert, J.W. 1975. Isolation and partial characterization of globoids from aleurone grains of *Arachis hypogaea* seed. *Physiol. Plant.* **33**: 1–7.

Shetty, K.J. and Rao, M.S.N. 1974. Studies on groundnut proteins. III. Physicochemical properties of arachin prepared by different methods. *Anal. Biochem.* **62**: 108–120.

Shetty, K.J. and Rao, M.S.N. 1977. Studies on groundnut proteins. VI. Isolation, characterization and hydrogen ion titration of conarachin II. *Int. J. Peptide Protein Res.* **9**: 11–17.

Shutov, A.D., Koroleva, T.N. and Vaintraub, I.A. 1978. Participation of protease of dormant vetch seeds in decomposition of storage proteins during germination. *Sov. Plant Physiol.* **25**: 573–578.

Singh, R., and Axtell, J.D. 1973. High lysine mutant gene (hl) that improves protein quality and biological value of grain sorghum. *Crop Sci.* **13**: 535–539.

Singh, U. and Jambunathan, R. 1980. Evaluation of rapid methods for the estimation of protein in Chickpea (*Cicer arietinum* L.). *J. Sci. Food Agric.* **31**: 247–254.

Smith, A.K. and Circle, S.J. 1972. Chemical composition of the seed. In: *Soybeans: Chemistry and Technology.* (Vol. I. Proteins). A.K. Smith and S.J. Circle, eds. pp. 61–92. AVI Publishing Co.

Sobolev, A.M., Buzulukova, N.P., Dmitrieva, M.E. and Barbashova, A.K. 1977. Structural-biochemical organization of aleurone grains in seeds of yellow lupine. *Sov. Plant Physiol.* **23**: 621–628.

Sodek, L., Vecchia, P.T.D. and Lima, M.L.G.P. 1975. Rapid determination of tryptophan in beans (*Phaseolus vulgaris*) by the acid ninhydrin method. *J. Agric. Food Chem.* **23**: 1147–1150.

Stockman, D.R., Hall, T.C. and Ryan, D.S. 1976. Affinity chromatography of the major seed protein of the bean (*Phaseolus vulgaris* L.). *Plant Physiol.* **58**: 272–275.

Sun, S.M. and Hall, T.C. 1975. Solubility characteristics of globulins from *Phaseolus* seeds in regard to their isolation and characterization. *J. Agric. Food Chem.* **23**: 184–189.

Sun, S.M., Ma, Y., Buchbinder, B.U. and Hall, T.C. 1979. Comparison of G 1 polypeptides synthesized *in vitro* and *in vivo* in the presence and absence of a glycosylation inhibitor. *Plant Physiol. Ann. Meeting Abstracts: Amer. Soc. Plant Physiol., Ohio State University Columbus.* 1979, **63**: (suppl. No. 5) 94.

Sun, S.M., MacLeester, R.C., Bliss, F.A. and Hall, T.C. 1974. Reversible and irreversible dissociation of globulins from *Phaseolus vulgaris* seed. *J. Biol. Chem.* **249**: 2118–2121.

Sun, S.M., Mutschler, M.A., Bliss, F.A. and Hall, T.C. 1978. Protein synthesis and accumulation in bean cotyledons during growth. *Plant Physiol.* **61**: 918–923.

Sund, H. and Weber, K. 1966. The quaternary structure of proteins. *Angew. Chem. Int. Ed.* **5**: 231–245.

Sussex, I.M., Dale, R.M.K. and Crouch, M.L. 1980. Developmental regulation of storage protein synthesis in seeds. In: *Genome Organization and Expression in Plants.* NATO Advanced Studies Institute, Series A. Vol. 29. C.J. Leaver, ed. pp. 283–290. Plenum Press, New York.

Thanh, V.H., Okubo, K. and Shibasaki, K. 1975a. Isolation and characterization of the multiple 7 S globulins of soybean proteins. *Plant Physiol.* **56**: 19–22.

Thanh, V.H., Okubo, K. and Shibasaki, K. 1975b. The heterogeneity of the 7 S soybean protein by sepharose gel chromatography and disc gel electrophoresis. *Agric. Biol. Chem.* **39**: 1501–1503.

Thanh, V.H. and Shibasaki, K. 1976a. Heterogeneity of beta-conglycinin. *Biochim. Biophys. Acta* **439**: 326–338.

Thanh, V.H. and Shibasaki, K. 1976b. Major proteins of soybean seeds. A straight forward fractionation and their characterization. *J. Agric. Food Chem.* **24**: 1117–1121.

Thanh, V.H. and Shibasaki, K. 1977. Beta-conglycinin from soybean proteins. Isolation and immunological and physicochemical properties of the monomeric forms. *Biochim. Biophys. Acta* **490**: 370–384.

Thanh, V.H. and Shibasaki, K. 1978a. Major proteins of soybean seeds. Subunit structure of beta-conglycinin. *J. Agric. Food Chem.* **26**: 692–695.

Thanh, V.H. and Shibasaki, K. 1978b. Major proteins of soybean seeds. Reconstitution of beta-conglycinin from its subunits. *J. Agric. Food Chem.* **26**: 695–698.

Thanh, V.H. and Shibasaki, K. 1979. Major proteins of soybean seeds. Reversible and irreversible dissociation of β-conglycinin. *J. Agric. Food Chem.* **27**: 805–809.

Thompson, J.F., Madison, J.T. and Muenster, A.M.E. 1977. *In vitro* culture of immature cotyledons of soybean (*Glycine max* L. Merr). *Ann. Bot.* **41**: 29–39.

Thomson, J.A. and Doll, H. 1979. Genetics and evolution of seed storage proteins. In: *Seed Protein Improvement of Cereals and Grain Legumes.* (Proc. Symp. Neuherberg 1978). I.A.E.A. Vienna, Vol. I. 109–123.

Thomson, J.A., Millerd, A. and Schroeder, H.E. 1979. Genotype-dependent patterns of accumulation of seed storage proteins in *Pisum*. In: *Seed Protein Improvement in Cereals and Grain Legumes.* (Proc. Symp. Neuherberg, 1978). I.A.E.A. Vienna, Vol. I. 231–240.

Thomson, J.A., Schroeder, H.E. and Dudman, W.F. 1978. Cotyledonary storage proteins in *Pisum sativum* I. Molecular heterogeneity. *Aust. J. Plant Physiol.* **5**: 263–279.

Thomson, W.W. 1979. Ultrastructure of dry seed tissue after a non-aqueous primary fixation. *New Phytol.* **82**: 207–212.

Tkachuk, R. 1969. Nitrogen to protein conversion factors for cereals and oilseed meals. *Cereal Chem.* **46**: 419–424.

Tombs, M.P. 1965. An electrophoretic investigation of groundnut proteins: The structure of arachins A and B. *Biochem. J.* **96**: 119–133.

Tombs, M.P. 1967. Protein bodies of the soybean. *Plant Physiol.* **42**: 797–813.

Tombs, M.P. 1978. Filaments from proteins. In: *Plant Proteins.* G. Norton, ed. pp. 283–288. Butterworths.

Tombs, M.P. and Lowe, M. 1967. A determination of the subunits of arachin by osmometry. *Biochem. J.* **105**: 181–187.

Tombs, M.P., Newson, B.F. and Wilding, P. 1974. Protein solubility: Phase separation in arachin-salt-water systems. *Int. J. peptide Protein Res.* **6**: 253–277.

Utsumi, S. and Mori, T. 1980. Heterogeneity of broad bean legumin. *Biochim. Biophys. Acta* **621**: 179–189.

Vaintraub, I.A. and Gofman, Y.Y. 1961. N-kontzerie aminokisloty legumina i vilzilina gorokha. *Biokhimiya.* **26**: 13–14.

Vaintraub, I.A., Seliger, P. and Shutov, A.D. 1979. The action of pepsin on the reserve proteins of some leguminous seeds. *Nahrung.* **23**: 15–21.

Vaintraub, I.A., Shutov, A.D. and Klimenko, V.G. 1962. On vetch seed globulins. *Biokhimiya.* **27**: 349–358.

Van Der Wilden, W., Herman, E.M. and Chrispeels, M.J. 1980. Protein bodies of mung bean cotyledons as autophagic organelles. *Proc. Natl. Acad. Sci. USA* **77**: 428–432.

Van Etten, C.H., Kwolek, W.F., Peters, J.E. and Barclay, A.S. 1967. Plant seeds as protein sources for food or feed. Evaluation based on amino acid composition of 379 species. *J. Agric. Food Chem.* **15**: 1077–1089.

Van Etten, C.H., Miller, R.W., Wolff, I.A. and Jones, Q. 1963. Nutrients in seeds. Amino acid composition of seeds from 200 angiospermous plant species. *Agric. Food Chem.* **11**: 399–410.

Varner, J.E. and Schidlovsky, G. 1963. Intracellular distribution of proteins in pea cotyledons. *Plant Physiol.* **38**: 139–144.

Vuyst, A. de, Vervack, W. Van Belle, M., Arnould, R. and Moreels, A. 1963. Les acides aminés des tourteaux. *Agricultura.* **11**: 385–390.

Weber, E., Manteuffel, R. and Neumann, D. 1978. Isolation and characterization of protein bodies of *Vicia faba* seeds. *Biochem. Physiol. Pflanzen.* **172**: 597–614.

Weber, E., Suess, K.H., Neumann, D. and Manteuffel, R. 1979. Isolation and partial characterization of the protein body membrane from mature and germinating seeds of *Vicia faba* L. *Biochem. Physiol. Pflanzen.* **174**: 139–150.

Wolf, W.J. 1976. Chemistry and technology of soybeans. *Adv. Cereal Sci. Technol.* **1**: 325–377.

Wolf, W.J. 1977. Legumes: seed composition and structure, processing into protein products and protein properties. In: *Food Proteins.* J.R. Whitaker and S.R. Tannenbaum, eds. pp. 291–314. AVI Publishing Co.

Wolf, W.J. and Baker, F.L. 1972. Scanning electron microscopy of soybeans. *Cereal Sci. Today* **17**: 125–130.

Wolf, W.J. and Briggs, D.R. 1959. Purification and characterization of the 11 S component of soybean proteins. *Arch. Biochem. Biophys.* **85**: 186–199.

Wolf, W.J. and Sly, D.A. 1967. Cryoprecipitation of soybean 11 S protein, *Cereal Chem.* **44**: 653–668.

Wolff, G. 1975a. Relation between seed distribution in the pod, protein content, and thousand-kernel weight of *Pisum sativum*. *Pisum Newsl.* **7**: 60.

Wolff, G. 1975b. Distribution of seeds in the pods of *Pisum sativum* and some of its mutants. *Pisum Newsl.* **7**: 61.

Wright, D.J. and Boulter, D. 1972. The characterisation of vicilin during seed development in *Vicia faba* (L.). *Planta.* **105**: 60–65.

Wright, D.J. and Boulter, D. 1973. A comparison of acid extracted globulin fractions and vicilin and legumin of *Vicia faba*. *Phytochemistry* **12**: 79–84.

Wright, D.J. and Boulter, D. 1974. Purification and subunit structure of legumin of *Vicia faba* L. (broad bean). *Biochem. J.* **141**: 413–418.

Yamada, T., Aibara, S. and Morita, Y. 1979a. Dissociation-association behaviour of arachin between dimeric and monomeric forms. *Agric. Biol. Chem.* **43**: 2549–2556.

Yamada, T., Aibara, S. and Morita, Y. 1979b. Isolation and some properties of arachin subunits. *Agric. Biol. Chem.* **43**: 2563–2568.

Yamauchi, F., Kawase, M., Kanbe, M. and Shibazaki, K. 1975. Separation of the B-aspartamido-carbohydrate fractions from soybean 7 S protein: Protein-carbohydrate linkage. *Agric. Biol. Chem.* **39**: 873–878.

Yamauchi, F. Thanh, V.H., Kawase, M. and Shibasaki, K. 1976. Separation of the glycopeptides from soybean 7 S protein: Their amino acid sequences. *Agric. Biol. Chem.* **40**: 691–696.

Yamauchi, F. and Yamagishi, T. 1979. Carbohydrate sequence of a soybean 7 S protein. *Agric. Biol. Chem.* **43**: 505–510.

Yarwood, A. 1978. Biosynthesis of legume seed proteins. In: *Plant Proteins.* G. Norton, ed. pp. 41–56. Butterworths.

Yokoyama, Z., Mori, T. and Matsushita, S. 1972a. Characterization of ribonucleic acids in the protein bodies from soyabean seeds. *Agric. Biol. Chem.* **36**: 33–41.

Yokoyama, Z., Mori, T. and Matsushita, S. 1972b. Some aspects of a template activity of high molecular weight RNA from soybean protein bodies. *Agric. Biol. Chem.* **36**: 2237–2240.

Yomo, H. and Varner, J.E. 1973. Control of the formation of amylases and proteases in the cotyledons of germinating peas. *Plant Physiol.* **51**: 708–713.

4

Non-Protein Nitrogenous Compounds with Particular Attention to Ureides

E. JOLIVET and J. MOSSÉ

Plant organs (roots, tubers, stems, leaves, flowers and seeds) contain numerous compounds which generally correspond to minor quantities. Therefore, they are often classified among "secondary plant products" as suggested by Bonner (1950), Paech (1950) and reminded by Mothes (1980) and Bell (1980a).

A significant fraction of these secondary products consists of non-protein nitrogenous compounds. Many of them have been recently reviewed. Such is the case of:

Non-protein or uncommon amino acids (Fowden 1976; Bell 1976 and 1980b),

Amines (Smith 1975 and 1980),

Alkaloids (Leete 1980; Fodor 1980a and b; Gröger 1980),

Purine and pyrimidine derivatives like favism agents (Mager et al. 1980),

Lathyrogens (Padmanaban 1980),

Glucid derivatives like cyanogenic glycosides (Conn 1980; Montgomery 1980) or osamines.

At first, we will only briefly remind a few general features of all these non-protein nitrogenous compounds, many of which are encountered in legumes. The second section deals with ureides which play a significant role as transport forms in Leguminosae in relation with nitrogen fixation by nodules.

4.1. Non-Protein Nitrogenous Compounds

4.1.1. NON-PROTEIN AMINO ACIDS

In addition to the twenty amino acids of the Nirenberg code which are constituents of proteins, more than three hundred naturally occurring amino acids have been isolated in plant kingdom and their structure determined. Bell

(1971) has reviewed their distribution in Leguminosae. Fowden (1976) has proposed to classify them into three classes according to the structural links which exist between them and their protein counterparts. He suggested these links may take the form of homology, of simple substitution of protons by larger groups, or of isosterism. Homology corresponds to amino acids which could derive one from another by addition of one carbon atom. For instance, homoserine OH $(CH_2)_2CH$ (NH_2) COOH when compared to serine; α-amino adipic acid HOOC $(CH_2)_3$ CH (NH_2) COOH when compared to glutamic acid. Substitution can be illustrated by derivatives of glutamic acid like γ-methyl, γ-methylene, γ-ethyl, γ-hydroxy-γ-methyl derivatives several of which have been studied by Dardenne (1976). Isosterism characterizes amino acid molecules of very similar size and spatial configuration. In this acceptation arginine and canavanine HN (NH_2) CNHO $(CH_2)_2$ CH (NH_2) COOH are isosteric, and also lysine with S-aminoethylcysteine.

It is generally agreed that canavanine, one of the most nitrogen-rich amino acids of the Leguminosae seems to have a nitrogen storage function: it is indeed present in large amounts in the seeds of several species (10 g per 16 g of N in *Canavalia ensiformis*) and it is utilized during germination. Another one, γ-methylene-glutamine is a form of transport: its amount can reach 95 per cent of total nitrogen of sap of *Arachis hypogaea*.

Some of them are toxic compounds. In some *Lathyrus* species, β-aminopropionitrile and also its γ-glutamyl peptide are lathyrogens (Padmanaban 1980) which cause osteolathyrism. *Vicia sativa* contains β-cyanoalanine and its γ-glutamyl derivative which both are neurotoxic. More than fifty of nonprotein amino acids are likewise toxic for insects or for mammals. Other ones can compete with bacterial permease enzymes, or act as competitive inhibitors, or due to isosterism, substitute to protein amino acids for aminoacyl-t-RNA synthetases. Lastly, several authors have evidenced different D-aminoacids. For example *Pisum sativum* seedlings contain D-aspartic acid and D-glutamic acid (Ogawa et al. 1977) and also D-alanine (Fukuda et al. 1973).

Therefore non-protein amino acids unquestionably play a role of defence of plant species.

4.1.2. Amines

Monoamines, diamines and polyamines are almost always derived from amino acids by decarboxylation. They can also result from aldehyde transamination. And they are often precursors of alkaloids (Smith 1975 and 1980). For instance, *Vicia faba* contains several aliphatic monoamines: dimethylamine (which is derived from alanine); diethylamine; n-propylamine (derived from γ-aminobutyric acid); n-hexylamine; isobutylamine (derived from valine); and isoamylamine (derived from lysine). *Glycine max* contains agmatine (derived from arginine). *Lupinus luteus* and *Pisum sativum* contain cadaverin (derived from lysine). *Vigna radiata* contains S-methyl-adenosyl-homo-cysteamine (derived from S-adenosyl-methionine).

Like a few amino acids, amines can play a nitrogen storage role. But they can have several other important and various functions. Monoamines may facilitate pollination by acting as attractive for insects, due to their particular smell. Polyamines are able to stimulate plant growth and their concentration depends on mineral nutrition: mineral cation (K^+ and Mg^{++}) deficiency causes a formation of amines which act as pH regulators and may contribute to the electric neutrality of plant cells. They are known to interact with nucleic acids in the double helical form, their positive charges allowing the association with phosphate groups; so that they seem to be involved in protein biosynthesis (Algranati and Goldenberg 1977). Some of them appear in plants in response to infractions by different agents: fungi or viruses. Then they accumulate as conjugates with phenolics like ferulic acid, caffeic acid or para-coumaric acid (Cabanne et al. 1977; Martin-Tanguy et al. 1978). The latter have suggested that caffeyl-putrescine may act as a plant hormone-like substance in floral induction. It is a fact that amines are often synthesized in flowers at anthesis and in developing seeds. On the other hand, one cannot refrain to remind that polyamines are shown to be involved in rapid growth and cell proliferation of animal tissues as reviewed by Jänne et al. (1978).

4.1.3. ALKALOIDS

If isoprenoid alkaloids, also called pseudoalkaloids, are excepted, almost all plant alkaloids are derived from amino acids which can be either proteic (isoleucine, phenylalanine, tyrosine, lysine, histidine, tryptophan) or non-proteic (ornithine, nicotinic acid, anthranilic acid). There are very few available data about their content in plants. This lack of information is partially due to the fact that the alkaloid content can vary for an individual plant, for populations, and also according to the physiological stage of development. Mears and Mabry (1971) have collected and reviewed their structure and distribution in the Leguminosae through which they are widespread. *Lupinus* seeds contain many kinds of alkaloids which have lysine as precursor: lupinine, lupanine, sparteine, angustifoline, anagyrine and multiflorine. They also con-tain vicine and convicine which are β-D-glucopyranosides conjugated with pyrimidine derivatives. Both are known as the agents of favism, a human disease linked to glucose-6-phosphate dehydrogenase deficiency. Trigonelline, which is derived from anthranilic acid occurs in *Glycine max*, *Pisum sativum* and *Trigonella foenum-graecum*. But many consumable legume seeds are poor in alkaloids or are devoid of them.

It is still difficult to understand the actual function of these substances in plant metabolism. Because of their low content, they can hardly act as nitrogen metabolism regulators. It could be that they play any role in defence, for instance by inhibiting feeding (Bell 1980a).

4.1.4. CYANOGENIC GLYCOSIDES

Until now, two dozen cyanogenic glycosides have been identified in higher plants (Conn 1980). The sugar moiety is very often D-glucose. The aglycones are derived from amino acids. For instance *Phaseolus*, like *Lotus* and *Trifolium* contain linamarin and lotaustralin of which valine and leucine are respectively the aglycone precursors. Their biosynthesis and degradation are more and more well known, but their possible role remains unclear.

4.1.5. POSSIBLE ROLES AND SIGNIFICANCE

As we have discussed above, it can be seen that many nitrogenous secondary compounds play various roles. They can have biochemical or physiological function and can be implicated in the control of different metabolism steps. They are sometimes nitrogen reserve compounds for the young seedling, or also nitrogen transport forms through the plant. They can play an attractive role or more frequently, a repulsive role, and they are therefore involved in species' defence against possible animal or human consumers.

Due to the fact that their occurrence implies particular metabolic pathways with different special enzymes and regulatory mechanisms (Luckner 1980), they are often worth having as markers in genetics, phylogeny and taxonomy. This had been already pointed out by Harborne et al. (1971) in their collective book on *Chemotaxonomy of the Leguminosae*: the occurrence of the same secondary compound in two or in several species, already known as being related in other respects enables to assume that these species result from the evolution of a common ancestral form, the genome of which controls its biosynthesis. In this acceptance, the study of such compounds progressively becomes of the first significance.

4.2. Ureides in Legumes

The glyoxylic ureides, allantoin and allantoic acid, are very important nitrogenous compounds which have been found in numerous living organisms, particularly in higher plants and especially in a number of species of the

Allantoin   Allantoic acid legume family. In the past, several reviews concerning biochemistry and metabolic study of these substances have been published (Brunel and Capelle 1947; Tracey 1955; Bollard 1959; Mothes 1961; Reinbothe and Mothes 1962; Durand et al. 1965) and they bring out the fact that our knowledge in the field related to these compounds is still largely insufficient.

Recently, studies about ureides in legumes have extensively been developed because these substances have been found to be related to symbiotic dinitrogen fixation in a lot of leguminous plants. Works carried out are dealing with three species: soybean (*Glycine max*) mainly studied by Japanese authors (Tajima and Yamamoto 1975, 1977; Tajima et al. 1977; Fujihara et al. 1977; Fujihara and Yamaguchi 1978a, b; Matsumoto et al. 1977a, b, c, 1978; Ohyama and Kumazawa 1978, 1979), cowpea (*Vigna unguiculata*) studied by an Australian team (Herridge et al. 1978; Atkins et al. 1980; Pate et al. 1980), French bean (*Phaseolus vulgaris*) studied by a Swiss group (Thomas et al. 1979, 1980). A short analysis of the results obtained by the Japanese and Australian workers has just been published (Rawsthorne et al. 1980).

The present section is dealing with biosynthesis, distribution and role of ureides in legumes.

4.2.1 UREIDE FORMATION IN LEGUMES

(1) BIOSYNTHETIC PATHWAYS

In animals, allantoin is produced through a biosynthetic pathway including the formation of purines followed by their degradation (Fig. 4.1). Bacteria, yeasts, mushrooms, algae are also able to degrade oxypurines and accordingly allantoin (Fujihara and Yamaguchi 1978b).

In higher plants, ureides formation is likewise assumed to take place through purine degradation (Fosse et al. 1933; Echevin and Brunel 1937; Barnes 1959); a direct biosynthesis of allantoic acid from urea and glyoxylic acid, already suggested by Brunel and Brunel-Capelle (1951) in the Basidiomycetacae, has been also considered in the banana leaf by Freiberg et al. (1957) who found a great proportion of radioactivity in allantoic acid when the plant was supplied with ^{14}C urea. According to Mothes' (1961) and Reinbothe and Mothes (1962) this direct pathway for ureide synthesis would be acting too in the case of allantoin rich plants, soybean for example, though among legumes like french bean (*Phaseolus vulgaris*) this direct synthesis has neither been proved nor has it been discarded (Thomas et al. 1980). Nowadays, the ureide formation in soybean, the most studied species, is supposed to take place through a biosynthesis of purines followed by the degradation of these newly formed substances. The following reasons are in favour of this assumption:

(1) Matsumoto et al. (1977b) supply nodulated intact soybean plants with

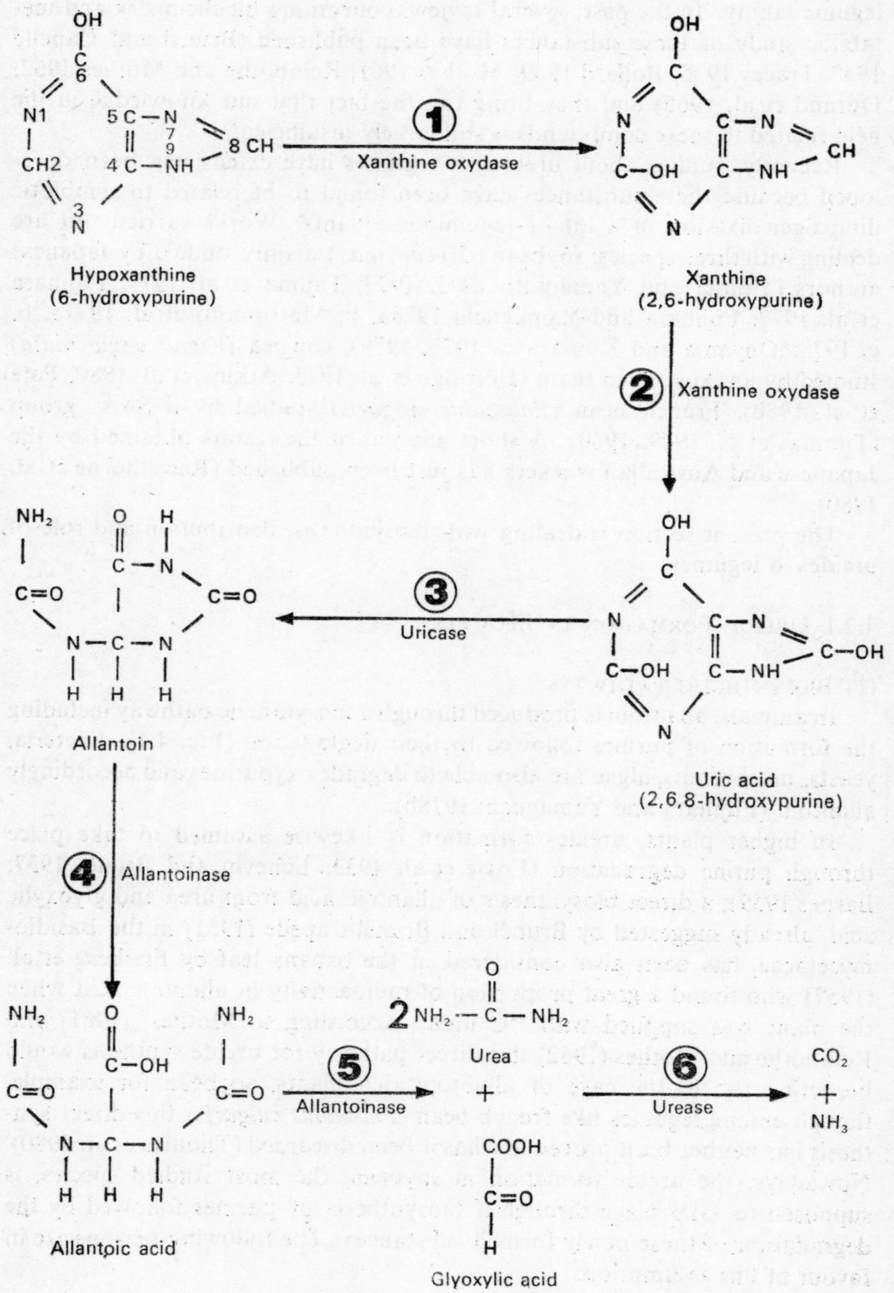

Fig. 4.1. Catabolism of purine compounds.

$^{15}N_2$ during 24 hours. They find that the level of ^{15}N in allantoin is higher than in allantoic acid, suggesting that allantoin has been synthesized before allantoic acid which would come from it.

(2) The activities of the three enzymes in the degradation pathway of xanthine into allantoic acid have been evidenced (Tajima and Yamamoto 1975, 1977; Tajima et al. 1977). These are xanthine oxidase (xanthine = O_2 oxidoreductase E.C.1.2.3.2.), uricase (urate = O_2 oxidoreductase E.C.1.7.3.3.) and allantoinase (allantoin amidohydrolase E.C.3.5.2.5.).

(3) When soybean seedlings, aseptically germinated upon agar, are supplied with allopurinol (4-hydroxypyrazolo (3.4 d) pyrimidine), a potent inhibitor of xanthine oxidase, a large decrease of allantoin and allantoic acid production is observed in cotyledons and axis with a high concomitant accumulation of xanthine (Fujihara and Yamaguchi 1978b).

Very recently, through the same approach based upon studies of activities of xanthine oxidase, uricase, allantoinase and by using allopurinol, Atkins et al. (1980) have drawn a conclusion identical to that of Fujihara and Yamaguchi (1978b). Then they have shown that in *Vigna unguiculata* ureide biosynthesis is performed through the pathway of purine degradation. By using allopurinol, Thomas et al. (1980) conclude that the same pathway is operating in *Phaseolus vulgaris*.

A ureide formation which would be based upon a previous purine biosynthesis requires two amide nitrogen atoms, originating from either glutamine or asparagine (Kapoor and Waygood 1962; Robern et al. 1965) and one nitrogen atom from aspartate to make up three of the four nitrogen atoms of allantoin and allantoic acid (Beevers 1976). The amide group of glutamine is the initial product of ammonium ion assimilation, following nitrate reduction or dinitrogen fixation (Miflin and Lea 1977). This amide nitrogen is directly used in the glutamate biosynthesis (Robertson et al. 1975) and in the synthesis of amide groups of asparagine by transamidation with aspartate (Scott et al. 1976). These facts explain that the biosynthesis of ureides can be considered as being competitive with the asparagine formation (Pate et al. 1980).

(2) SITE OF FORMATION

The allantoin production is carried out in the underground parts of the plant, as is demonstrated by experiments of reciprocal grafting carried out by Matsumoto et al. (1978) between two soybean varieties, the nodulated A62-1 and genetically non-nodulated A62-2. They have observed a high accumulation of allantoin in stems of A62-2 plants grafted onto the A62-1 nodulated variety but they have found a very reduced production in stems of A62-1 plants grafted onto the A62-2 non-nodulated variety (Fig. 4.2). The results of these grafting experiments, in agreement with other results from Matsumoto et al. (1977a and b) show that the aerial parts of the plant have no direct effect upon allantoin production. The formation of this compound depends

ratio is 10 to 50 times higher than any other compound ratio, which suggests that allantoin is actively synthesized in nodules.

Other results arising from enzyme studies are also in favour of ureide formation in nodules. Indeed enzymes implicated in allantoin biosynthesis from xanthine are almost exclusively found in nodules (Tajima et al. 1977). In the four weeks old nodulated soybean variety A62-1, xanthine oxidase is found solely in nodules. The localization of this enzyme explains the high accumulation of xanthine which is accompanied by a reduced content of allantoin in nodules, when plants have absorbed allopurinol (Fujihara and Yamaguchi 1978b). An unusual uricase having chemical properties different from typical seedling uricase, is localized in nodules and its activity is very high. However, after induction of this particular uricase, allantoin formation in nodules is probably a specific characteristic of symbiosis, for this uricase is not found in cells of a pure cultivation of *Rhizobium japonicum* 002 bacteria, supplied with RNA, nucleotides or urate (Tajima and Yamamoto 1975). Uricase has been evidenced in the bacteroids during symbiosis (Tajima et al. 1977), suggesting that purines are degraded inside these bacteroides. The mechanism of allantoin biosynthesis is likely more complicated in nodules than in young germinated plants because of the complex nature of nodules composed of host cells and bacteria (Tajima et al. 1977).

The high allantoinase activity exhibited not only in nodules, but also in roots, stems and leaves, shows that allantoin is translocated from nodules towards different parts of the plant where it is at first converted into allantoic acid by this allantoinase.

4.2.2. UREIDE DISTRIBUTION IN LEGUMES

(1) UREIDE REPARTITION IN LEGUME FAMILY

Ureides are widespread in Aceraceae (Mothes and Engelbrecht 1952; Barnes 1959) and Borraginaceae, for example, *Symphytum officinale* (Mothes and Engelbrecht 1954; Butler et al. 1961). As mentioned above, in Leguminosae, they have been essentially studied in *Glycine max*, *Vigna unguiculata*, *Phaseolus vulgaris*. Pate et al. (1980) have evidenced these compounds in other tropical series: mung bean (*Vigna radiata*) in which the presence of allantoin and allantoic acid had been previously noticed (Nagai and Funahashi 1961), adzuki bean (*Vigna angularis*), rice bean (*Vigna umbellata*), black gram (*Vigna mungo*), cluster bean (*Cyamopsis tetragonoloba*), horse gram (*Macrotyloma uniflorum*), wing bean (*Psophocarpus tetragonolobus*). They can be found as allantoin and/or as allantoic acid. This latter compound is particularly abundant in *Phaseolus vulgaris* (Pate et al. 1980). Ureides however have not been found in all legume species. For instance *Lupinus albus* or *Pisum sativum* which contain only small amounts (Fosse et al. 1933) are on the contrary very rich in amides, especially asparagine (Pate et al. 1980).

(2) REPARTITION INSIDE UREIDE-RICH LEGUME

The repartition of enzymes acting in ureide metabolism among the different parts of soybean plant, shows that allantoin is mostly synthesized in nodules (Ishizuka 1972; Matsumoto et al. 1977a, b, c). It is translocated towards roots in which it is also lightly synthesized, then translocated towards stem internodes, leaves, pods and seeds as far as nodules are expanding (Table 4.2). As a matter of fact, in nodulated four weeks old A62-1 soybean plants, xanthine

TABLE 4.2. Variation in allantoin concentration of each organ of nodulating soybean plants

Organs	Allantoin concentration (mg/100 g *fresh weight*)									
	Weeks after sowing									
	3	4	5	6	7	8	9	10	11	12
Seeds					149	125	53.6	14.8	15.8	16.0
Pods					951	678	676	214	175	54.8
Stems	4.7	51.3	342	408	734	853	1010	913	287	16.3
Roots	2.8	4.9	7.9	20.3	26.0	73.1	95.0	11.9	6.0	—
Nodules	34.7	42.2	54.2	54.8	57.6	92.6	112	77.2	14.8	—

Source: Matsumoto et al. 1977a, modified.

oxidase is found only in nodules, uricase is present in nodules and roots, while allantoinase occurs in nodules, roots, stems and leaves. With regards to allantoicase (allantoate amidohydrolase E.C. 3.5.3.4.) nodules and roots are devoid of it, its level is low in stems and it is mainly located in leaves (Tajima and Yamamoto 1975; Tajima et al. 1977). Thus, allantoin-producing enzymes are chiefly localized in nodules and enzymes breaking it down occur in stems and leaves (Matsumoto et al. 1978). The non-nodulated A62-2 genotype always contains only low amounts of allantoin, and differences between allantoin contents in the nodulated A62-1 variety and the non-nodulated A62-2 one increase during growth (Table 4.1).

In A62-1 variety, allantoin is stored in roots and stems. Its level is higher in the basal part of roots on which the nodules are mainly growing. As far as nodules are developing, allantoin accumulates in internodes of the stems where its content reaches a maximum following flowering, at the green pod stage, when it can amount to 70 to 80 per cent of total nitrogen (Matsumoto et al. 1977c). The allantoin level increases in leaves during leaf expansion and can reach 40 per cent of soluble nitrogen; thereafter it slows down, fully developed leaves containing only small amounts of it. This variation of allantoin amounts in leaves suggests that this compound is used for the development of new leaves. Allantoin occurs in as much important amount in young pods, then it disappears quickly during seeds maturation (Fujihara et al. 1977). Nevertheless, *Phaseolus mungo* and *Dolichos sinensis* seeds respectively contain 0.18 and

0.34 per cent of allantoin which is also present in dry soybean seeds (Fosse et al. 1931). Its amount gradually increases during germination (Matsumoto et al. 1977c) which confirms the results obtained by Fosse et al. (1931) and Van der Drift and Vogels (1966).

(3) Ureide transport through sap

Allantoin is exported from *Phaseolus vulgaris* (Pate 1973), *Glycine max* (Matsumoto et al. 1977a) and *Vigna unguiculata* (Herridge et al. 1978) nodules. This nitrogenous exportation form remained unknown for a long time in studies concerning dinitrogen fixation by nodules since it was assumed that the only nitrogenous exported compounds were amino acids and amides. The presence of ureides, allantoin and allantoic acid, had however been pointed out in the xylem sap of many plants (Bollard 1957; Pate 1971). The ureide concentration of the xylem sap changes according to legume species. In soybean Streeter (1979) has found that the ureide level of exudate collected from nodules can be as high as 5.3 nitrogen mg per ml, that is to say, 94 allantoin micromoles per ml. In *Vigna unguiculata*, during the developing cycle, the average ureide amounts account for 70 per cent, of soluble nitrogen whereas in *Pisum sativum*, they represent only 10 per cent, asparagine accounting for 63 per cent (Atkins et al. 1978).

Therefore from the viewpoint of nitrogen metabolism in relation to xylem sap composition, we suggest that two categories could be distinguished in legumes:

(i) Legume producing and exporting amides (asparagine, glutamine). The contents of ureides in these species (*Lupinus albus, Pisum sativum*) are low or very low. Asparagine is found in greater quantities than glutamine, and the xylem sap composition does not appreciably change in function of the nature of supplied nitrogen whatever it could be: dinitrogen, nitrates, ammonium ions or urea (Atkins et al. 1978; Pate et al. 1980).

(ii) Legume producing and exporting ureides (*Glycine max, Vigna unguiculata, Phaseolus vulgaris*) which on the contrary show variations in the xylem sap composition according as the plant, only fixes dinitrogen or is supplied in addition with combined nitrogen. Plants having only fixed N_2 contain higher amounts of ureides and glutamine and lower quantities of asparagine, as compared to plants furnished with nitrates or other nitrogenous forms and *vice versa*.

It results that the exportations of nitrogenous compounds by the xylem sap in ureide-producing legumes and in ureide non-producing ones are similar if the various plants are furnished with combined nitrogen—e.g. nitrates—but these exportations are quite different if nitrogen only arises from air dinitrogen (Pate et al. 1980).

The ureide biosynthesis during the process of ammonium assimilation and the utilization of these compounds as nitrogen form of transport expresses

Nitrogen compound	Formula	C/N Ratio
Allantoin	NH_2 $\mid$ CO $CO-NH$ $\mid$ $\mid$ $\diagdown CO$ $NH-CH-NH \diagup$	1.0
Allantoic acid	NH_2 NH_2 $\mid$ $\mid$ CO $COOH$ CO $\mid$ $\mid$ $\mid$ $NH-CH-NH$	1.0
Arginine	$NH=C-NH-(CH_2)_3-CH-COOH$ $\mid$ $\mid$ NH_2 NH_2	1.5
Citrulline	$NH_2-C-NH-(CH_2)_3-CH-COOH$ $\parallel$ $\mid$ O NH_2	2.0
Asparagine	$CO-CH_2-CH-COOH$ $\mid$ $\mid$ NH_2 NH_2	2.0
Ornithine	$CH_2-(CH_2)_2-CH-COOH$ $\mid$ $\mid$ NH_2 NH_2	2.5
Glutamine	$CO-(CH_2)_2-CH-COOH$ $\mid$ $\mid$ NH_2 NH_2	2.5
Methylene glutamine	$CO-C-CH_2-CH-COOH$ $\mid$ $\parallel$ $\mid$ NH_2 CH_2 NH_2	3.0
Homoserine	$CH_2OH-CH_2-CH-COOH$ $\mid$ NH_2	4.0
Aspartic acid	$COOH-CH_2-CH-COOH$ $\mid$ NH_2	4.0
Glutamic acid	$COOH-(CH_2)_2-CH-COOH$ $\mid$ NH_2	5.0

Fig. 4.3. C/N ratio of nitrogen-rich compounds transported in xylem of plants (*Source:* Pate 1973, modified).

a better economy, as calculated in term of ATP utilization, than in case of asparagine involvement (Atkins et al. 1978). In the same way, if we consider the C/N ratio in the different transported nitrogen forms (ureides, amides, other compounds) one can see that the transfer of a given nitrogen quantity requires less carbon with ureides (Fig. 4.3).

4.2.3 Ureide Role in Legumes

(1) Implication of ureides in dinitrogen fixation

Results from Matsumoto et al. (1978) show that well-nodulated soybean plants contain large amounts of ureides whereas non-nodulated isolines contain only little amounts (Fujihara et al. 1977). The ureide formation reaches a maximum at the term of maximum nitrogen fixation rate, when the reproductive phaseb egins (Fujihara and Yamaguchi 1978a) and it is more important in nodulated plants having fixed N_2 than in plants previously supplied with nitrates or ammonium salts (Matsumoto et al. 1977c; McClure and Israel 1979). Nitrogenous manuring is indeed known to reduce the allantoic nitrogen content and on the contrary to increase the nitrate and alpha-amino nitrogen content in nodulated plants. Alpha-amino nitrogen depends upon nitrogen fertilizers and allantoic nitrogen depends upon the nodulation degree (Ishizuka 1972). Ureide formation is therefore in connection with dinitrogen fixation, but the exact nature of relationships between nodule formation and allantoin accumulation is not known (Matsumoto et al. 1977c). On the other hand, it is known that in ureide producing plants the level of ureides in tissues or in transported fluids, such as xylem sap, varies according to the plant which has been or has not been supplied with combined nitrogenous forms, fertilizers or soil mineral nitrogen. This fact allows to detect if plant-*Rhizobium* symbiosis is well developing: it requires only a determination of ureide level (Pate et al. 1980).

(2) Ureides as nitrogen storage substances

Ureides are known to be nitrogen storage forms in some plants (Mothes and Engelbrecht 1952; Butler et al. 1961; Mothes 1961). In legumes they represent a nitrogen storage form characteristic of the reproductive stage whereas amino nitrogen characterizes the vegetative stage (Ishizuka 1972). In soybean, allantoin and allantoic acid are the main nitrogen forms transported from the nodulated roots towards the aerial parts of the plant. Stems, petioles and pods, at the beginning of the reproductive stage, are storage organs for ureic nitrogen. Indeed, before flowering, nitrate nitrogen and amino nitrogen on the whole exceed ureic nitrogen in stem exudates, but after flowering, ureic nitrogen is 2 to 6 times higher than nitrate and amino nitrogen all together. During pod and seed development, ureic nitrogen amounts to 2.3, 37.7 and 15.8 per cent of total nitrogen of respectively leaf blades, stems,

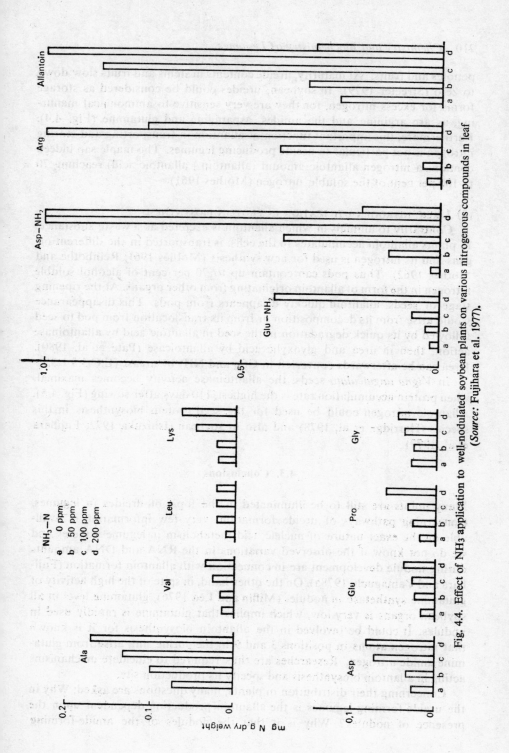

Fig. 4.4. Effect of NH₃ application to well-nodulated soybean plants on various nitrogenous compounds in leaf (*Source*: Fujihara et al. 1977).

petioles and fruits. At maturity, ureide contents in stems and fruits slow down to zero (Streeter 1979). In soybean, ureides could be considered as storage forms for excess nitrogen, for they are very sensitive to ammoniacal manuring, as are arginine and the amides, asparagine and glutamine (Fig. 4.4). Furthermore, this important function of allantoin in transporting and storing nitrogen is not particular to ureide producing legumes. The maple sap indeed contains a nitrogen allantoic amount (allantoin + allantoic acid) reaching 70 to 100 per cent of the soluble nitrogen (Mothes 1961).

(3) UREIC NITROGEN UTILIZATION IN PROTEIN BIOSYNTHESIS

Contrarily to animals in which allantoin is excreted as a waste substance, in plants allantoin accumulates in the cells, is transported in the different organs and its nitrogen is used for new synthesis (Mothes 1961; Reinbothe and Mothes 1962). Thus pods can contain up to 70 per cent of alcohol soluble nitrogen in the form of allantoin originating from other organs. At the ripening stage of seeds, allantoin quickly disappears from pods. This disappearance would arise from its decomposition or from its translocation from pod to seed followed by its quick degradation in the seed in allantoic acid by allantoinase action, then in urea and glyoxylic acid by allantoicase (Pate et al. 1980). Urea can be afterwards converted in CO_2 and NH_3 by urease (E.C.3.5.1.5.).

In *Vigna unguiculata* seeds, the allantoinase activity becomes maximal, when protein accumulation rate is the highest, 110 days after sowing (Fig. 4.5). Allantoic nitrogen could be used for the seed protein biosynthesis in this species (Herridge et al. 1978) and also in soybean (Ishizuka 1972; Fujihara et al. 1977).

4.3. Conclusions

Many points are still to be illuminated in the topic of ureides in legumes. Concerning pathways of ureide formation, very few information is available on the exact nature of nucleic acid metabolism in legume nodules and we do not know if the observed variations in the RNA and DNA amounts during nodule development are in connection with allantoin formation (Fujihara and Yamaguchi 1978a). On the other hand, in spite of the high activity of glutamine synthetase in nodules (Miflin and Lea 1976), glutamine level in all soybean organs is very low, which implies that glutamine is rapidly used in nodules. It could be involved in the allantoin biosynthesis for it is known that nitrogen atoms in positions 3 and 9 in the purine ring arise from glutamine amide nitrogen. Researches are thus required to elucidate mechanisms acting in allantoin biosynthesis and specify its production site.

Concerning their distribution in plants, many questions are asked: Why in the ureide forming legumes is the allantoin production dependent upon the presence of nodules? Why is it that the nodules of the amide-forming

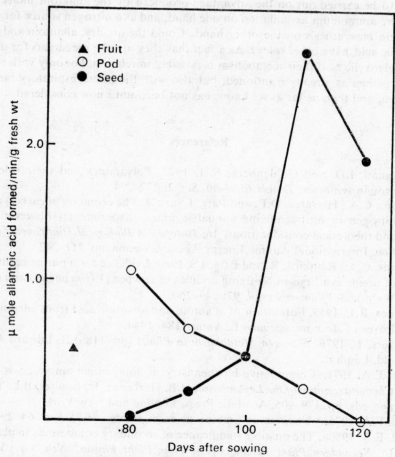

Fig. 4.5. Activity of allantoinase in extracts of fruits of nodulated cowpea.
Plant flowered 61 days after sowing, fruits were all fully ripened by 120
days. Time of highest enzyme activity in seeds (110–120 days after sowing)
was when highest rates of accumulation of seed storage protein were recorded (*Source:* Herridge et al. 1978).

legumes (Lupin) do not form ureides? Why are other vegetables species,
non-dinitrogen fixing synthesizing ureides in their roots?

With the intention of understanding their role, it is necessary to have a
better knowledge of the relations existing between nodule formation and
allantoin accumulation (Matsumoto et al. 1977c) just as the influence of the
aerial part upon the ureide formation in the nodule. It is also necessary to
specify the site for allantoin degradation and determine the localization of
enzymes acting in this degradation (Fujihara et al. 1977). Works have

also to be carried out on the advantage produced by the allantoin molecule for the ammonium assimilation on one hand, and as a nitrogen source for the protein biosynthesis on the other hand. Could the ureides, allantoin and allantoic acid have other roles? As a fact that they are the precursors for urea and glyoxylic acid their metabolism is possibly in relation, not only with that of arginine, as already mentioned, but also with the photorespiratory metabolism, and that, as far as we know, has not been until now considered.

References

Algranati, I.D. and Goldenberg, S.H. 1977. Polyamines and their role in protein synthesis. *Trends Biochem. Sci.* **2**: 272–274.

Atkins, C.A., Herridge, D.F. and Pate, J.S. 1978. The economy of carbon and nitrogen in nitrogen-fixing annual legumes. Experimental observations and theoretical considerations. In: *Isotopes in Biological Dinitrogen Fixation*. International Atomic Energy Agency, Vienna. pp. 211–242.

Atkins, C.A., Rainbird, R. and Pate, J.S. 1980. Evidence for a purine pathway of ureide synthesis in N_2-fixing nodules of cowpea (*Vigna unguiculata* (L.) Walp.). *Z. Pflanzenphysiol.* **97**: 249–260.

Barnes, R.L. 1959. Formation of allantoin and allantoic acid from adenine in leaves of *Acer saccharinum* L. *Nature* **184**: 1944.

Beevers, L. 1976. *Nitrogen Metabolism in Plants.* pp. 118–121. Edward Arnold, London.

Bell, E.A. 1971. Comparative biochemistry of non-protein amino acids. In: *Chemotaxonomy of the Leguminosae.* J.B. Harborne, D. Boulter, B.L. Turner, eds. pp. 179–206. Acadmic Press, London and New York.

Bell, E.A. 1976. "Uncommon" amino acids in plants. *FEBS Lett.* **64**: 29–35.

Bell, E.A. 1980a. The possible significance of secondary compounds in plants. In: *Secondary Plant Products (Encyclop. Plant Physiol. New Ser.)* V. 8. pp. 11–22. E.A. Bell and B.V. Charlwood, eds. Springer-Verlag.

Bell, E.A. 1980b. Non-protein amino acids in plants. In: *Secondary Plant Products (Encyclop. Plant Physiol. New Ser.)* V. 8. pp. 403–432. E.A. Bell and B.V. Charlwood, eds. Springer-Verlag.

Bollard, E.G. 1957. Translocation of organic nitrogen in the xylem. *Austral. J. Biol. Sci.* **10**: 292–301.

Bollard, E.G. 1959. Urease, urea and ureides in plants. *S.E.B. Symposia* **13**: 304–329.

Bonner, J. 1950. *Plant Biochemistry.* 537 pp. Academic Press, New York.

Brunel, A. and Capelle, G. 1947. Sur l'importance biologique des uréides glyoxyliques chez les êtres vivants. I. L'allantoïne et l'acide allantoïque chez les vègétaux. *Bull. Soc. Chim. Biol.* **29**: 427–444.

Brunel, A. and Capelle, G. 1951. Synthèse de l'acide allantoïque chez les champignons Basidiomycètes. *C.R. Acad. Sci.*, Paris **232**: 1130–1132.

Butler, G.W., Ferguson, J.D. and Allison, R.M. 1961. The biosynthesis of allantoin in *Symphytum. Physiol. Plant.*, **14**: 310–321.

Cabanne, F., Martin-Tanguy, J. and Martin, C. 1977. Phénolamines associées à l'induction florale et ā l'état reproducteur du *Nicotiana tabacum* var., *Xanthi n. c. Physiol. vég.* **15**: 429–433.

Conn, E.E. 1980. Cyanogenic glycosides. In: *Secondary Plant Products* (*Encyclop. Plant Physiol. New Ser.*). V. 8. pp. 461–492. E.A. Bell and B.V. Charlwood, eds. Springer-Verlag.

Dardenne, G.A. 1976. Recherche, isolement et structure de nouveaux acides aminés libres dans les vègétaux. *Mémoire Agr. Ens. Sup.* (*Fac. Sci. Gembloux*). 129 pp.

Durand, G., Guitton, Y. and Brunel, A. 1965. Le mètabolisme de l'azote chez les microorganismes et les végétaux supérieurs. II. *Les uréides. Bull. soc. fr. Physiol. vég.* **11**: 15–35.

Echevin, R. and Brunel, A. 1937. Uréides et urée libre, dégradation des purines chez le *Soja hispida Mnch. C.R. Acad. Sci.*, *Paris* **205**: 294–296.

Fodor, G.B. 1980a. Alkaloids derived from phenylalanine and tyrosine. In: *Secondary Plant Products* (*Encyclop. Plant Physiol. New Ser.*). V. 8. pp. 92–127. E.A. Bell and B.V. Charlwood, eds. Springer-Verlag.

Fodor, G.B. 1980b. Alkaloids derived from histidine and other precursors. In: *Secondary Plant Products* (*Encyclop. Plant Physiol. New Ser.*). V. 8. pp. 160–166. E.A. Bell and B.V. Charlwood, eds. Springer-Verlag.

Fosse, R., Brunel, A. and Thomas, P.E. 1931. Application de l'analyse quantitative spectrophotométrique de l'allantoïne au sang de quelques mammifères et ā la graine de nombreux végétaux, *C.R. Acad. Sci.*, *Paris* **193**: 7–11.

Fosse, R., de Graeve, P. and Thomas, P.E. 1933. Rôle de l'acide allantoïque chez les végétaux supérieurs. *C.R. Acad. Sci.*, *Paris* **196**: 1264–1267.

Fowden, L. 1976. Amino acids: occurrence, biosynthesis and analogue behaviour in plants. In: *Perspectives in Experimental Biology T. 2: Botany.* pp. 263–272. Pergamon Press, Oxford.

Freiberg, S.R., Bollard, E.G. and Hegarty, M.P. 1957. The natural occurrence of urea and ureides in the soluble nitrogen of the banana plant. *Plant Physiol.* **32**: suppl. iii.

Fujihara, S., Yamamoto, K. and Yamaguchi, M. 1977. A possible role of allantoin and the influence of nodulation on its production in soybean plants. *Plant Soil* **48**: 233–242.

Fujihara, S. and Yamaguchi, M. 1978a. Probable site of allantoin formation in nodulating soybean plants. *Phytochem.* **17**: 1239–1243.

Fujihara, S. and Yamaguchi, M. 1978b. Effects of allopurinol (4-hydroxypyrazolo (3, 4 d) pyrimidine) on the metabolism of allantoin in soybean plants. *Plant Physiol.* **62**: 134–138.

Fukuda, M., Tokumura, A. and Ogawa, T. 1973. D-alanine in germinating *Pisum sativum* seedlings. *Phytochem.* **12**: 2593–2595.

Gröger, D. 1980. Alkaloids derived from tryptophan and anthranilic acid. In: *Secondary Plant Products* (*Encyclop. Plant Physiol. New Ser.*). V. 8. pp. 128–159. E.A. Bell and B.V. Charlwood, eds. Springer-Verlag.

Harborne, J.B., Boulter, D. and Turner, B.L. 1971. *Chemotaxonomy of the Leguminosae.* 612 pp. Academic Press, London and New York.

Herridge, D.F., Atkins, C.A., Pate, J.S. and Rainbird, R.M. 1978. Allantoin and allantoic acid in the nitrogen economy of the cowpea (*Vigna unguiculata* (L.) Walp.). *Plant Physiol.* **62**: 495–498.

Ishizuka, J. 1972. Physiological roles of soluble nitrogenous components on vegetative growth and seed protein formation of soybean plants. *Res. Bull. Hokkaido National Agricult. Exp. Station* **101**: 51–121 (in Japanese); English summary. pp. 119–121.

Jänne, J., Pösö, H. and Raina, A. 1978. Polyamines in rapid growth and cancer. *Biochim. Biophys. Acta* **473**: 241–293.

Kapoor, M. and Waygood, E.R. 1962. Initial steps of purine biosynthesis in wheat germ. *Biochem. biophys. Res. Commun.* **9**: 7–19.

Leete, E. 1980. Alkaloids derived from ornithine, lysine, and nicotinic acid. In: *Secondary Plant Products* (*Encyclop. Plant Physiol. New Ser.*). V. 8. pp. 65–91. E.A. Bell and B.V. Charlwood, eds. Springer-Verlag.

Luckner, M. 1980. Expression and control of secondary metabolism. In: *Secondary Plant Products* (*Encyclop. Plant Physiol. New Ser.*). V. 8. pp. 23–64. E.A. Bell and B.V. Charlwood, eds. Springer-Verlag.

Mager, J., Chevion, M. and Glaser, G. 1980. Favism. In: *Toxic Constituents of Plant Foodstuffs.* pp. 266–294. I.E. Liener, ed. (2nd).

Martin-Tanguy, J., Cabanne, F., Perdrizet, E. and Martin, C. 1978. The distribution of hydroxycinnamic acid amides in flowering plants. *Phytochemistry* **17**: 1927–1928.

Matsumoto, T., Yatazawa, M. and Yamamoto, Y. 1977a. Distribution and change in the contents of allantoin and allantoic acid in developing nodulating and non-nodulating soybean plants. *Plant Cell Physiol.* **18**: 353–359.

Matsumoto, T., Yatazawa, M. and Yamamoto, Y. 1977b. Incorporation of ^{15}N into allantoin in nodulated soybean plants supplied with $^{15}N_2$. *Plant Cell Physiol.* **18**: 459–462.

Matsumoto, T., Yatazawa, M. and Yamamoto, Y. 1977c. Effects of exogenous nitrogen-compounds on the concentration of allantoin and various constituents in several organs of soybean plants. *Plant Cell Physiol.* **18**: 613–624.

Matsumoto, T., Yatazawa, M. and Yamamoto, Y. 1978. Allantoin metabolism in soyabean plants as influenced by grafts, a delayed inoculation with *Rhizobium*, and a late supply of nitrogen-compounds. *Plant Cell Physiol.*

19: 1161–1168.

McClure, P.R. and Israel, D.W. 1979. Transport of nitrogen in the xylem of soybean plants. *Plant Physiol.* **64**: 411–416.

Mears, J.A. and Mabry, T.J. 1971. Alkaloids in the Leguminosae. In: *Chemotaxonomy of the Leguminosae.* pp. 73–178. J.B. Harborne, D. Boulter, B.L. Turner, eds. Academic Press, London and New York.

Miflin, B.J. and Lea, P.J. 1976. The pathway of nitrogen assimilation in plants. *Phytochemistry* **15**: 873–885.

Miflin, B.J. and Lea, P.J. 1977. Amino acid metabolism. *Ann. Rev. Plant Physiol.* **28**: 299–329.

Montgomery, R.D. 1980. Cyanogens. In: *Toxic Constituents of Plant Foodstuffs.* pp. 143–160. I.E. Liener, ed. (2nd).

Mothes, K. 1961. The metabolism of urea and ureides. *Canad. J. Bot.* **39**: 1785–1807.

Mothes, K. 1980. Historical introduction. In: *Secondary Plant Products* (*Encyclop. Plant Physiol. New Ser.*). V. 8. pp. 1–10. E.A. Bell and B.V. Charlwood, eds. Springer-Verlag.

Mothes, K. and Engelbrecht, L. 1952. Über Allantoinsäure und Allantoin. I. Ihre Rolle als Wanderform des Stickstoffs und ihre Beziehungen zum Eiweissstoffwechsel des Ahorns. *Flora* **139**: 586–616.

Mothes, K. and Engelbrecht, L. 1954. Über Allantoinsäure und Allantoin. II. Ihr Verhalten in den Speicherwurzeln von *Symphytum officinale. Flora* **141**: 356–378.

Nagai, Y. and Funahashi, S. 1961. Allantoinase from mung bean seedlings. *Agric. Biol. Chem.* **25**: 265–268.

Ogawa, T., Kimoto, M. and Sasaoka, K. 1977. Identification of D-aspartic acid and D-glutamic acid in pea seedlings. *Agric. Biol. Chem.* **41**: 1811–1812.

Ohyama, T. and Kumazawa, K. 1978. Incorporation of ^{15}N into various nitrogenous compounds in intact soybean nodules after exposure to ^{15}N$_2$ gas. *Soil Sci. Plant Nutr.* **24**: 525–533.

Ohyama, T. and Kumazawa, K. 1979. Assimilation and transport of nitrogenous compounds originated from ^{15}N$_2$ fixation and ^{15}NO$_3$ absorption. *Soil Sci. Plant Nutr.* **25**: 9–19.

Padmanaban, G. 1980. Lathyrogens. In: *Toxic Constituents of Plant Foodstuffs.* pp. 239–265. I.E. Liener, ed. (2nd).

Paech, K. 1950. *Biologie und Physiologie der sekundären Pflanzenstoffe.* Springer, Berlin. 268 pp.

Pate, J.S. 1971. Movement of nitrogenous solutes in plants. In: *Nitrogen-15 in Soil-Plant Studies.* International Atomic Energy Agency, Vienna. pp. 165–187.

Pate, J.S. 1973. Uptake, assimilation and transport of nitrogen compounds by plants. *Soil Biol. Biochem.* **5**: 109–119.

Pate, J.S., Atkins, C.A., White, S.T., Rainbird, R.M. and Woo, K.C. 1980. Nitrogen nutrition and xylem transport of nitrogen in ureide-producing grain legumes. *Plant Physiol.* **65**: 961–965.

Rawsthorne, S., Minchin, F.R., Summerfield, R.J., Cookson, C. and Coombs J. 1980. Carbon and nitrogen metabolism in legume root nodules. *Phytochemistry* **19**: 341–355.

Reinbothe, H. and Mothes, K. 1962. Urea, ureides and guanidines in plants. *Ann. Rev. Plant Physiol.* **13**: 129–150.

Robern, H., Wang, D. and Waygood, E.R. 1965. Biosynthesis of nucleotides in wheat. I. Purines from ^{14}C-labelled compounds. *Can. J. Biochem.* **43**: 225–235.

Robertson, J.G., Warburton, M.P. and Farnden, K.J.F. 1975. Induction of glutamate synthase during nodule development in lupin. *FEBS Lett.* **55**: 33–37.

Scott, D.B., Farnden, K.J.F. and Robertson, J.G. 1976. Ammonia assimilation in lupin nodules. *Nature* **263**: 703–708.

Smith, T.A. 1975. Recent advances in the biochemistry of plant amines, *Phytochemistry* **14**: 865–890.

Smith, T.A. 1980. Plant amines. In: *Secondary Plant Products* (*Encyclop. Plant Physiol. New Ser.*). V. 8. pp. 433–460. E.A. Bell and B.V. Charlwood, eds. Springer-Verlag.

Streeter, J.G. 1979. Allantoin and allantoic acid in tissues and stem exudate from field-grown soybean plants. *Plant Physiol.* **63**: 478–480.

Tajima S. and Yamamoto, Y. 1975. Enzymes of purine catabolism in soybean plants. *Plant Cell Physiol.* **16**: 271–282.

Tajima S. and Yamamoto, Y. 1977. Regulation of uricase activity in developing roots of *Glycine max*, non-nodulating variety A62-2. *Plant Cell Physiol.* **18**: 247–253.

Tajima S., Yatazawa M. and Yamamoto, Y. 1977. Allantoin production and its utilization in relation to nodule formation in soybeans—Enzymatic studies. *Soil Sci. Plant Nutr.* **23**: 225–235.

Thomas R.J., Feller, U. and Erismann, K.H. 1979. The effect of different inorganic nitrogen sources and plant age on the composition of bleeding sap of *Phaseolus vulgaris*. *New Phytol.* **82**: 657–669.

Thomas, R.J., Feller, U. and Erismann, K.H. 1980. Ureide metabolism in non-nodulated *Phaseolus vulgaris* L. *J. Exp. Bot.* **31**: 409–417.

Tracey, M.V. 1955. Urea and ureides. In: *Modern Methods of Plant Analysis*. Vol. IV. pp. 119–141. K. Paech and M.V. Tracey, eds. Springer-Verlag. Berlin, Gottingen and Heidelberg.

Van der Drift, C. and Vogels, G.D. 1966. Allantoin and allantoate in higher plants. *Acta Botan. Neerland* **15**: 209–214.

5

Toxic Constituents in Legumes

IRVIN E. LIENER

5.1. Introduction

For reasons which scientists have yet to fathom, Nature has seen fit to endow many plants with the genetic capacity to synthesise a wide variety of chemical substances that are known to exert a deleterious effect when ingested by animals or man. Although these substances are frequently referred to by such non-descript terms as "toxic factors" or "toxicants", these terms may be misleading. Strictly speaking, they imply that the substance in question is lethal beyond a given level of intake, and the toxicologist may in fact assess its toxicity in terms of its LD_{50}, that is, that dose which causes the death of 50 per cent of the animals tested. In fact, although some plants are known to produce a violent expression of poisoning, much more subtle effects, produced only by prolonged ingestion of a given plant, are more commonly observed. Such effects might include an inhibition of growth, a decrease in food efficiency, a goiterogenic response, pancreatic hypertrophy, hypoglycemia, and liver damage. Other factors that should be taken into consideration include the species of animal, its age, size, and sex, its state of health and plane of nutrition, and any stress factors that might be superimposed on these variables. The reader will therefore be asked to give the term "toxic" as used throughout this chapter, a most liberal interpretation to mean nothing more or less than an adverse physiological response produced in man or animals by a particular food or a substance derived therefrom.

It should perhaps be emphasized that the evidence that a particular food constitutes a hazard to the health of man is frequently only presumptive. As might be expected, much of the research relating to the toxicity of foodstuffs has been done with plants commonly used as food by farm animals. Under these conditions the animals consume large quantities of a particular plant over a long period of time, a situation quite foreign from the normal eating patterns of man. Thus, a toxic substance, if present, might produce symptoms of poisoning that might not otherwise be apparent. If the

217

causative factor has been isolated, its toxicity is frequently evaluated by using a route of administration (such as intraperitoneal or subcutaneous injection) that is physiologically unrelated to the normal mode of ingestion. The final link in the chain of evidence, namely, the demonstration that the ingestion of the purified toxin will produce some physiological damage to man at a level comparable to that which would be present in the quantity of food that he normally consumes, must be left undone. In effect, therefore, the only evidence for toxicity to man is often only the knowledge that a substance known to be toxic to animals under a given set of conditions is present in a food that he consumes. This information is nevertheless of sufficient importance to at least alert one to the possible hazards involved in the consumption of such foods by man.

5.2. Protease Inhibitors

Substances that have the ability to inhibit the proteolytic activity of certain enzymes are found throughout the plant kingdom, particularly among the legumes. These so-called protease inhibitors have attracted the attention of scientists in many disciplines—nutritionists, because of their possible effect on the nutritive value of plant proteins; protein chemists because the reaction of these inhibitors, which are proteins, with enzymes provides a simple model system for studying the protein-protein interaction; and members of the medical profession because the unique pharmacological properties of these inhibitors hold considerable promise for clinical application in the field of medicine.

Read and Hass (1938) appear to have been the first to recognize the presence of an inhibitor in plant material. They reported that an aqueous extract of soybean flour inhibited the ability of trypsin to liquefy gelatin. The fraction of soybean protein responsible for this effect was partially purified by Bowman (1944) and Ham and Sandstedt (1944) and subsequently isolated in crystalline form by Kunitz (1945). The existence of an inhibitor of trypsin in soybeans that could be inactivated by heat seemed to offer, at the time, a reasonable explanation for the observation made many years before by Osborne and Mendel (1917) that the heat treatment improved the nutritive value of soybean protein. The realization that protease inhibitors might be of nutritional significance in soybeans stimulated a search for similar factors in other legumes. The list of protease inhibitors found in various legumes is now a long and growing one. Table 5.1 is a summary of our current knowledge of the distribution of these inhibitors among the legumes.

5.2.1. BIOCHEMICAL PROPERTIES

The protease inhibitors that have been isolated from soybeans fall into two

TABLE 5.1. Distribution of protease inhibitors present in legumes[a]

Botanical name	Common name	Proteases inhibited[b]
Arachis hypogaea	peanut, groundnut	T, C, Pl, K
Cajanus cajan	pigeon pea, red gram	T
Canavalia ensiformis	jack bean, sword bean	T, C, S
Chamacrista fasiculata	partridge pea	T
Cicer arietinum	chick pea, Bengal gram, garbanzo	T, C
Clittoria ternatea	butterfly pea	T, C, S
Cyamopsis tetragonoloba	cluster bean	T, C, S
Dolichos biflorus	horse gram	T
Dolichos lablab	hyacinth bean, field bean, Hakubenzu bean	T, C, Th
Faba vulgaris	double bean	T
Glycine max	soybean	T, C
Lathyrus odoratus	sweet pea	T
Lathyrus sativus	chickling vetch	T, C
Lens esculenta (culinaris)	lentil	T
Lupinus albus	lupine	T
Mucana deeringianum	Florida velvet bean	T
Phaseolus aconetifolius	moth bean	T
Phaseolus angularis	Adzuki bean	T, C
Phaseolus aureus	mung bean, green gram	T, endopeptidase
Phaseolus coccineus	scarlet runner bean	T, C
Phaseolus lunatus	lima bean, butter bean	T, C
Phaseolus mungo (radiatus)	black gram	T, C, S
Phaseolus vulgaris	navy bean, kidney bean, pinto bean, French bean, white bean, wax bean, haricot bean, garden bean	T, C, E, S
Pisum sativum	field bean, garden pea	T
Psophocarpus tetragonolobus	winged bean, Gao bean	T
Stizobolium deeringianum	velvet bean	T
Vicia faba	broad bean, field bean, faba bean	T, C, Th, Pr, Pa
Vigna unguiculata (sinensis)	Cowpea, black-eyed pea, Southern pea, serido pea	T, C
Voandzeia subterranea	Bambara beans	T

[a]Data taken from Liener and Kakade (1980).
[b]Key to abbreviations: C, chymotrypsin; E, elastase; K, kallikrein; Pa, papain; Pl, plasmin; Pr, pronase; S, subtilisin; T, trypsin; Th, thrombin.

main categories: those that have a molecular weight of 20,000–25,000 with relatively few disulfide bonds and a specificity directed primarily toward trypsin, and those that have a molecular weight of only 6,000–10,000 with a high proportion of disulfide bonds and capable of inhibiting trypsin and chymotrypsin at independent binding sites. The most thoroughly character-

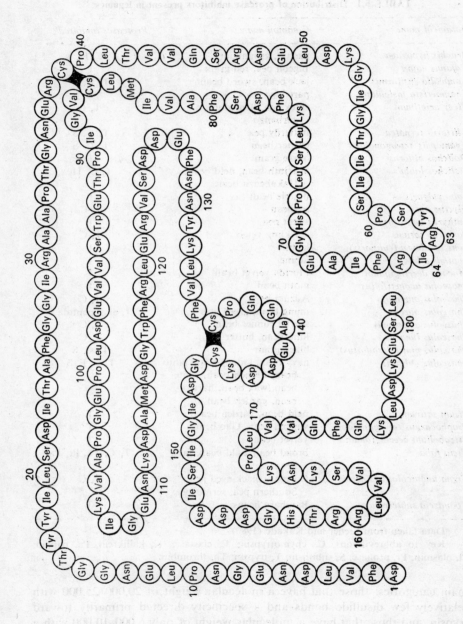

Fig. 5.1. Amino acid sequence of the Kunitz soybean trypsin inhibitor (Koide et al. 1973).

ized examples of these two classes of inhibitors are the so-called Kunitz and Bowman-Birk inhibitors derived from the soybean.

The complete amino acid sequence of the Kunitz inhibitor is shown in Fig. 5.1. It consists of 181 amino acids with the reactive site (site directly involved in its interaction with trypsin) being located at residues Arg 63 and Ile 64. This molecule combines with trypsin in a stoichiometric fashion, i.e. 1 molecule of the inhibitor inactivates 1 molecule of trypsin. X-ray crystallography has given a better insight into the nature of the inhibitor-enzyme complex (Fig. 5.2).

The amino acid sequence of the Bowman-Birk inhibitor is shown in Fig. 5.3. A unique feature of this molecule is the fact that it possesses two inde-

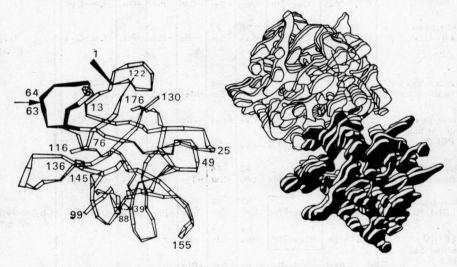

Fig. 5.2. Folding of the polypeptide backbone chain of the Kunitz inhibitor is shown on the left. Amino residues in intimate contact with trypsin shown in black. Shown on the right is a model of the Kunitz inhibitor trypsin complex. That part representing trypsin is less heavily shaded. Taken from Sweet et al. (1974).

pendent binding sites—a trypsin-reactive site (Lys 16–Ser 17) and a chymotrypsin-reactive site (Leu 44–Ser 45). The sequences of amino acids surrounding these two reactive sites are remarkably similar to each other, and a high degree of homology exists between the Bowman-Birk inhibitor and the inhibitors which have been isolated from other legumes (Table 5.2).

5.2.2. NUTRITIONAL SIGNIFICANCE

The fact that protease inhibitors are so widely distributed among those very plants that constitute an important source of dietary protein throughout the

TABLE 5.2. Sequence homology around reactive sites of double-headed inhibitors in legumes[1]

Reactive site ↓

				16	17					
Soybeans Bowman-Birk	Cys	-Ala	-Cys-Thr	-Lys (49)	-Ser (50)	-Asn	-Pro	-Pro	-Gln-Cys	Trypsin inhibitors
CII	Cys	-Ala	-Cys-Thr	-Arg (24)	-Ser (25)	-Met	-Pro	-Gly	-Gln-Cys	
DII	Cys	-Met	-Cys-Thr	-Arg (51)	-Ser (52)	-Met	-Pro	-Pro	-Gln-Cys	
	Cys	-Met	-Cys-Thr	-Arg	-Ser	-Gln	-Pro	-Gly	-Gln-Cys	
Lima bean LBI-I, IV	Cys	Leu -Ala	-Cys-Thr	-Lys (26)	-Ser (27)	-Ile	-Pro	-Pro	-Gln-Cys	
Garden bean Isoinhibitor II	Cys	-Met	-Cys-Thr	-Arg (53)	-Ser (54)	-Met	-Pro	-Gly-Lys	-Cys	
Chick pea **Runner bean** PCI-3	Cys	-Val	-Cys-Thr	-Lys (14)	-Ser (15)	-Ile	-Pro	-Pro	-Gln-Cys	
			Ile- Tyr	-Lys	-Ser	-Gln	-Pro.			
Soybeans Bowman-Birk	Cys	-Ile	-Cys-Ala	-Leu (43)	-Ser (44)	-Tyr	-Pro	-Ala	-Gln-Cys	Chymotrypsin inhibitors
Lima bean LBI-I, IV	Cys	-Ile	-Cys-Thr	-Leu (53)	-Ser (54)	-Ile	-Pro	-Ala	-Gln-Cys	
LBI-IV **Runner bean** PCI-3	Cys	-Ile	-Cys-Thr	-Phe (53)	-Ser (54)	-Ile	-Pro			
		Asp	-Val-Ala	-Leu	-Ser	-Pro				
Soybean CII	Cys	-Met	-Cys-Thr	-Ala (22)	-Ser (23)	-Met	-Pro	-Pro	-Gln-Cys	Elastase inhibitors
Garden bean Isoinhibitor II			Cys-Thr	-Ala (26)	-Ser (27)	-Ile	-Pro	-Pro	-Gln-Cys	

[1]Data taken from Liener and Kakade (1980).

world has stimulated a vast amount of research regarding their possible nutritional significance. Because of the important role that the soybean plays in animal feeding and its potential contribution to human nutrition, the protease inhibitors of this plant have received particular attention.

With the demonstration of a heat-labile trypsin inhibitor in soybeans and its subsequent crystallization, it was generally assumed that the benefici-

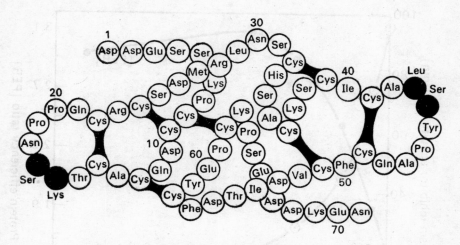

Fig. 5.3. Amino acid sequence of the Bowman-Birk inhibitor. The reactive sites of trypsin (Lys 16-Ser 17) and chymotrypsin (Leu 44-Ser 45) are shown in black. Taken from Odani and Ikenaka (1973).

al effect of heat treatment could be ascribed to the destruction of this inhibitor. The inactivation of the trypsin inhibitor does in fact appear to parallel the improvement in nutritive value effected by heat as demonstrated with rats (Fig. 5.4). Further evidence came from experiments in which it was shown that the addition of purified preparations of the trypsin inhibitor to heated soybeans, so as to provide the same inhibitory activity as raw soybeans, caused a significant reduction in growth.

With the recognition of the presence of a trypsin inhibitor in soybeans, it was tempting to conclude that the growth inhibition which it evoked in animals was simply due to an inhibition of digestion of dietary protein by proteolytic enzymes present in the intestinal tract. The most destructive blow to this theory was the observation that preparations of trypsin inhibitor were capable of inhibiting growth even when incorporated into diets containing the predigested protein or free amino acids (Liener et al. 1949). Such experiments obviously rule out an inhibition of proteolysis as the sole factor responsible for growth inhibition and thus served to focus attention on some alternative mode of action of the trypsin inhibitor.

Perhaps the most significant observation which has ultimately led to a better understanding of the mode of action of the soybean inhibitor was the finding that raw soybeans and the trypsin inhibitor itself could cause hypertrophy of the pancreas, an effect which is accompanied by an increase in the secretory activity of the pancreas (Chernick et al. 1948). This led to the suggestion that the growth depression caused by the trypsin inhibitor might be

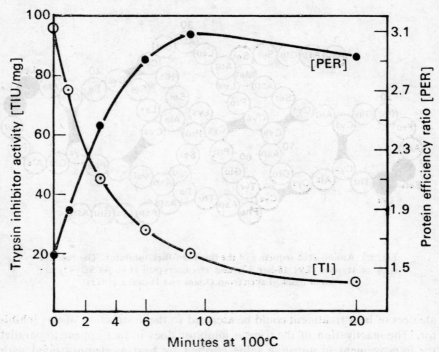

Fig. 5.4. Effect of heat treatment on the trypsin inhibitory activity and nutritive value, as measured by the protein efficiency ratio (PER), of raw soybean meal. Taken from Rackis (1974).

the consequence of an endogenous loss of essential amino acids being secreted by a hyperactive pancreas (Lyman and Lepkovsky 1957; Booth et al. 1960). Since pancreatic enzymes such as trypsin and chymotrypsin are particularly rich in the sulphur-containing amino acids, pancreatic hypertrophy causes a drain on the body tissue of these particular amino acids in order to meet an increased need for the synthesis of these enzymes. This loss in sulphur-containing amino acids serves to accentuate an already critical situation with respect to soybean protein, which is inherently deficient in these amino acids.

The mechanism whereby the trypsin inhibitor induces pancreatic enlargement is still not fully understood. Lyman and co-workers (Green and Lyman 1972; Lyman et al. 1974) have shown that pancreatic secretion is controlled by a mechanism of feedback inhibition which depends upon the level of trypsin and chymotrypsin present at any given time in the small intestine. When the level of these enzymes falls below a certain critical threshold value, the pancreas is induced to produce more enzyme. The suppression of negative feedback inhibition can thus occur if the trypsin is complexed with the inhibitor. It is believed that the mediating agent between trypsin and the pancreas is the hormone cholecystokinin (CCK), which is released from the intestinal mucosa when the level of trypsin in the intestine falls below its threshold level. These relationships are illustrated in Fig. 5.5.

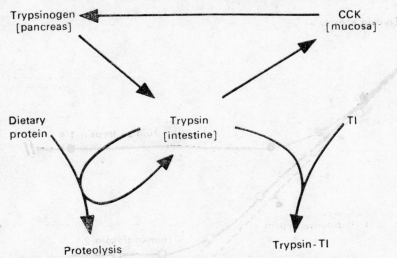

Fig. 5.5. Mode of action of soybean trypsin inhibitors on pancreas. Taken from Anderson et al. (1979).

In view of the increasing interest in the use of soybean products for the human diet, it becomes important to assess, if possible, the risk to human health that might be associated with the consumption of soybean preparations in which the protease inhibitors may not have been completely inactivated. Actually, the only report in which human subjects were fed raw soybean flour is that of Lewis and Taylor (1947), who found that the unheated soybean flour could support a positive nitrogen balance, albeit not as efficiently as autoclaved flour. A much more basic question that should be answered is whether the soybean inhibitor is in fact capable of inhibiting human trypsin. Trypsin inhibitor activity in soybeans is invariably measured *in vitro* on the basis of the extent to which soybean preparations can inhibit bovine or porcine trypsin, since the latter are readily available commercially in a pure crystalline form. Human trypsin is known to exist in two forms, a cationic species (trypsin 1) that constitutes the major component of human pancreatic juice (70–80 per cent) and an anionic species (trypsin 2) that accounts for the remainder of the tryptic activity (Figarella et al. 1974). The inhibition of these two forms of human trypsin is shown in Figure 5.6. Although trypsin 2, the minor trypsin component, is inhibited in a 1:1 molar ratio by the soybean inhibitor (as is bovine trypsin), trypsin 1, the major component, is very weakly inhibited. From these considerations one is tempted to conclude that, despite the considerable body of evidence that implicates the trypsin inhibitors as a factor contributing to the poor nutritive value of raw legumes in animals, their relevance to human nutrition seems dubious.

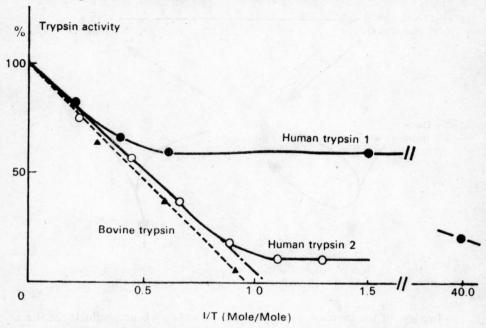

Fig. 5.6. Inhibition of human trypsins with Kunitz soybean inhibitor. Ordinate, remaining trypsin activity as measured on tosyl-DL-arginine methyl ester; abscissa, molar ratio of inhibitor to trypsin. Taken from Figarella et al. (1974).

5.2.3. PHYSIOLOGICAL ROLE IN PLANT

The physiological role of the protease inhibitors in the plant has sometimes been presumed to be one that favours protein catabolism during germination or prevents the degradation of storage protein during seed maturation. Attractive as this hypothesis may appear, there are a number of observations that do not support the generality of this concept:

1. In some cases, the protease inhibitors isolated from plants are in fact incapable of inhibiting endogenous proteases of the same plant. This was shown to be true in the case of soybeans (Birk 1968) and peas (Hobday et al. 1973). Moreover, no change in the protease inhibitor activity of soybeans occurs during germination (Collins and Sanders 1976), and, in the case of kidney beans (Palmer et al. 1973), there may be an actual increase.

2. In some seeds, such as the pea (Hobday et al. 1973) and mung bean

the leaves being fed. The nutrient absorption capacity of young cowpea roots has been found to decrease rapidly with age (Adepetu and Akapa 1977). Similarly, the nutrient content of plant tissues decreased progressively with the age of pea crop (Fageria 1977).

PHOSPHORUS

Phosphorus plays an important role in energy transfer in both respiration and photosynthesis. The plant tissues characteristically maintain a high inorganic phosphate content. The organic forms of the phosphate occur in great variety in metabolites and macro-molecules such as nucleic acid polymers and phospholipids. The proportion of organophosphorus compounds in tissues is not affected significantly by the nutrient status in soil and the stage of growth of the plant. Leguminous seeds are rich in phytic acid and the phosphorus is stored in seeds. Grain legumes have been found to respond to phosphatic fertilizers. An increased yield has been reported in black gram (Rajendran and Krishnamoorthy 1975), soybean (Robertson and Yuan 1973; Kumar 1978), pea (Fageria 1977), green gram (Aulakh and Pasricha 1977), pigeon pea (Dalal and Quilt 1977) and *dhaincha (Sesbania aculeata)*, (Chattopadhyay et al. 1976) with the application of phosphorus.

Fageria (1977) found a depression in the growth of peas due to the phosphorus deficiency and its application increased the uptake of both phosphorus and potassium. The increase in the supply caused an increase in the content of calcium, magnesium, potassium and sodium, and a decline in the concentration of iron in chick pea (Maftoun and Bassiri 1975). The uptake of phosphorus is affected by the type of the soil (Sheldrake and Narayanan 1979).

In black gram the nitrogen, phosphorus and potassium uptake increased with the application of phosphatic fertilizers (Rajendran and Krishnamoorthy 1975) and the highest amounts of nitrogen and phosphorus were recorded between 30 and 50 days of potassium and calcium at 30 days (Shanker and Kushwaha 1971) but the magnesium content remained constant throughout the growing season. Aulakh and Pasricha (1977) found a negative significant interaction between phosphorus and sulphur on the concentration and uptake of both these elements in green gram. In *dhaincha (Sesbania aculeata)*, Chattopadhyay et al. (1976) recorded an increased nitrogen content due to the application of phosphorus.

The uptake in the peanut was found to be 13.5 kg P/ha, of which 64 per cent was in the pods (Bromfield 1973). In soybean, the phosphorus, magnesium, iron and sulphur contents in the seed increased with phosphorus fertilization (Robertson and Yuan 1973; Kapoor and Gupta 1977a; Kumar 1978).

The decreased concentration and uptake of zinc has been noted due to a higher dose of phosphatic fertilizers in groundnut (Quintana 1972), soybean (Paulsen and Rotini 1968; and Keefer Singh 1969), chick pea (Yadav

6.2.1. ESSENTIAL ELEMENTS

NITROGEN

The symbiotic nitrogen fixation system of legumes in general excludes the type of the beneficial response that cereals give to the added nitrogen. Still its deficiency has been reported in peanut and soybeans. Blondel (1969) observed its deficiency in the peanuts in the absence of normal rhizobial population in acidified sandy soils. In soybean, its removal in the early stages resulted in earlier leaf and petiole abscission, early maturity, smaller seeds and reduced yield (Egli et al. 1978). Streeter (1978) found that nitrogen stress (87 days after planting) resulted in a marked decrease in the seed weight and nitrogen concentration in the seed, but did not result in premature leaf senescence or in a decline in the nitrogen concentration of vegetative organs or a decline in the number of fruits and seeds per plant, but the longer periods of nitrogen stress (after 65 days) caused major reductions in pod and seed numbers. The application of nitrogenous fertilizers resulted in a reduced nodulation and grain yield in *Phaseolus vulgaris* (Sanchez et al. 1977).

Peanuts were found to contain 157 kg N/ha in the harvested crop, of which 62 per cent was in the pods (Bromfield 1973). Williams (1979) observed that the peanuts accumulated 2.39 kg N/ha/day during vegetative growth and 3.77 during the first half of the reproductive growth, after which the nitrogen accumulation ceased and a total of 30 g/m² was accumulated by the crop. Its redistribution from the leaves was a substantial contribution to the kernel nitrogen. Similarly, Egli et al. (1978) found that 20 to 60 per cent of this element in seeds was translocated from the soybean leaves before leaf abscission could take place. Leaves and podwalls were the major sources of redistributed nitrogen. Its remobilization from stems and leaves has also been observed by Sheldrake and Narayanan (1979) in pigeon pea. The uptake of nitrogen is also affected by the type of the soil. The mean uptake by Cv. ICP-1 was 89 kg/ha on Vertisols and 79 kg/ha on Alfisols. In *Vicia-faba*, the youngest vegetative tissue has been found initially to be the most powerful sink for the recently absorbed and redistributed nitrogen among the above ground parts (Cooper et al. 1976). However, after the fruit had set, competition developed between the growing tip and the developing beans until at the later stages the beans became the more powerful. Quantitatively, redistribution made only a small contribution to the bean nitrogen.

Oghoghorie and Pate (1972) studied the nitrogen transport system of a nodulated legume using ^{15}N. They found that the nitrogen assimilated by roots or nodules moved preferentially toward the shoot through the xylem. A considerable proportion of the nitrogen received by a shoot was apparently returned to the root in the translocation stream. Nitrogen from $^{15}NO_3$ fed to the mature leaf was distributed to all parts of the plant except older regions of the shoot. The maximum share of this nitrogen was in roots, nodules and detailed account of mineral absorption and its mechanism has been dealt with in the recent review (Clarkson and Hanson 1980).

The surface of young roots appears to have a coating of mucilaginous material along with other plant secretions which modify the environment at and near the absorption sites. The site of the most active absorption is often 1 to 2 mm behind the root apex which varies along with the species and the periodic effect of time. Different theories to account for the mineral absorption have been put forward from time to time, but it is clear that no single mechanism can be responsible for all the cases of elemental uptake. Roots tend to retain metals such as lead, cobalt, nickel, zinc, iron, copper, manganese, chromium and vanadium. Chloroplasts are rich in copper, iron and magnesium compared to cytoplasm. Calcium may be accumulated in mitochondria, whereas these organelles lack in boron.

The infection of plant roots by symbiotic fungi and the transformation of these roots into unique morphological structures called mycorrhyza constitute one of nature's widespread, persistent and interesting examples of parasitism. The mycorrhyzal association is normally associated by penetration of hyphae produced from germinating spores of fungal endophytes present in soils. The fungus colonises the root cortex and forms the external mycelium in the soil. Most of the minerals absorbed by the plants pass first through the fungus. Recent findings suggest that the fungal hyphae may be highly effective in promoting the mineral absorption by roots. The ectotrophic mycorrhyza increase the area of the absorbing surface of the root system by increasing its diameter, promoting its branching and prolonging the life of the short roots.

The processes governing the movement of minerals to the root surface are the subject of recent research. There are two processes, namely, the movement by convective flow in the soil water to the root, and the diffusion in response to the concentration gradient. Both these processes are strongly influenced by soil properties. The third process called root-interception also helps in the absorption of some plant nutrients.

6.2. Mineral Nutrition and Its Effect on Growth

Less attention is being paid to the mineral nutrition of grain legumes in comparison to cereals like wheat and paddy in tropics and subtropics. Although the low yield potential and lack of an assured yield and income are the constraints in using higher quantities of fertilizers, the yield, composition and quality of these crops may be improved to certain limits with adequate mineral fertilization. Few recent references for each element have been taken and there is no intention to ignore anybody's work, except to keep brevity.

inhibitor (Kunitz) at 2.6 Å resolution. *Biochemistry* **13**: 4212–4228.

Torquati, T. 1913. Presence of a nitrogenous substance in the seeds of *Vicia faba*. *Arch. Farm. Sper*. **15**: 213–223.

Turner, R.H. and Liener, I.E. 1975. The effect of the selective removal of hemagglutinins on the nutritive value of soybeans. *J. Agr. Food Chem*. **23**: 484–487.

Van Veen, A.G. 1973. In "Toxicants Occurring Naturally in Foods". *Nat. Acad. Sci*. pp. 464–476.

Van Veen, A.G. and Hyman, A.J. 1933. On the toxic component of the Djenkol bean. *Genecsk. Tydschr. Nederl. Indie* **73**: 991–1001.

Van Veen, A.G. and Latuasan, H.E. 1949. The state of djenkolic acid in the plant. *Chronica Naturae* **105**: 288–290.

Van Wyk, J.J., Arnold, M.B., Wynn, J. and Pepper, F. 1959. The effects of a soybean product on thyroid function in humans. *Pediatrics* **24**: 752–760.

Vilhjalmsdottir, L. and Fisher, H. 1971. Castor bean meal as a protein source for chickens: detoxification and determination of limiting amino acids. *J. Nutr*. **101**: 1185–1192.

Warden, C.J.H. and Waddell, L.A. 1884. The non-bacillar nature of abrus poison with observations on its chemical and physiological properties. Bengal Secretarial Press, Calcutta, India, 76.

Yoshida, R.K. 1944. A chemical and physiological study on the nature and properties of the toxic principle in *Leucaena glauca* (koa haole). Ph.D. thesis, U. Minnesota.

(Baumgarten and Chrispeels 1976), most of the protease inhibitor is located outside the protein bodies. This would appear to preclude the inhibitor from playing a role in regulating the proteolysis of storage protein in these organelles during germination.

3. A kinetic study of the rise in the endopeptidase activity and decline in inhibitory activity in the mung bean revealed no causal relationship between these two phenomena (Baumgarten and Chrispeels 1976). That the inhibitors might function to protect the cytoplasm from accidental rupture of the protease-containing bodies was suggested as a possibility.

It has also been postulated that the protease inhibitors may serve as a defence mechanism against insect predators. A soybean fraction has been described that inhibits the larval growth and digestive proteases of the flour beetles *Tribolium confusum* (Lipke et al. 1954) and *Tribolium castaneum* (Birk and Applebaum 1960). Birk et al. (1963) were able to purify the larval growth inhibitor by chromatography on hydroxylapatite. Although this growth inhibitor fraction was also capable of inhibiting the activity of larval proteases, it was devoid of trypsin and chymotrypsin inhibitor activity. It is obvious that more work is needed to establish the role of protease inhibitors as a defence mechanism against insects.

5.3. Lectins

That the seeds of certain plants are highly toxic to man and animals has been known for a long time. During the latter part of the 19th century, when the science of bacteriology was still in its infancy, it was widely believed that the toxicity of such seeds was due to a bacterial toxin. This theory was disproved, however, when, in 1884, Warden and Waddell observed that the toxicity of the jequirity bean, *Abrus precatorius*, resided in fractions that could be precipitated by alcohol from an aqueous extract of the bean. Stillmark (1889) was the first to observe that a protein fraction of the castor bean, *Ricinus communis*, which he called ricin, was capable of agglutinating red blood cells, a property that led to the term "phytohemagglutinin", which is still used today in referring to this class of substances. The work of Stillmark attracted the attention of Ehrlich (1891), who chose to work with ricin rather than the bacterial toxins that were then so popular among the bacteriologists of that time. The use of these substances led Ehrlich to the discovery of the most fundamental principles of immunology.

Landsteiner and Raubitcheck (1908) showed for the first time that even the seeds of the edible species of some common legumes such as navy beans, lentils, and garden peas also contained these so-called phytohemagglutinins. Landsteiner pointed out that the relative hemagglutinating activities of various seed extracts were quite different when tested with erythrocytes from different animals. It was, in fact, this specificity toward specific types of blood

cells that led Boyd and Sharpleigh (1954) to coin the word "lectin" (Latin, *legere = to choose*), a term that is today used interchangeably with phytohemagglutinin.

It has since been found that the lectins are very widely distributed in the plant kingdom, particularly among the legumes, which are known to be very low in nutritive value unless subjected to some form of heat treatment. It is indeed curious that one finds so little information in the literature regarding what would seem to be a possible relationship between the lectins and the poor nutritive value of raw legumes. Perhaps one explanation for this vacuum lies in the fact that the discovery of protease inhibitors in soybeans and other legumes served to direct attention away from the lectins, and nutritionists became preoccupied with studying the role of protease inhibitors, particularly in the soybean.

5.3.1. BIOCHEMICAL PROPERTIES

In addition to being able to agglutinate red blood cells, the lectins exhibit a host of other interesting and unusual biological and chemical properties, including the interaction with specific blood groups, mitogenesis, agglutination of tumor cells, and toxicity toward animals. All of the effects exhibited are believed to be a manifestation of the ability of lectins to bind to specific kinds of sugars present on the surface of cells. By virtue of this property, the lectins have provided investigators with an extremely useful tool, not only for the isolation and characterization of polysaccharides and glycoproteins, but also for probing the molecular structure of cell surfaces and the changes induced therein by chemical and biological agents. For a most recent comprehensive review of the lectins, see Goldstein and Hayes (1978). Although hemagglutinating activity has been detected in over 800 different plant species, of which over 600 are in the family *Leguminosae*, relatively few of these have been purified to the point of meaningful characterization. Those which have been purified and at least partially characterized are listed in Table 5.3. Included in this table is the sugar specificity which these lectins display. It is evident from this compilation that, despite a wide diversity in the properties of lectins, certain generalizations can be made. Most lectins appear to have molecular weights ranging from 100,000 to 150,000 and are composed of four subunits which may be identical. A smaller number of lectins, such as those obtained from the lentil and lima bean, appear to be dimers having about one-half of the molecular weight of the lectins which are tetramers. With few exceptions (such as soybean), each subunit has a sugar binding site. It is this feature of multivalency which accounts for the ability of lectins to agglutinate cells or precipitate glycoproteins, a property which is lost if the molecule is dissociated into subunits. Some lectins with reduced valence, although incapable of causing agglutination, sometimes exhibit toxicity (abrin and ricin).

TABLE 5.3. Physico-chemical properties and sugar specificity of phytolectins

Plant source	Sugar specificity[a]	Molecular weight[b]	Subunits	Carbohydrate content, %
Abrus precatorius (jequirity bean)	α-D-Gal	65,000[c] 134,000	2 4	
Arachis hypogeae (peanut)	α-D-Gal	110,000	4	0
Bandiera simplicifolia	α-D-Gal	114,000	4	
Bauhinia purpurea alba	α-D-GalNAc	195,000		11
Canavalia ensiformis (jack bean)	α-D-Man, α-D-Glc	55,000 110,000	2 4	0
Dolichos biflorus (horse gram)	α-D-GalNAc	113,000 109,000	4 4	3.8
Glycine max (soybean)	D-Gal, α-D-GalNAc	122,000	4	6
Lens culinaris[d] (lentil)	α-D-Man, α-D-Glc	52,000	2	2
Phaseolus coccineus (scarlet runner bean)	GlcNAc	120,000	4	
Phaseolus lunatus[e] (lima bean)	D-GalNAc	124,000 247,000	2 4	4 4
Phaseolus vulgaris (black bean)		128,000		5.7
Phaseolus vulgaris (red kidney bean)	D-GalNAc	120,000	4	4.1
Phaseolus vulgaris (wax bean)	D-GalNAc	120,000	4	10.4
Phaseolus vulgaris (navy bean)	D-GalNAc	128,000	4	4
Pisum sativum (garden pea)	α-D-Man, α-D-Glc	53,000	4	0.3
Psophocarpus tetragonolobus (winged bean)	α-L-Fuc	120,000 58,000	4 2	9.4 4.8
Ricinus communis (castor bean)	D-Gal, D-GalNAc D-Gal	60,000[c] 120,000	2 4	
Vicia faba (field bean)	D-Man, D-GlcN	50,000	4	

[a]Glc=glucose; GlcN=glucosamine; GlcNAc=N-acetylglucosamine; Gal=galactose; GalN=galactosamine; GalNAc=N-acetylgalactosamine; Fuc=fucose; Man=mannose.

[b]Where more than one value is given for molecular weight, there is evidence for the existence of multiple forms of the lectins (isolectins).

[c]Nonhemagglutinating toxins.

[d]Also known as *L. esculenta*.

[e]Also known as *P. limensis*.

Most of the lectins are glycoproteins containing 4 to 10 per cent carbohydrates. Notable exceptions are concanavalin A of the jack bean and the lectin from the peanut which are devoid of carbohydrate. The lectins from

the soybean, peanut, and wax bean do not contain any cystine residues. In a few cases which have been examined, namely, concanavalin A and the lectins from the pea and lima bean, metal ions are required for activity. How general this metal requirement remains to be established.

Of all the lectins which have thus far been isolated, only the structure of concanavalin A has been established with respect to amino acid sequence and three-dimensional structure by x-ray crystallography (Edelman et al. 1972). Much remains to be learnt about the structure of the other lectins. Particularly pertinent are such questions as whether there is any homology of structure among the lectins, and in what manner the sugar specificity of lectins is governed by their structural features.

5.3.2. NUTRITIONAL SIGNIFICANCE

(1) SOYBEANS

Liener et al. (1949) presented evidence that indicated that the trypsin inhibitor did not account for all of the growth inhibition observed with rats on a raw soybean diet. This observation led to the subsequent isolation from raw soybeans of a protein fraction that had hemagglutinating activity and was toxic when injected into rats (Liener 1951). Since the destruction of this phytohemagglutinin by heat paralleled the improvement in the nutritive value of soybean meal (1953), the possibility was considered that this protein might be responsible for at least part of the poor nutritive value of raw soybeans. Feeding experiments in which the purified lectin was added to heated soybean meal revealed that this protein was capable of inhibiting growth under conditions of *ad libitum* feeding (Liener 1953). However, when the food intake of a control group of animals not receiving the lectin was restricted to that of animals fed diets containing the lectin, no significant growth inhibition was observed. It thus appeared that the growth inhibitory effect of the soybean lectin may have been due to a depressing effect on the appetite. A more definitive answer to this problem came from experiments in which a crude soybean extract from which the lectin had been removed by affinity chromatography was fed to rats and compared with a similar preparation from which the lectin had not been removed (Turner and Liener 1975). No significant difference in the growth response was observed, from which it was concluded that the soybean lectin has little if any direct effect on the nutritive properties of soybean protein.

(2) *Phaseolus vulgaris*

Although the lectins appear to have little significance as far as the soybean is concerned, the situation with respect to some other legumes is quite different. The common bean, *P. vulgaris*, constitutes an important source of dietary protein for large segments of the world's population, and numerous

reports may be found in the literature concerning the toxic effects that have sometimes accompanied the ingestion of raw or inadequately cooked beans (see, for example, Griebel 1950). Although the presence of phytohemagglutinins in *P. vulgaris* had been reported as early as 1908 (Landsteiner and Raubitcheck 1908), the toxicity of partially purified preparations of the lectin from this bean was first reported by Jaffé and co-workers (1955). Honavar et al. (1962) fed rats purified preparations of lectins from two varieties of *P. vulgaris*, black bean and kidney bean, and observed that levels as low as 0.5 per cent of the diet caused a definite inhibition of growth; higher levels of these lectins hastened the onset of death. These results have more recently been confirmed by Pusztai and Palmer (1977) for the kidney bean lectin. These authors also reported that kidney bean protein preparations from which the lectin has been removed by affinity chromatography are nontoxic.

Preliminary soaking prior to autoclaving seemed to be required for the complete elimination of the toxicity of the kidney bean (Jaffé 1949), although autoclaving alone for five minutes served to eliminate the toxicity of finely ground navy bean meal (Kakade and Evans 1965). However, autoclaving for 30 minutes was necessary to destroy the hemagglutinating activity of certain African varieties of *P. vulgaris* (DeMuelenaere 1964). Of particular significance was the observation that the hemagglutinating activity was still detectable after 18 hours of dry heat.

One of the complicating factors involved in relating hemagglutinating activity to toxicity is the fact that there are hundreds of different strains and cultivars of *P. vulgaris*. The hemagglutinins present in legumes are known to exhibit different degrees of specificity depending on the species of animal from which the red blood cells have been derived, their blood group in the case of human bloods, and whether or not the cells have been pretreated with proteolytic enzymes or neuraminidase (Sharon and Lis 1972). Jaffé and his colleagues (Brücher et al. 1969; Jaffé and Brücher 1972; Jaffé et al. 1972) have made a systematic study of the hemagglutinating activity of a large number of different varieties and cultivars of *P. vulgaris* with respect to their action on the blood from different animals, with and without trypsinization, and the toxicity of these extracts when injected into rats. As shown in Table 5.4, only those extracts that agglutinated trypsinated cow cells were toxic when injected into rats. Feeding tests confirmed the fact that those varieties that exhibited agglutinating activity toward trypsinated cow cells supported very poor growth when fed to rats (Jaffé and Lette 1968). Those varieties that were non-agglutinating or agglutinated only rabbit cells were nontoxic when fed. These results serve to emphasize the importance of testing the hemagglutinating activity of seed extracts against several species of blood cells before one is justified in concluding that a particular bean is toxic or not. The use of trypsinated cow cells would appear to be the most useful system for detecting potentially toxic beans.

TABLE 5.4. Correlation of specific hemagglutinating activity with the intra-peritoneal toxicity in rats of extracts of different varieties and cultivars of *Phaseolus vulgaris*[a]

Variety	Rabbit blood	Trypsinated cow blood	Toxicity[b]
Valin de Albenga	+	+	5/4
Merida	+	+	9/9
Negro Nicoya	+	+	5/4
Saxa	+	+	5/5
Peruvita	+	−	5/0
Palleritos	+	−	6/0
Juli	+	−	5/0
Cubagua	+	−	5/0
Porillo	−	−	5/0
Negra No. 584	−	+	5/5
Varnica Saavegra	−	+	5/3
Hallado	−	+	10/6
Madrileno	−	−	5/0
Alabaster	−	−	5/0
Triguito	−	−	5/0
			6/0

[a]From Jaffé and Brücher (1972).
[b]Number of rats receiving injections per number of dead rats.

Other species of *Phaseolus* that have demonstrated hemagglutinating activity are the lima bean (*P. lunatus*), mung bean (*P. aureus*), white tipary bean (*P. acutifolius*), and the scarlet runner bean (*P. multiflorus*). The lectins from the lima bean and mung bean, however, have bean reported to be nontoxic (DeMuelenaere, 1965; Manage et al. 1972). Despite its lack of toxicity, the oral administration of the lima bean lectin severely restricted the growth of rats (Manage et al. 1972). Similar studies with the mung bean lectin have not been reported. Extracts of the white tipary bean and the scarlet runner bean do display some degree of toxicity when injected into rats (DeMuelenaere 1965), but the extent to which they influence the nutritive properties of these beans is not known.

(3) *Ricinus communis*

Ricin, the lectin of the castor bean, *R. communis*, was one of the first lectins to attract the attention of investigators, presumably because of its extreme toxicity; its minimum lethal dose is about 0.001 μg/g (mice), which makes it about 1000 times more toxic than any of the other bean lectins (Jaffé 1969). This toxicity persists after oral ingestion, and for this reason the detoxification of castor pomace is essential for its safe handling and for its use for animal feeding. Steam heating, as used for the recovery of the solvents employed for the extraction of castor oil, has been found to produce a 1000-fold reduction in the toxicity and to render the pomace harmless for sheep,

rabbits and rats when used in the respective diets in a proportion of not more than 10 per cent (Clemens 1963). The steaming of castor bean meal for 1 h at 15 pounds of pressure reduces the toxicity of the meal to 1/2000th of its original value (Jenkins 1963). Rats fed 23.9 per cent of the autoclaved meal were in good health after four weeks, although growth and food conversion were lower than in the casein controls. Effective detoxification can also be achieved by extraction with hot water (Vilhjalmsdottir and Fischer 1971) or treatment with dilute alkali or formaldehyde (Gardner et al. 1960).

In most of the earlier work dealing with the toxicity of ricin it was assumed that toxicity and hemagglutinating activity were associated with the same protein. It is now clear that these are two distinct components—a toxic protein, for which the name ricin has been reserved and which is devoid of hemagglutinating activity, and the so-called castor bean agglutinin, which exhibits agglutinating activity but is nontoxic (Olsnes and Pihl 1978). It may be reasonably assumed that most of the toxic effects observed with the castor bean preparations in the past have been due to the toxic component rather than to the agglutinin that may have accompanied it. This, of course, raises the question as to whether the toxicity observed with other lectins may be due to a toxic protein that is not a hemagglutinin but is strongly associated with another protein having hemagglutinating activity. Thus far, however, no evidence for the existence of such a toxin in other legumes (except for *A. precatorius*) has been reported.

Not to be confused with the toxic effects of ricin is the presence of the castor bean allergen and the alkaloid ricinine (Jenkins 1963). The latter is generally considered harmless and is growth inhibitory to chicks only when fed in large amounts. Individuals handling castor bean pomace that has not been properly processed are known to develop severe symptoms of irritation of eyes, nose, and throat frequently accompanied by asthma, nausea, vomiting, weakness, and pain (Perlman 1969). Although most of these symptoms are most likely due to the castor bean allergen, Cooper et al. (1964) attribute at least some of these reactions to ricin.

(4) *Canavalia ensiformis*

Concanavalin A is the name given to the lectin first isolated from the jack bean (*Canavalia ensiformis*) by Sumner (1919). In recent years this protein has been the object of considerable study because of the many and varied biological effects that it induces in cells because of its interaction with glycoprotein receptor sites on the surface of cell membranes (Bittiger and Schnebli 1976). Despite the attention that concanavalin A has received in this regard, very little is known regarding its toxicity. It has been noted that the direct injection of concanavalin A into animals caused the agglutination of red blood cells, followed by hemolysis, and finally death (Dameshek and Miller 1943; Ham and Castle 1940). Jack bean meal is of poor nutritive value unless

heated (Borchers and Ackerson 1950), and consumption of the raw bean has been reported to cause a variety of pathological lesions in rats (Orru and Demel 1941) and cattle (Shone 1961). Japanese quails raised under germ-free conditions are able to tolerate the toxic effects of raw jack bean meal or concanavalin A itself much better than conventional birds (Jayne-Williams 1973). There appears to be some doubt, however, as to whether the harmful effects that accompany the ingestion of raw jack bean meal are entirely due to concanavalin A. Dennison et al. (1971) removed the agglutinating activity of a crude extract of jack bean meal by selective absorption onto Sephadex and observed that the unabsorbed fraction still retained some toxicity, albeit less than the original extract, when injected into rats. This would indicate that a portion of the toxicity of the jack bean meal may reside in a fraction devoid of hemagglutinating activity.

(5) OTHER EDIBLE LEGUMES

Although the extracts of peas (*P. sativum*) exhibit hemagglutinating activity, the isolated lectin does not produce any toxic effects when injected into rats or when incorporated into their diet at a level of 1 per cent (Manage et al. 1972). This may perhaps explain why the nutritive value of peas is not enhanced by heat treatment (Liener 1962).

The lectin from the field bean (*D. lablab*) is toxic when injected into rats (Manage et al. 1972) and, when fed at a level of 2.5 per cent in the diet, inhibits the growth of rats and causes zonal necrosis of the liver (Salgarkar and Sohonie 1965). The lectin presumably accounts for only part of the growth depression and toxicity of the raw field bean, since the effects seen with the purified lectin are less than that noted with raw field bean meal containing an equivalent level of hemagglutinating activity (Salgarkar and Sohonie 1965). Although the hemagglutinin of the horse gram (*D. biflorus*) is non-toxic when injected, it retarded the growth of rats when administered orally (Manage et al. 1972).

The broad bean (*V. faba*) is known to contain hemagglutinins, but this activity would appear to have little relationship to the disease in humans known as "favism", which sometimes accompanies the ingestion of this bean by humans (Mager et al. 1969). Although lectins have been isolated from other edible legumes such as the lentil (*L. esculenta*) and peanuts (*Arachis hypogaea*) (Goldstein and Hayes 1978), their influence on the nutritive value of these legumes is not known.

5.3.3. SIGNIFICANCE IN HUMAN DIET

It is difficult, of course, to assess the significance of the lectins in the human diet based on experiments in animals alone. As long as sufficient heat treatment has been applied to ensure destruction of the lectins, there would appear

to be little cause for concern. Nevertheless, it should be recognized that conditions may prevail wherein complete destruction of the lectins may not always be achieved. For example, a massive outbreak of poisoning occurred in Berlin in 1948 after the consumption of partially cooked bean flakes (Griebel 1950). Mixtures of ground beans and cereals have been recommended in child-feeding programmes in more primitive countries (King et al. 1966). Such mixtures can be prepared locally from easily available foodstuffs and can be formulated in proportions to give an amino acid pattern comparable to that of milk protein. However, cooking such mixtures requires a relatively short heating time to become palatable, so that the lectin may not be completely destroyed (Korte 1972). Furthermore, primitive cooking is often done in earthen pots on a wood fire, so that with a tough viscous mass like cooked beans, heat transfer may be imperfect, and, in the absence of constant and vigorous stirring, the temperature reached in parts of the preparation may well be inadequate for the destruction of the lectins (Jaffé 1973). A reduction in the boiling point of water such as would be encountered in certain mountainous regions of the world might also result in the incomplete elimination of lectins.

The marked resistance of lectins to inactivation by dry heat (DeMuelenaere 1964) deserves special emphasis. Thus, the addition of raw bean flour to wheat in bread formulations and in other baked goods should be viewed with caution.

5.3.4. MODE OF ACTION IN ANIMALS

Jaffé and co-workers (Jaffé et al. 1955; Jaffé 1960) had proposed many years ago that the toxic effect of lectins when ingested orally may be due to their ability to bind to specific receptor sites on the surface of intestinal epithelial cells, which thus cause a nonspecific interference with the absorption of nutrients across the intestinal wall. This could be reflected *in vivo* by a decrease in the digestibility of the protein, as evidenced by the fact that the addition of the lectin purified from *P. vulgaris* markedly reduced protein digestibility when added to diets containing casein (Jaffé and Camejo 1961). *In vitro* experiments with isolated intestinal loops taken from rats fed the purified lectin showed a 50 per cent decrease in the rate of absorption of glucose across the intestinal wall as compared with control animals that had not received the lectin (Jaffé 1960). Support for this hypothesis comes from the studies of Etzler and Branstrator (1974), who found that a number of different lectins react with the crypts and/or villi of the intestine, but at different regions of the intestine, depending on the specificity of the lectin. Since surface-bound lectins are known to produce profound physiological effects on the cells with which they interact (Lis and Sharon 1973), one of these effects could be a serious impairment in the ability of these cells to absorb nutrients from

the gastrointestinal tract, thus causing inhibition of growth, and in extreme cases, death.

An alternative effect on the intestinal cells is suggested by the studies of Jayne-Williams and Burgess (1974), who observed that the germ-free Japanese quail was much better able to tolerate the toxic effects of concanavalin A and the navy bean lectin than conventional birds. It was theorized that the binding of lectins to cells lining the intestine may interfere with the normal defence mechanism of these cells whereby normally innocuous intestinal bacteria are prevented from passing from the lumen of the gut into the lymph, blood, and other tissues of the animal body.

5.3.5. PHYSIOLOGICAL FUNCTION IN THE PLANT

The fact that lectins make up 2 to 10 per cent of the total protein in most leguminous seeds would suggest that they must play some important physiological role in the plant. Speculations in this regard in the past have been many and varied (see review by Liener 1976), and have included such possibilities as: they act as antibodies to counter the soil bacteria; serve to transport or store sugars; act as glue for the attachment of glycoprotein enzymes in organized multi-enzyme systems; and play a key role in the development and differentiation of embryonic cells.

More recently, however, strong evidence has been obtained which suggests that the lectins may be the key mediator in determining the specificity between nitrogen-fixing bacteria and their host plants. Thus, specific interactions have been demonstrated between the lectins of the soybean, pea, red kidney bean, jack bean, and clover and the corresponding *Rhizobia* which are known to the specific symbionts with each of these plants. The exact manner in which these lectins act as mediators in this symbiotic relationship is not known, although it has been suggested that the lectins may serve to crosslink sites (lipopolysaccharides) on the bacterial cell surface with specific antigenic sites on the surface of the root hairs of the host.

There is also some evidence to indicate that the lectins may serve to protect plants against certain pathogenic fungi and insect predators. A simulation of pollen germination and seedling growth and a regulation of cell wall extension are other physiological roles which have been postulated for phytolectins on the basis of recent studies.

If the lectins are indeed involved in so many diverse reactions, as would seem likely from the above evidence, it is clear that they represent a most remarkable class of substances which should merit the increased attention of plant physiologists in the future.

5.4. Goitrogens

Goitrogenic substances are most commonly found in plants of the cabbage

family which includes such common edible plants as cabbage, turnip, cauli-flower, kale, brussel sprouts, rapeseed, mustard seed, etc. Among the legumes, however, only the soybeans and peanuts have been reported to produce goitrogenic effects in animals. Unheated soybeans, for example, cause a marked enlargement of the thyroid gland of the rat and chick, an effect which can be counteracted by the administration of iodide or partially eliminated by heat (Patton et al. 1939; Block et al. 1961). Several workers (Van Wyk et al. 1959; Hydowitz 1960) have reported a number of cases of goitre in human infants fed soybean milk. Apparently the heat treatment employed for sterilizing these particular soybean preparations was not sufficient to destroy the goitrogenic agent. Iodine supplementation, however, alleviated this goitre condition in human infants (Van Wyk et al. 1959). The goitrogenic principle from soybeans has been recently partially purified and characterized as a low-molecular-weight oligopeptide composed of two or three amino acids or a glycopeptide consisting of one or two amino acids and a sugar (Konijn et al. 1972, 1973).

Rats fed ground nuts also develop enlarged thyroids, but in this instance the goitrogenic principle has been identified as a phenolic glycoside which resides in the skin (Mougdal et al. 1951; Sreenivasan et al. 1957). It has been suggested that the phenolic metabolites formed from this glycoside are pre-ferentially iodinated and thereby deprive the thyroid of available iodine. Thus the goitrogenic effect of ground nuts is effectively counteracted by iodine supplementation but not by heat treatment.

5.5. Cyanogenic Glycosides

It has been known for a long time that a wide variety of plants are potential-ly toxic because they contain glycosides from which HCN may be released by hydrolysis (Conn 1979). It will be noted in Table 5.5 that the legumes predominate in their cyanide-producing potential. In the years immediately following the turn of 20th century and again during World War I lima beans imported into Europe from tropical countries (Java, Puerto Rico, and Burma) were responsible for serious outbreaks of cyanide poisoning, and cases of human intoxication from the consumption of certain varieties of lima beans are not uncommon today in some of the tropical countries. Most of the lima beans consumed in the U.S.A. and Europe at the present time are well below the toxic levels implicated in fatal cases of poisoning.

Cyanide in the form of HCN is released from a glycoside (*phaseolunatin* in the case of lima beans) through the action of an enzyme present in the plant tissue (see Fig. 5.7). Hydrolysis occurs quite rapidly when the ground bean meal is cooked in water, and the most of the liberated HCN is lost by volatilization. Further cooking also leads to the eventual destruction of the enzyme. Yet many cases of human intoxication have occurred even with cook-

TABLE 5.5. Cyanide content of certain plants

Plant	HCN yield, mg/100 g
Lima bean (*Phaseolus lunatus*)	
Samples incriminated in fatal human poisoning	210.0–312.0
Normal levels	14.4–16.7
Sorghum	250.0
Cassava	113.0
Linseed meal	53.0
Black-eyed pea (*Vigna sinensis*)	2.1
Garden pea (*Pisum sativum*)	2.3
Kidney bean (*Phaseolus vulgaris*)	2.0
Bengal gram (*Cicer arietinum*)	0.8
Red gram (*Cajanus cajan*)	0.5

[a]Data taken from Montgomery (1969).

ed lima beans. For example, it has been reported (Gabel and Kruger 1920) that when lima beans, which had been cooked so as to destroy the enzymes responsible for cyanide formation were fed to human subjects, cyanide could be detected in the urine. This has led to the supposition that perhaps enzymes secreted in the intestinal tract, or by the microflora of the colon, may be responsible for releasing HCN after ingestion of the cooked beans.

5.6. Anti-Vitamin Factors

The inclusion of unheated soybean meal, or the protein isolated therefrom, in the diet of chicks may cause rickets unless higher than normal levels of

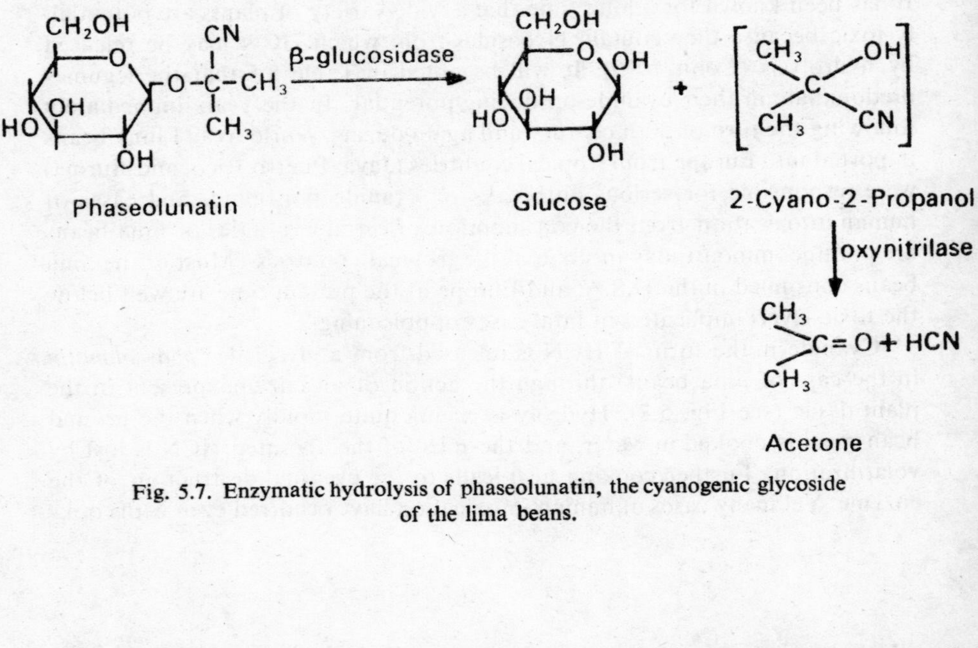

Fig. 5.7. Enzymatic hydrolysis of phaseolunatin, the cyanogenic glycoside of the lima beans.

vitamin D_3 are added to the diet (Carlson et al. 1964). The rachitogenic effect could be eliminated by autoclaving the soybean meal, but supplementation with calcium or phosphorus was ineffective. Along the same lines, raw kidney beans are believed to contain an antagonist of vitamin E as evidenced by liver necrosis in rats and muscular dystrophy and low levels of plasma tocopherol in chicks (Hintz and Hogue 1964; Desai 1966). The anti-vitamin E effect of raw kidney beans can be partially eliminated by heat treatment. Fisher et al. (1969) have reported that a soybean protein isolate increased the chick's requirement for α-tocopherol as measured by growth, mortality, exudative diathasis and encephalomalacia. Whether this factor is the α-tocopherol oxidase described by Murillo and Gaunt (1975) remains to be demonstrated. Neither the identity of the anti-vitamin D factor of soybeans nor of the anti-vitamin E factor of kidney beans is known. Edelstein and Guggenheim (1970a, b) demonstrated that unheated soy flour is deficient in vitamin B_{12} and contains a heat-labile substance that increases the requirement for vitamin B_{12}. The identity of this anti-vitamin B_{12} factor remains to be established.

5.7. Metal-Binding Constituents

The inclusion of isolated soybean protein in animal diets has been noted to lead to a decrease in the availability of certain trace minerals such as zinc, manganese, copper, and iron (Erdman 1979). In fact, an anaemia in monkeys due to a deficiency in iron may be induced in monkeys on a soybean protein diet, an effect which may be eliminated by heat treatment or chelating agents such as EDTA (Fitch et al. 1964). Peas (*Pisum sativum*) have also been shown to contain a factor which interferes with the availability of zinc for chicks (Kienholz et al. 1962). Autoclaving the peas eliminated the requirement for supplemental zinc. Since zinc supplementation was only one-third as effective as autoclaving, the presence of an additional heat-labile growth inhibitor in peas was postulated.

The exact mechanism whereby certain plant proteins exhibit this ability to interfere with the availability of metals is not known, although it may be

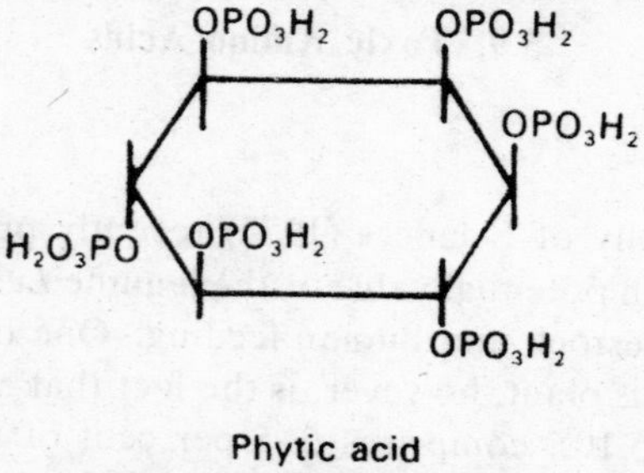

Phytic acid
Fig. 5.8. Structure of phytic acid.

related to the observation that a soy protein-phytic acid complex (Fig. 5.8) has a special affinity for metal ions (O'Dell and Savage 1960).

5.8. Estrogenic Factors

Factors capable of eliciting an estrogenic response in animals have been shown to occur naturally in certain legumes (Stob 1973). Chemically, these have been identified as derivatives of isoflavone, combined with a sugar residue (Fig. 5.9), the latter being attached to one of the hydroxyl groups. Genistein and

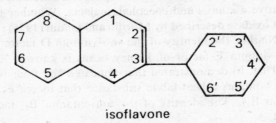

isoflavone

Genistein	4',5,7-trihydroxyisoflavone
Daidzein	4',7-dihydroxyisoflavone
Isogenistein	2',5,7-trihydroxyisoflavone
Tatoin	8-methyl-4',5-dihydroxyisoflavone
Methylgenistein	8-methyl-4',5,7-trihydroxyisoflavone
Methylisogenistein	8-methyl-2',5,7-trihydroxyisoflavone

Fig. 5.9. Estrogenic isoflavones found to occur in legumes.

daidzein have been isolated from the soybean and shown to have estrogenic activity when tested on female rats and mice. It is doubtful, however, that the normal consumption of legumes would provide sufficient amounts of these substances to elicit a physiological response in humans.

5.9. Toxic Amino Acids

5.9.1. MIMOSINE

The National Academy of Sciences (1977) recently published a monograph that describes the high potential value of the legume *Leucaena leucocephala* as a forage crop for livestock and human feeding. One of the principal factors limiting the use of this plant, however, is the fact that an unusual amino acid, mimosine (see Fig. 5.10), comprises 3–5 per cent of the dry weight of the protein. This amino acid is responsible for the poor growth performance of

cattle when *leucaena* makes up more than one-half of the diet. This adverse effect on growth has been traced to the underproduction of thyroxine presumably due to the fact that the rumen bacteria convert mimosine to 3,4-dihydroxypyridine, which acts as a goitrogenic agent (Hegarty et al. 1976). In nonruminants, such as the horse, pig, and rabbit, the goitrogenic effect is

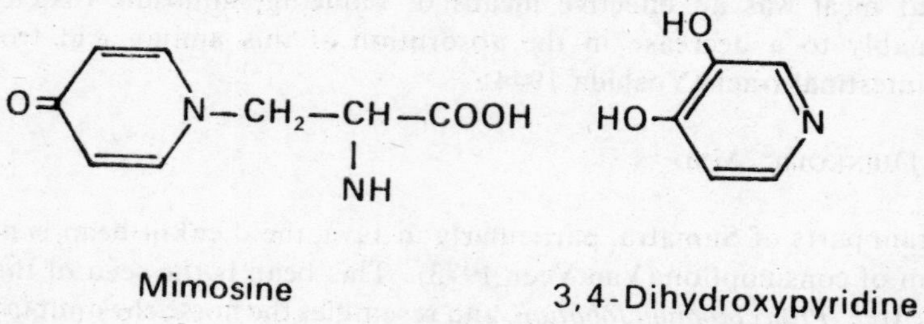

Mimosine   3,4-Dihydroxypyridine

Fig. 5.10. Structure of mimosine and its goitrogenic metabolite, 3,4-dihydroxypyridine.

not very marked, but the animals nevertheless do very poorly on diets containing *leucaena*, one of the characteristic symptoms being a loss of hair (Owen 1958; Hegarty et al. 1964). In fact, it has even been suggested that mimosine might be used as a defleecing agent in sheep (Reis et al. 1975). Certain segments of the human population, particularly in Indonesia, are known to consume portions of the *leucaena* in their diet, and a loss of hair has been frequently observed among those individuals who have eaten the leaves, pods, and seeds in the form of a soup (Van Veen 1973). As far as is known, mimosine has no effect on the meat or milk of ruminants that can be detrimental to human beings.

Although the goitrogenic effect of mimosine in ruminants seems to be well established, the precise mechanism of toxicity in other animals remains obscure. It can act as an inhibitor of pyridoxal-containing transaminases (Lin et al. 1965), tyrosine decarboxylase (Crounze et al. 1962), and both cystathionine synthetase and cystathionase (Hylin 1969). The latter may be of particular significance, since an inhibition of the conversion of methionine to cysteine, a major component of hair protein, could account for the hair loss that is so characteristic of mimosine toxicity. Mimosine may also exert a more direct effect on the hair growth since Montagna and Yun (1963) showed that *leucaena* extracts destroyed the matrix of the cells of the hair follicles of mice. This effect was reversible, however, since hair growth returned to normal once the animals were taken off the mimosine-containing diet.

The structural resemblance of mimosine to tyrosine would suggest that it might function *in vivo* as an antagonist of this amino acid. Lin et al. (1964) in fact reported that the growth inhibition produced by feeding 0.5 per cent mimosine to rats could be partially reversed with phenylalanine and wholly

reversed with tyrosine.

Matsumoto et al. (1951) reported that the mimosine content of the seeds and leaves of *leucaena* could be decreased by storing the plant at temperatures in excess of 70°C in the presence of moisture. These authors also showed that ferrous sulphate added to the rations of rats containing unheated *leucaena* leaf meal was an effective means of reducing mimosine toxicity, due presumably to a decrease in the absorption of this amino acid from the gastrointestinal tract (Yoshida 1944).

5.9.2. DJENKOLIC ACID

In certain parts of Sumatra, particularly in Java, the djenkol bean is a popular item of consumption (Van Veen 1973). This bean is the seed of the leguminose tree, *Pithecolobium lobatum,* and resembles the horse chestnut in colour and size. The bean is not particularly toxic except that it sometimes leads to kidney failure, which is accompanied by the appearance of blood and white needlelike clusters in the urine. The latter substance is a sulfur-containing amino acid known as djenkolic acid (Fig. 5.11), which is present in the free

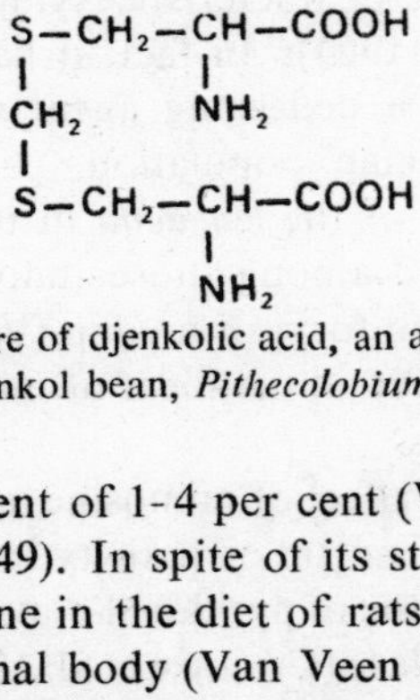

$$S-CH_2-CH-COOH$$
$$\quad\quad\quad\quad\; |$$
$$CH_2 \quad\quad NH_2$$
$$|$$
$$S-CH_2-CH-COOH$$
$$\quad\quad\quad\quad\; |$$
$$\quad\quad\quad\quad NH_2$$

Fig. 5.11. Structure of djenkolic acid, an amino acid present in the djenkol bean, *Pithecolobium lobatum.*

state in the bean to the extent of 1-4 per cent (Van Veen and Hyman 1933; Van Veen and Latuasan 1949). In spite of its structural resemblance to cystine, it cannot replace cystine in the diet of rats, although it can apparently be metabolized by the animal body (Van Veen and Hyman 1933). Because of its relative insolubility, and djenkolic acid that escapes metabolic degradation tends to crystallize out in the kidney tubules and urine, hence the observations in human beings noted above.

5.9.3. DIHYDROXYPHENYLALANINE

The amino acid 3,4-dihydroxyphenylalanine (dopa) (Fig. 5.12) is present in fairly high concentrations in the fava bean, *Vicia faba* (Torquati 1913; Guggenheim 1913; Nagasawa et al. 1961) and the velvet bean, *Stizolobium decringianum* (Miller 1920). Since the consumption of the fava bean is frequently associated with a disease in human beings known as favism (see section 5.11 below), the question has been raised as to whether dopa might not play a

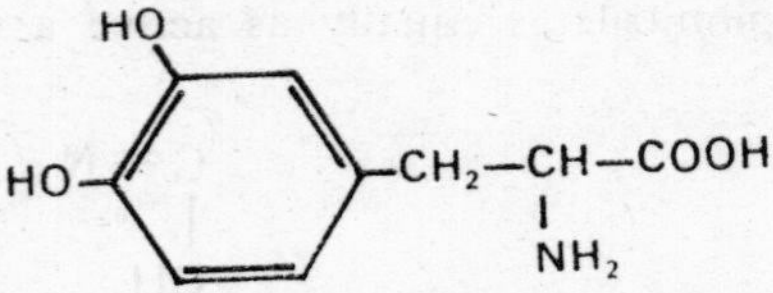

Fig. 5.12. Structure of 3,4-dihydroxyphenylalanine (dopa), an amino acid present in the fava been (*Vicia faba*) and velvet bean (*Stizolobium decringianum*).

causative role in this disease (Kosower and Kosower 1967). Persons genetically deficient in the enzyme glucose-6-phosphate dehydrogenase appear to be particularly susceptible to favism, and one of the characteristic clinical features of this disease is a hemolytic type of anaemia. The hemolysis of the red blood cells is believed to be due to a marked lowering of the glutathione content of the erythrocytes (Davies and Gower 1964). In view of these facts, it may be pertinent to note that the *in vitro* addition of dopa to red blood cells from individuals deficient in glucose-6-phosphate dehydrogenase produced a significant lowering of glutathione content of such cells (Kosower and Kosower 1967). Because of its high content of dopa, it has been suggested that *Vicia faba* might be of therapeutic value in the treatment of Parkinson's disease (Natelson 1969).

5.10. Lathyrogens

Lathyrism, as it is known to occur in humans, is a disease associated with the consumption of a legume or peas known scientifically as *Lathyrus sativus*, or by its common name as chicking vetch or kesari dal. This disease is particularly prevalent in India, especially during periods of famine resulting from droughts when the field crops become blighted, and, as an alternate crop, this particular legume is cultivated. We are not dealing here with an occasional case of poisoning but with a disease which can almost reach epidemic proportions. As recently as 1975 over 100,000 cases of lathyrism in men between the ages of 15 and 45 years were reported (Natarajan 1976). Lathyrism seems to affect only males, particularly young adults. This disease is characterized by a nervous paralysis of the lower limbs which forces the victim to walk with short, jerky steps; death may result in extreme cases (Attal et al. 1978).

Attempts to identify the causative agent of human lathyrism have been complicated by the fact that the sweet pea (*Lathyrus odoratus*) produces another type of lathyrism (osteolathyrism) which is characterized in rats by skeletal deformities. In contrast to this, rats thrive quite well on *Lathyrus sativus* and do not display the nervous disorder associated with the consumption of this species in man. Historically, the lathyrogen of the sweet pea was the first to be isolated and was identified as β-(N-γ-glutamyl)-aminopropionitrile

although β-aminopropionitrile is equally as active as an osteolathyrogen (Fig. 5.13).

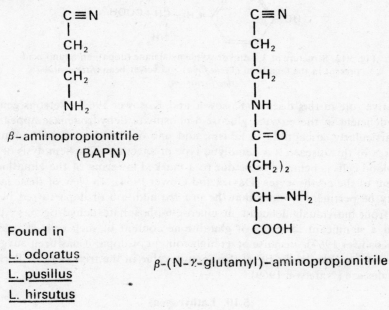

Found in
L. odoratus
L. pusillus
L. hirsutus

β-(N-γ-glutamyl)-aminopropionitrile

Fig. 5.13. Structures of compounds found in certain *Lathyrus* spp. which cause osteolathyrism.

Several groups of workers in India (Adiga et al. 1962; Murti et al. 1964; Rao et al. 1964) have succeeded in isolating a compound from *Lathyrus sativa* which may very well be the causative principle of human neurolathyrism. This compound, identified as β-N-oxalyl-α, β-diaminopropionic acid (Fig. 5.14), produced severe neurotoxic symptoms when injected into young chicks, rats, and monkeys. This compound as well as α, γ-diaminobutyric acid and β-cyano-L-alanine have been isolated from other *Lathyrus* spp. as well as *Vicia sativa* and have been shown to produce neurotoxic effects when injected into animals (Fig. 5.14). Roy (1973) has shown that the oral administration of β-N-oxalyl-α, β-diaminopropionic acid to baby chicks can induce neurological symptoms but at a much higher dose than that required by the intraperitoneal route. Because of its structural similarity to glutamic acid, it is not surprising that this compound has been found to interfere with the role of glutamic acid as an excitory neurotransmitter in brain tissue (Lakshaman and Padmanabhan 1974).

Assuming that β-N-oxalyl-α, β-diaminopropionic acid is the causative factor of human lathyrism, much of the misery in the past associated with the consumption of *Lathyrus sativus* could have been avoided by a relatively

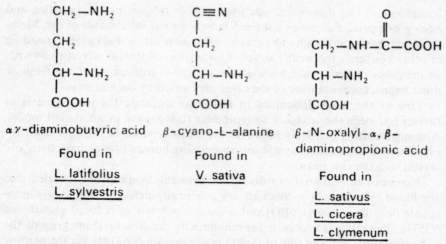

Fig. 5.14. Structures of compounds found in certain *Lathyrus* spp. and *Vicia* spp. which act as neurotoxins.

simple detoxification procedure involving steeping the dehusked seeds overnight, followed by steaming for 30 minutes or roasting at 150°C for 20 min. Amino acid analysis of the protein of *Lathyrus sativus* indicates that it is rich in lysine and is therefore potentially useful as a lysine supplement.

The breeding of species and varieties of *Lathyrus* which are genetically deficient in the neurotoxin is also a possibility. In extensive surveys of different *Lathyrus* spp. and common vetch (*Vicia sativa*) a number of samples were found to have little or no β-N-oxalyl-α, β-diaminopropionic acid (Bell 1964; Jeswani et al. 1970). Unfortunately, many of these samples contain other compounds which produce neurotoxic effects when injected into animals. It is obvious that any serious breeding programme involving *Lathyrus* spp. must take into account the possible role of these compounds as well in the pathogenesis of lathyrism.

5.11. Favism

The field or broad bean (*Vicia faba*) has been extensively used as a source of good quality protein for feeding livestock and poultry (Eppendorfer 1971; Nitsan 1971). However, its use for human consumption has been tempered by the fear that its consumption could lead to a disease known as favism in some susceptible individuals. This would be particularly true in the Middle East, where the field bean is a major good staple and where the genetic defect associated with favism is more prevalent.

Favism is a disease characterized by hemolytic anaemia which affects certain individuals following the ingestion of the field bean (Belsey 1973).

Symptoms of this disease include weakness or fatigue, pallor, jaundice and hemoglobinuria. Favism is confined largely to the inhabitants of the Mediterranean basin, although individuals of the same ethnic background residing in other countries frequently suffer from favism. Although about two-thirds of the cases of favism are associated with the consumption of the fresh or dried beans, the remainder of the cases are caused by cooked beans.

One of the main difficulties in trying to elucidate the pathogenesis of favism has been the inability to reproduce this disease in an animal model. Although heating definitely improves the nutritive value of the field bean for experimental animals, no symptoms resembling human favism have been observed with the raw bean.

Extensive clinical studies with favism-prone individuals has revealed that the blood cells of such individuals are genetically deficient in glucose-6-phosphate dehydrogenase (G6PD) and contain low levels of reduced glutathione (GSH). The latter is necessary for maintaining the structural integrity of the cell membrane, and the role of G6PD is to generate NADPH via the pentose phosphate shunt. NADPH is necessary for the action of glutathione reductase which serves to reduce oxidized glutathione to GSH. These relationships are shown in Fig. 5.15.

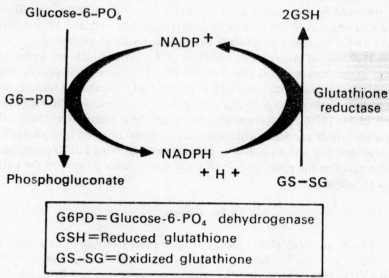

Fig. 5.15. Metabolic reactions governing the level of reduced glutathione (GSH) in the blood.

It follows that any factors which lead to a decrease in GSH, particularly in the absence of G6PD, might be expected to cause hemolysis of red blood cells. The field bean is known to contain the glucosides, vicine and convicine

(Fig. 5.16), the aglycone components of which (divicine and isouramil respectively) have been shown to cause a rapid oxidation of GSH in G6PD-deficient erythrocytes but not in normal cells (Mager et al. 1969) (Fig. 5. 17). If vicine

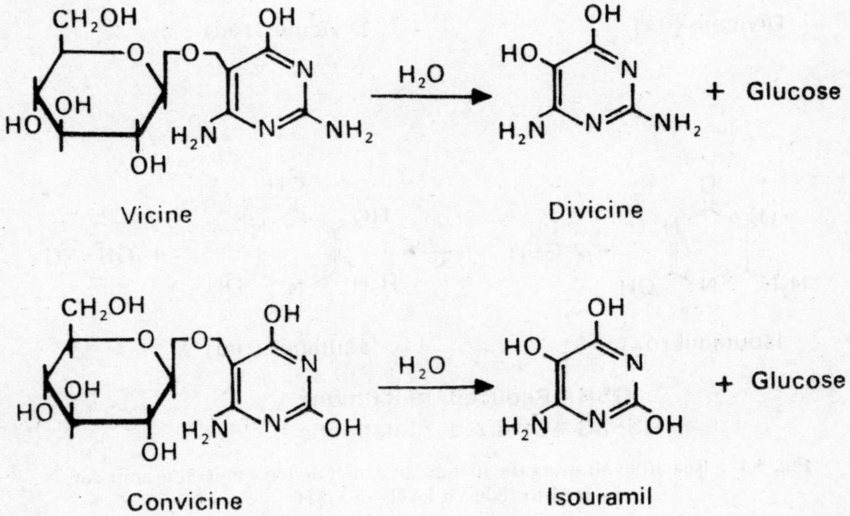

Fig. 5.16. Structure of vicine and convicine, the favism-producing factors in *Vicia faba*.

and convicine are in fact the causative agents of favism, it would appear that the only way of diminishing the risk of this disease is to effect their removal by genetic breeding or by some form of processing. With the availability of relatively simple chemical tests for vicine and convicine (Brown and Roberts 1972; Higazi and Read 1974; Jamalian et al. 1976), it should be possible to screen cultivars of *V. faba* and to assess various processing techniques for the effectiveness in eliminating these compounds. Although protein concentrates and isolates have been prepared from the field bean and have been found to be of good nutritional quality in animal tests (Duthrie et al. 1972, 1977), the extent to which they might still be contaminated with vicine or convicine does not appear to have received the attention this problem deserves if such products are to be used in the human diet. It is evident that in order for divicine and isouramil to function as oxidizing agents *in vivo*, they must first be released from their parent glycosides and converted to their corresponding orthoquinones. How this is accomplished is not known; this presumably could take place through hydrolysis by β-glycosidases in the intestinal tract (Flohé et al. 1971).

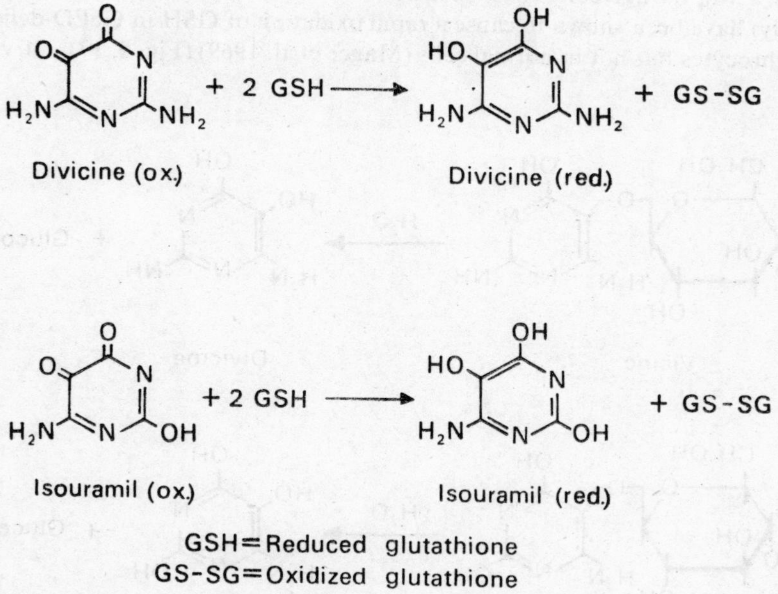

Fig. 5.17. Reactions showing the manner in which divicine and isouramil can lead to reduced levels of GSH.

5.12. Conclusions

It should be apparent that, although there are numerous examples of so-called toxic constituents in legumes, they have nevertheless provided man over the centuries with a valuable source of protein. This can be attributed in part to the fact that man has learned how to detoxify them by suitable preparative measures. The varied nature of our diet also minimizes the contribution of a toxicant from any one foodstuff. Nevertheless, there is the ever present possibility that the prolonged consumption of a particular legume which may be improperly processed could bring to the surface these toxic effects which would otherwise not be apparent. As the shortage of protein becomes more acute, it is not unlikely that in the future much of the population of the world will be faced with a more limited selection of protein foods, most of which will be of plant origin and hence potential carriers of toxic constituents. The nutritionist, food scientist, and plant breeder should all be at least cognizant of such a possibility and prepared to apply their knowledge and skill to meeting this challenge.

References

Adiga, P.G., Padmanaban, G., Rao, S.L.N. and Sarma, P.S. 1962. The isolation of a toxic principle from *Lathyrus sativus*. *J. Sci. Ind. Res.*, India **21**: 284–286.

Anderson, R.L., Rackis, J.J. and Tallent, W.H. 1979. In: *Soy Protein and Human Nutrition*. H.L. Wilcke, D.T. Hopkins and D.H. Waggle, eds. pp. 209–233. Academic Press, New York.

Attal, H.C., Kulkarni, S.W., Choubey, B.S., Palkar, N.D. and Deotale, P.G. 1978. A field study of lathyrism—some clinical aspects. *Ind. J. Med. Res.* **67**: 608–615.

Baumgarten, B. and Chrispeels, M.J. 1976. Partial characterization of a protease inhibitor which inhibits the major endopeptidase in cotyledons of the navy bean. *Plant Physiol.* **58**: 1–6.

Bell, E.A. 1964. Relevance of biochemical taxonomy to the problem of lathyrism. *Nature* **203**: 378–380.

Belsey, M.A. 1973. The epidemiology of favism. *Bull. W.H.O.* **48**: 1–13.

Birk, Y. 1968. Chemistry and nutritional significance of proteinase inhibitors from plant sources. *Ann. N.Y. Acad. Sci.* **146**: 388–399.

Birk, Y. and Applebaum, S.W. 1960. Effect of soybean trypsin inhibitors on the development and midgut proteolytic activity of *Tribolium castaneum* larvae. *Enzymologia* **22**: 318–326.

Birk, Y., Gertler, A. and Khalef, S. 1963. Separation of a *Tribolium* protease inhibitor from soybeans on a calcium phosphate column. *Biochim. Biophys. Acta* **67**: 326–328.

Bittiger, H. and Schnebli, H.P. 1976. *Concanavalin A as a Tool*. John Wiley and Sons, New York.

Block, R.J., Mandl, R.H., Howard, H.W., Bauer, C.D. and Anderson, D.W. 1961. The curative action of iodine on soybean goiter and the changes in the distribution of iodiamino acids in the serum and in thyroid gland digests. *Arch. Biochem. Biophys.* **93**: 15–24.

Booth, A.N., Robbins, D.J., Ribelin, W.E. and DeEds, F. 1960. Effect of raw soybean meal and amino acids on pancreatic hypertrophy in rats. *Proc. Soc. Exp. Biol. Med.* **104**: 681–683.

Borchers, R. and Ackerson, C.W. 1950. The nutritive value of legume seeds. X. Effect of autoclaving and the trypsin inhibitor test for 17 species. *J. Nutr.* **41**: 339–345.

Bowman, D.E. 1944. Fractions derived from soybeans and navy beans which retard the tryptic digestion of casein. *Proc. Soc. Exptl. Biol. Med.* **57**: 139–140.

Boyd, W.C. and Sharpleigh, E. 1954. Specific precipitating activity of plant agglutinins. *Science* **119**: 419.

Brown, E.G. and Roberts, F.M. 1972. Formation of vicine and convicine by

Vicia faba. Phytochemistry **11**: 3203–3206.

Brücher, O., Wecksler, M., Levy, A., Palozza, A. and Jaffé, W.G. 1969. Comparison of phytohemagglutinin in wild beans (*Phaseolus aborigineus*) and in common beans (*Phaseolus vulgaris*) and their inheritance. *Phytochemistry* **8**: 1739–1743.

Carlson, C.W., McGinnis, J. and Jensen, L.S. 1964. The antirachitic effects of soybean preparations for turkey poults. *J. Nutr.* **82**: 366–370.

Chernick, S.S., Lepkovsky, S. and Chaikoff, I.L. 1948. A dietary factor regulating the enzyme content of the pancreas. Changes induced in size and proteolytic activity of the chick pancreas by the ingestion of raw soybean meal. *Am. J. Physiol.* **155**: 33–41.

Clemens, E. 1963. Toxicity and tolerance of Ricinus seed oilmeal by different animals. *Landwitsch. Forsch.* **17**: 202–207.

Collins, J.L. and Sanders, G.G. 1976. Changes in trypsin inhibitor activity in some soybean varieties during maturation and germination *J. Food Sci.* **41**: 168–172.

Conn, E.E. 1979. In: *Biochemistry of Nutrition*. A. Neuberger and T.H. Jukes, eds. Vol. 27, pp. 21–43. University Park Press, Baltimore, MD.

Cooper, W.C., Perone, V.B., Scheel, L.D. and Keenan, R.G. 1964. Occupational hazards from castor bean pomace: Tests for toxicity. *Am. Ind. Hyg. Assoc. J.* **25**: 431–438.

Crounze, R.G., Maxwell, J.D. and Blank, H. 1962. Inhibition of growth of hair by mimosine. *Nature* **194**: 694–695.

Dameshek, W. and Miller, E.B. 1943. Pathogenic mechanisms in hemolytic anemias. *Arch. Int. Med.* **72**: 1–7.

Davies, J.P.H. and Gower, D.B. 1964. Conditions affecting the glucose-6-phosphate dehydrogenase activity and glutathione stability of human erythrocytes. *Nature* **203**: 310–311.

DeMuelenaere, H.J.H. 1964. Effect of heat treatment on the hemagglutinating activity of legumes. *Nature* **201**: 1029–1030.

DeMuelenaere, H.J.H. 1965. Toxicity and hemagglutinating activity of legumes. *Nature* **206**: 827–828.

Dennison, C., Stead, R.H. and Quicke, G.V. 1971. A non-hemagglutinating toxic factor from the jack bean (*Conavalia ensiformis*). *Agroplantae* **3**: 27–30.

Desai, I.D. 1966. Effect of kidney beans (*Phaseolus vulgaris*) on plasma tocopherol levels and its relation to nutritional muscular dystrophy in the chick. *Nature* **209**: 810.

Duthrie, I.F., Owen, E., Miller, E.L., Laws, B.M. and Owers, M.J. 1977. Preparation of field bean (*Vicia faba*) cotyledons as a substitute for dried skim-milk in calf feeding. *Proc. Nutr. Soc.* **35**: A115–116.

Duthrie, I.F., Porter, P.D. and Gadaly, B. 1972. Nutritional value of a protein isolate from the Throws M.S. variety of field beans (*Vicia faba* L.)

grown in the U.K. *Proc. Nutr. Soc.* **31**: 1980–1981.

Edelman, G.M., Cunningham, B.A., Reeke, G.N. Jr., Becker, J.W., Waxdal, M.J. and Wang, J.L. 1972. The covalent and three-dimensional structure of concanavalin A. *Proc. Nat. Acad. Sci. USA* **69**: 2580–2584.

Edelstein, S. and Guggenheim, K. 1970a. Causes of the increased requirement for vitamin B_{12} in rats subsisting on an unheated soybean flour diet. *J. Nutr.* **100**: 1377–1382.

Edelstein, S. and Guggenheim, K. 1970b. Changes in the metabolism of vitamin B_{12} and methionine in ratsfe d unheated soyabean flour. *Brit. J. Nutr.* **24**: 735–740.

Ehrlich, P. 1891. Experimentelle Untersuchungen über Immunität 1. Über Ricin. *Deut. Med. Wochschr.* **17**: 976.

Eppendorfer, W.H. 1971. Effects of S, N, and P on amino acid composition of field beans (*Vicia faba*) and responses of the biological value of seed protein to S-amino acid content. *J. Sci. Food Agr.* **22**: 501–510.

Erdman, J.W. Jr. 1979. Oil seed phytates: nutritional implications. *J. Am. Oil Chem. Soc.* **56**: 736–741.

Etzler, M.E. and Branstrator, M.L. 1974. Differential localization of cell surface and secretory components in rat intestinal epithelium by use of lectin. *J. Cell Biol.* **62**: 329–343.

Figarella, C., Negri, G.A. and Guy, O. 1974. Studies on inhibition of two human trypsins. In: *Bayer Symp. V on Proteinase Inhibitors.* H. Fritz, L. Tochesche, L.J. Greene and E. Truscheit, eds. pp. 213–222. Springer-Verlag, Berlin.

Fisher, H., Griminger, P. and Budowski, P. 1969. Anti-vitamin E activity of isolated soybean protein for the chick. *Z. Ernachrungswiss.* **9**: 271–278.

Fitch, C.D., Harville, W.E., Dinning, J.S. and Porter, F.S. 1964. Iron deficiency in monkeys fed diets containing soybean protein. *Proc. Soc. Exp. Biol. Med.* **116**: 130–133.

Flohé, L., Niebeh, G. and Reiber, H. 1971. The effect of divicine on human erythrocytes. *Z. Klin. Chem. Klin. Biochem.* **9**: 431–437.

Gabel, W. and Kruger, W. 1920. The toxic action of Rangoon beans. *Muensch. Med. Woehschr.* **67**: 214–215.

Gardner, H.K. Jr., D'Aquin, E.L., Kottun, S.P., McCourtney, E.J., Vix, H.L.E. and Gastroch, E.A. 1960. Detoxification and deallergenization of castor beans. *J. Am. Oil Chem. Soc.* **37**: 142–148.

Goldstein, I.J. and Hayes, C.E. 1978. The lectins: carbohydrate-binding proteins of plants and animals. *Adv. Carbohyd. Chem. Biochem.* **35**: 128–340.

Green, G.M. and Lyman, R.L. 1972. Feedback regulation of pancreatic enzyme secretion as a mechanism for trypsin-induced hypersecretion in rats. *Proc. Soc. Exp. Biol. Med.* **140**: 6–12.

Griebel, C. 1950. Erkrankungen durch Bohnenflochen (*Phaseolus vulgaris* L.) and Plattererbsen (*Lathyrus tingitanus* L.). *Z Lebensm. Untersuch. Forsch.*

90: 191–195.

Guggenheim, M. 1913. Dihydroxyphenylalanine, a new amino acid from *Vicia faba*. *Z. Physiol. Chem.* **88**: 276–284.

Ham, T.H. and Castle, W.B. 1940. Relation of increased hypotonic fragility and of erythrostasis to the mechanism of hemolysis in certain animals. *Trans. Assoc. Am. Phys.* **55**: 127–132.

Ham, W.E. and Sandstedt, R.M. 1944. A proteolytic inhibitory substance in the extract from unheated soybean meal. *J. Biol. Chem.* **154**: 505–506.

Hegarty, M.P., Court, R.D., Christie, G.S. and Lee, C.P. 1976. Mimosine in *Leucaena leucocephala* is metabolized to a goitrogen in ruminants. *Austral. Vet. J.* **52**: 490.

Hegarty, M.P., Schinckel, P.G. and Court, R.D. 1964. Reaction of sheep to the consumption of *Leucaena glauca* Benth and to its toxic principle. *Austral. J. Agric. Res.* **15**: 153–167.

Higazi, M.I. and Read, W.W.C. 1974. Method for the determination of vicine in plant material and in blood. *J. Agr. Food Chem.* **22**: 570–571.

Hintz, H.F. and Hogue, D.E. 1964. Kidney beans (*Phaseolus vulgaris*) and the effectiveness of vitamin E for prevention of nutritional muscular dystrophy in the chick. *J. Nutr.* **84**: 283–287.

Hobday, S.M., Thurman, D.A. and Barber, D.J. 1973. Proteolytic and trypsin inhibitory activities in extracts of germinating *Pisum sativum* seeds. *Phytochemistry* **12**: 1041–1046.

Honavar, P.M., Shih, C.-V. and Liener, I.E. 1962. Inhibition of the growth of rats by purified hemagglutinin fractions isolated from *Phaseolus vulgaris*. *J. Nutr.* **77**: 109–114.

Hydowitz, J.D. 1960. Occurrence of goiter in an infant on soy diet. *N.E.J. Med.* **262**: 351–353.

Hylin, J.W. 1969. Toxic peptides and amino acids in foods and feeds. *J. Agr. Food Chem.* **17**: 492–496.

Jaffé, W.G. 1949. Toxicity of raw kidney beans. *Experientia* **5**: 81–83.

Jaffé, W.G. 1969. Über Phytotoxine aus Bohnen (*Phaseolus vulgaris*). *Anneim. Forsch.* **12**: 1012–1016.

Jaffé, W.G. 1969. In: *Toxic Constituents of Plant Foodstuffs*. I.E. Liener, ed. pp. 69–101. Academic Press, New York, NY.

Jaffé, W.G. 1973. In: *Nutritional Aspects of Common Beans and Other Legume Seeds as Animal and Human Foods*. W.G. Jaffé, ed. p. 199. Arch. Latinoamer. Nutr., Caracas, Venezuela.

Jaffé, W.G. and Brücher, O. 1972. Toxicity and specificity of different phytohemagglutinins of beans (*Phaseolus vulgaris*). *Arch. Latinoamer. Nutr.* **22**: 267–281.

Jaffé, W.G. and Camejo, G. 1961. The action of a toxic protein from the black bean (*Phaseolus vulgaris*) on intestinal absorption in rats. *Acta Cient. Venez.* **12**: 59–61.

Jaffé, W.G. and Lette, C.L.V. 1968. Heat-labile growth-inhibiting factors in beans (*Phaseolus vulgaris*). *J. Nutr.* **94**: 203–210.

Jaffé, W.G., Brücher, O. and Palozza, A. 1972. Detection of four types of specific phytohemagglutinins in different lines of beans (*Phaseolus vulgaris*). *Z. Immunitat.-Forsch.* **142**: 439–447.

Jaffé, W.G., Planchert, A., Páez-Pumar, J.I., Torrealba, R. and Franceshi, D.N. 1955. New studies on the toxic factor of raw beans (*Phaseolus vulgaris*). *Arch. Venez. Nutr.* **6**: 195–205.

Jamalian, J., Aylward, F. and Hudson, B.J.F. 1976. Favism-inducing toxins in broad beans (*Vicia faba*). Examination of bean extracts for pyrimidine glucosides. *Qual. Plant., Pl. Fds. Human Nutr.* **26**: 331–339.

Jayne-Williams, D.J. 1973. The influence of dietary jack beans (*Canavalia ensiformis*) and of concanavalin A on the growth of conventional and gnobotic quail (*Coturnix coturnix japonica*). *Nature* **243**: 150–151.

Jayne-Williams, D.J. and Burgess, C.D. 1974. Further observations on the toxicity of navy beans (*Phaseolus vulgaris*) for Japanese quail (*Coturnix coturnix japonica*). *J. Appl. Bacteriol.* **37**: 149–169.

Jenkins, F.P. 1963. Allergenic and toxic components of castor bean meal: review of the literature and studies of the inactivation of these components. *J. Sci. Food Agr.* **14**: 773–780.

Jeswani, L.M., Lal, B.M. and Prakash, S. 1970. Studies on the development of low neurotoxin (β-N-oxalyl-α, β-diaminopropionic acid) lines in *Lathyrus sativus* (Khesari). *Current Sci.* **39**: 518–520.

Kakade, M.L. and Evans, R.J. 1965. Nutritive value of navy beans (*Phaseolus vulgaris*). *Brit. J. Nutr.* **19**: 269–276.

Kunitz, M. 1945. Crystallization of a trypsin inhibitor from soybeans. *Science* **101**: 668–669.

Kienholz, E.W., Jensen, L.S. and McGinnis, J. 1962. Evidence for chick growth inhibitors in several legume seeds. *Poultry Sci.* **41**: 367–371.

King, K., Fourgere, W., Frecald, J., Dominique, G. and Beghin, I.D. 1966. Response of pre-school children to high intake of Haitian cereal-bean mixture. *Arch. Latinoamer. Nutr.* **16**: 53–64.

Koide, T., Tsunasawa, S. and Ikenaka, T. 1973. Studies on soybean trypsin inhibitors. *Eur. J. Biochem.* **32**: 408–416.

Konijn, A.M., Edelstein, S. and Guggenheim, K. 1972. Separation of a thyroid-active fraction from unheated soya bean flour. *J. Sci. Food Agr.* **23**: 549–555.

Konijn, A.M., Gershon, B. and Guggenheim, K. 1973. Further purification and mode of action of a goitrogenic material from soybean flour. *J. Nutr.* **103**: 378–383.

Korte, R. 1972. Heat resistance of phytohemagglutinins in weaning food mixtures containing beans (*Phaseolus vulgaris*). *Ecol. Food Nutr.* **1**: 303–307.

Osborne, T.B. and Mendel, L.B. 1917. The use of soybean as food. *J. Biol. Chem.* **32**: 369–387.

Owen, L.N. 1958. Hair loss and other toxic effects of *Leucaena glauca*. *Vet. Rec.* **70**: 454–456.

Palmer, R., McIntosh, A. and Pusztai, A. 1973. The nutritional evaluation of kidney beans (*Phaseolus vulgaris*). The effect on nutritional value of seed germination and changes in trypsin inhibitor content. *J. Sci. Food Agr.* **24**: 937–944.

Patton, A.R., Wilgus, H.S. Jr. and Hershfield, G.S. 1939. The production of goitre in chickens. *Science* **89**: 162.

Perlman, F. 1969. In: *Toxic Constituents of Plant Foodstuffs*. I.E. Liener, ed. pp. 348–349. Academic Press, New York.

Pusztai, A. and Palmer, R. 1977. Nutritional evaluation of kidney beans (*Phaseolus vulgaris*): the toxic principle. *J. Sci. Food Agr.* **29**: 620–623.

Rackis, J.J. 1974. Biological and physiological factors in soybeans. *J. Am. Oil Chem. Soc.* **51**: 161A–174A.

Rao, S.L.N., Adiga, P.R. and Sarma, P.S. 1964. The isolation and characterization of β-N-oxalyl-α, β-diaminopropionic acid: a neurotoxin from the seeds of *Lathyrus sativus*. *Biochemistry* **3**: 432–437.

Read, J.W. and Haas, L.W. 1938. Studies on the baking quality of flour as affected by certain enzyme actions. V. Further studies concerning potassium bromate and enzyme activity. *Cereal Chem.* **15**: 59–68.

Reis, P.J., Tunks, D.A. and Hegarty, M.P. 1975. Fate of mimosine administered orally to sheep and its effectiveness as a defleecing agent. *Austral. J. Biol. Sci.* **28**: 495–498.

Roy, D.N. 1973. Effect of oral administration of β-(N)-oxalyl-L-alanine (BOAA) with or without *Lathyrus sativus* trypsin inhibitor (LS-T1) in chicks. *Environ. Physiol. Chem.* **3**: 192–195.

Salgarkar, S. and Sohonie, K. 1965. Hemagglutinins of field bean (*Dolichos lablab*). II. Effect of feeding field bean haemagglutinin A on rat growth. *Indian J. Biochem.* **2**: 197–199.

Sharon, N. and Lis, H. 1972. Lectins: Cell-agglutinating and sugar-specific proteins. *Science* **177**: 949–959.

Shone, D.K. 1961. Toxicity of the jack bean. *Rhodesia Agr. J.* **58**: 18–20.

Sreenivasan, V., Mougdal, N.R. and Sarma, P.S. 1957. Goitrogenic agents in food. I. Goitrogenic action of groundnut. *J. Nutr.* **61**: 87–96.

Stillmark, H. 1889. Über Ricin. *Arch. Pharmakol. Inst. Dorpat* **3**: 59–72.

Stob, M. 1973. In "Toxicants Occurring Naturally in Foods", *Nat. Acad. Sci.* 550–557.

Sumner, J.B. 1919. The globulins of the jack bean, *Canavalia ensiformis*. *J. Biol. Chem.* **37**: 137–144.

Sweet, R.M., Wright, H.T., Janin, J., Clothis, C.H. and Blow, D.M. 1974. Crystal structure of the complex of porcine trypsin and soybean trypsin

1979), pigeon pea (Dalal and Quilt 1977) and *Phaseolus vulgaris* (Burleson et al. 1961; Boawn and Brown 1968; Ambler and Brown 1969; Christensen 1972; Wallace et al. 1974). The various reasons attributed for this behaviour are the reduced rate of absorption, dilution effect, less translocation from roots to shoots, etc. The high levels of phosphorus decreased the concentration of copper, iron, strontium and barium in *Phaseolus vulgaris* (Wallace et al. 1974). Tiffin (1970) also observed a decrease in the iron translocation due to the high percentage of phosphorus in the growing medium and this was due to a reduced citrate level in the plant.

POTASSIUM

Potassium is present in plant tissues as free ions or in readily exchangeable combinations and is the most mobile of the essential elements. In peanuts, the potash deficiency was evident when a high proportion of the pods produced only one bold seed (Rachie and Roberts 1974). Potassium was transported mainly in cowpea stems during the vegetative growth but during the reproductive phase this was translocated to the seed (Jacquinot 1967). Tremblay and Bauer (1948) observed that either leaves or the petiole selected at the third node from the top of the pea plant, when it has reached an eight- or nine-node stage, appeared to be the most indicative of its potassium status. Its highest amount has been recorded at 30 days after sowing in black gram (Shanker and Kushwaha 1971).

Bray (1961) suggested that when the soil test shows 25 ppm of available potassium the soybean yield will be approximately 50 per cent of the maximum, whereas, with a soil test of 100 ppm, the soybean yield will be at 97 per cent of the maximum. The highest uptake was 1.7 kg/ha/day, during the 87th to 94th day of growth (Hammond et al. 1951). The extreme potassium concentrations reported in the reference are 0.30 to 5.7 per cent (Ohlrogge 1960). Graves et al. (1978) obtained an increased yield of soybean cultivars with the application of potassium.

CALCIUM

Calcium is important in the nutrition of grain legumes for the formation and development of the seed. In addition, on acid soils, liming is done for the proper growth and nutrient availability to the crops. But in alkaline soils an excess of calcium may reduce the availability of certain micronutrients, viz. iron and zinc. A deficiency of calcium decreased the shelling per cent and resulted in a high proportion of 'pops' or empty pods (Rachie and Roberts 1974) of peanuts. An increased uptake of nitrogen, calcium, sulphur and boron has been noticed with the addition of gypsum (Sankaran et al. 1977). Most of the calcium was taken up in the first 40 days of growth, and afterwards its translocation to pods took place in cowpeas (Jacquinot 1967).

Singh and Dahiya (1976) found a decrease in the concentration and uptake of phosphorus, iron and manganese in pea with the application up to 8 per cent of $CaCO_3$. The highest amount of calcium has been observed at 30 days in black gram (Shanker and Kushwaha 1971). The calcium concentration in the soybean plant at the pre-bloom stage ranges from 0.26 to 2.8 per cent (Ohlrogge 1960). Hammond et al. (1951) observed an initial low rate of calcium uptake which reached a maximum of 2.7 kg/ha/day, after 73–80 days of growth. Calcium is relatively less mobile in plants.

MAGNESIUM

Magnesium concentration ranging from 0.092 to 1.49 per cent has been reported in soybean at the pre-bloom stage (Ohlrogge 1960). Webb et al. (1954) reported a peak uptake rate of almost 5 mg of Mg/plant/day for a five-day period of full bloom with the lower pods developing. Hammond et al. (1951) observed a gradual increase reaching a maximum, in the 73 to 80 days period, of 1.4 kg Mg/ha/day by soybean. The uptake rate at podfilling stage was less, leading to its less content in seed. The maximal uptake of magnesium was found during the last stage of growth and foliar concentrations were slightly higher than in other plant organs of cowpeas (Jacquinot 1967).

The omission of magnesium from the nutrient solution did not retard the phosphorus absorption but did have a significant effect upon the movement and final location of phosphorus in the plants. The magnesium deficient plants contained a higher phosphorus percentage in the vegetative organs than in the seeds as compared to normal plants. Magnesium deficient plants absorbed slightly larger amounts of calcium and potassium on percentage basis (Ohlrogge 1960).

Carter and Hartwig (1962) considered a level of 150 kg/ha of exchangeable soil magnesium to be adequate for field legumes on soils of low to moderate exchange capacity.

SULPHUR

Sulphur is present as the inorganic sulphate ion and in various organic compounds, most of which are present as sulphydryl compounds and their derivatives. Prominent amongst these is protein which contains the sulphur amino acids, methionine, cystine and cysteine. Sulphur is also present in the sulphate ester linkage in sulpholipids and in a number of compounds of vital importance in the intermediary metabolism, including coenzyme A, thiamine and biotin, etc. The normal requirement of peanuts for sulphur has been reported to be 10 to 15 kg/ha (Rachie and Roberts 1974). Sulphur can be taken up by roots as well as by fruits. In the pot experiments on peanuts, uptake of fertilizer sulphur was 77 per cent by roots and 23 per cent by fruits in sierozem sandy soils (Chahal and Virmani 1974). The maximum uptake was during the eighth week, with a daily average of 0.22 kg/ha. Harvested plants contained 11.6 kg S/ha, of which 5.4 was in the haulms, 0.8 in the shells and

5.4 in the kernels (Bromfield 1973).

Yadav and Singh (1970) reported that sulphur application increased the uptake and percentage content of nitrogen, phosphorus, potassium, calcium, magnesium and sulphur in peanuts. A decrease in the content of iron and zinc has been found due to the sulphur application (Dungarwal et al. 1974; Shukla and Prasad 1979).

The appearance of sulphur deficiency in young leaves indicated little mobility of sulphur (Eaton 1935). Sulphur deficiency decreased the growth and concentration of nonsoluble nitrogen in soybean plants (Wooding et al. 1970). Austin (1930) found that the sulphur concentration in soybean at the various stages of growth ranged from 0.22 to 0.32 per cent. Saggar and Dev (1974) studied the effect of sulphur fertilization in soybean. It ranged from 0.205 to 0.352 per cent without sulphur and 0.308 to 0.37 per cent with sulphur at the flowering, and from 0.102 to 0.128 per cent without sulphur and 0.142 to 0.215 per cent with sulphur at the pod formation. Most of the varieties removed more fertilizer sulphur before the flowering stage. The plants removed a higher fraction of sulphur from the soil source than from the applied fertilizer. Whereas the increasing levels of sulphur significantly raised its uptake and increased the amount of sulphur derived from fertilizers (Dhillon and Dev 1978), the utilization of the applied source was found to decrease when the dose was raised to 20 ppm S.

The application of 15 to 45 ppm sulphur to soybean increased the content of manganese in plants but could not affect potassium, calcium, magnesium, zinc and copper concentrations (Robertson and Yuan 1973), whereas the uptake of phosphorus and zinc increased when applied at the rate of 40 and 80 ppm (Kumar 1978).

The molybdenum uptake in pea plants was markedly reduced by the applications of sulphate fertilizer. The application at higher rates also reduced the apparent efficiency of utilization of molybdenum by the plant. Maximum pea yields were associated with plant sulphur content of 200 to 600 ppm of sulphate-sulphur (Reisenauer 1963). A decrease in the phosphorus content in straw as well as in the seed of green gram was noted with the sulphur application at the rate of 5 to 40 ppm sulphur (Aulakh and Pasricha 1977).

IRON

The chlorosis due to iron deficiency has been observed both in peanuts (Rachie and Roberts 1974) and soybean (Ohlrogge 1960). Iron is required for chlorophyll synthesis and respiration. In peanuts, the most effective way to control iron chlorosis was one or two foliar sprays of 10 kg iron chelate per hectare, applied 3 to 6 weeks after planting (Hartzook et al. 1971, 1972).

The iron transport in pea shoots has been found to be dependent upon the metabolic activity of the root cells (Branton and Jacobson 1962). The root cells of iron-starved plants absorbed iron first to meet their own requirements before transferring it to the transpiration stream. In beans the iron was translocated by phloem and xylem and moved particularly to the actively growing young developing tissues (Brown et al. 1965).

Ambler et al. (1971) found that the reduction of iron was the most pronounced between the regions of root elongation and root maturation at both the epidermis and endodermis in soybean roots. The reducing capacity was the greatest in the young lateral diarch roots indicating that these roots contribute significantly to the ability of the plant to take up iron. The movement of iron from the external solution into root cells facilitated the release of citrate to the stem exudates. Iron was chelated with citrate and moved as iron citrate in the plant (Brown and Chaney 1971; Tiffin and Chaney 1974).

The application of iron aggravated zinc deficiency in the sanilac cv. of soybean (Ambler and Brown 1969). Iron not only depressed zinc absorption but also its translocation to shoots (Reddy et al. 1978). Similarly, it decreased the concentration of phosphorus and manganese in peas (Dahiya and Singh 1976) and phosphorus uptake in beans (Biddulph and Woodbridge 1952). The reason for reduced uptake has been assigned to the formation of Fe-P precipitate on the absorbing surfaces of roots and in the conducting tissues of stem and leaf.

ZINC

Viets et al. (1954 a, b) found that zinc is readily redistributed in the plant as the most severe symptoms occur on the older soybean leaves. Deficiency symptoms were associated with about 20 ppm or less zinc in the mature leaves or the total top. Deficient plants showed a light yellowish green colour with stunted growth (Hutton and Fiskell 1963). Yellow colour gradually changed into necrotic spots. The veins were sometimes green but not as distinctive as they were in the case of potassium deficiency. In severe deficiency symptoms complete defoliation occurred prior to maturity with no pod being formed. The zinc deficiency in soybean reduced its concentration, nodule weights and amount of nitrogen fixed (Demeterio et al. 1972). Ohki (1977) observed a reduction in the top and root dry weight, plant height, flower number and branching in soybean. Under zinc stress, however, nodule development was not affected. High manganese concentrations were associated with low zinc in blade 3 of plants. The critical zinc levels in blade 1, 2, 3, 4, and 5 were 15, 15, 14, 17, and 21 ppm of dry weight, respectively. In a study carried by Sudia and Green (1972), 23.9 per cent of the ^{65}Zn contained in the first generation soybean seeds was translocated to second generation. Zinc content in soybean decreased in order root > leaves > seeds > stem > pod (Bilteanu et al. 1974). It was concluded that the optimum conditions for its uptake by soybeans could be given by adjusting the rates of the applied zinc, phosphorus and potassium in relation to the soil moisture content. Soybean was found to be relatively tolerant for higher dose (11.1 kg Zn/ha) and six annual applications did not increase the concentration in seed to level that would preclude its use as animal feed (Martens et al. 1974). Lingle et al. (1963) observed the less uptake and translocation of iron to soybean tops with zinc application. The concentrations of phosphorus, potassium, calcium, magnesium and copper were not significantly affected in beans (Viets et al. 1954b).

Peanut fruits and the main stem contained the highest zinc concentration (Martens et al. 1969). Of the total ^{65}Zn in the whole plant 56.7 per cent has been found to be absorbed by roots and remaining through fruiting organs. It translocated from the vegetative parts to the fruits when absorbed through the roots and vice versa when absorbed through the fruits (Singh 1975). Takkar et al. (1975) concluded that zinc application at the rate of 11, 22, 33, 44 and 55 kg/ha to peanut may provide residual zinc for at least 5, 7, 9, 11 and 13 crops, respectively. The deficiency and toxicity limits for peas and bean had been suggested to be < 20 ppm and > 50 ppm zinc, respectively, in plant tissues (Melton et al. 1970). In *Vicia faba*, most of the ^{65}Zn entering the seed remained bound to the seed coat, the remainder entered the embryo and cotyledons and was concentrated around the abaxial part of the latter, lessening towards the adoxial region (Polar 1970).

MANGANESE

Manganese is a constituent of chlorophyll. In its deficiency soybeans may show light green to yellow mottling between leaf veins, which generally occur when soybean is grown on near neutral soils (Carter and Hartwig 1962). The lower critical manganese contents in 1st (young), 3rd (recently mature) and 5th (old) blade have been found to be 9, 10 and 11 ppm of the tissue sample (Ohki 1976). Blade 3 was suggested for plant analysis.

The toxicity of Mn has also been observed in peanuts (Boyd 1971), beans (Dobereiner 1966) and soybean (Parker et al. 1969; Carter et al. 1975; Ohki 1976). The nodule number and the nitrogen fixation was less. The leaves were wrinkled and had interveinal chlorosis, necrotic spots were associated with decrease in soil pH. Soybean Cv. Hill, Dare, Davis, Delmer and Bragg have been found to be very sensitive, whereas, Hawkeye, Custer, Lee and Amredo were tolerant ones. The critical toxic limits for the maximum growth of soybean have been found to be 100, 160 and 250 ppm in blade 1, 3, and 5 respectively.

Snider (1943) observed the following range of manganese concentration (ppm) in different plant parts of soybean; roots, 12 to 220; stems, 7 to 69; leaves, 98 to 825; hulls, 13 to 161; and beans, 14 to 85. Higher amount of Mn in 3 Cv. of peanut was found in mainstem than the hypocotyl-crown, the stemmy leafy portions of the first and second lateral branches, or the fruit (Martens et al. 1969). The manganese reduced the uptake of zinc due to the reduction both in absorption (Hauf and Schmid 1967) and in translocation (Reddy et al. 1978) in soybean.

MOLYBDENUM

Molybdenum is known to help in nitrogen fixation by legumes. This element becomes increasingly available as the soil pH is raised (Reid et al. 1960). It is needed in the symbiotic N fixation or in nitrate reduction (Anderson and Spencer 1950, Evans 1956).

An increase of 30 to 55 per cent in soybean yields has been recorded with the application of 0.2 kg Mo/ha, which increased the nitrogen and molybdenum contents in soybean (Parker and Harris 1962; Boswell and Anderson 1969) and *Sesbania aculeata* (Chattopodhyay et al. 1976). Widdowson (1965) reported that on saline soils approximately 60 per cent of the applied molybdenum was recovered in bean tops compared with 40 per cent on the more acid soils. Gurley and Giddens (1969) observed that soybean seeds having high molybdenum content (48.4 ppm) supplied the molybdenum needs of soybean grown on the deficient soils. There was little carryover into second generation seed. Molybdenum accumulated in the cotyledons of seed and there was a higher molybdenum content in the seed on the lower part of the plant than on the upper part.

BORON

Boron, a micronutrient, has been found to be essential for the normal cell division and growth and the general metabolism of the plants (Russell 1957). Boron deficiency has been found to be associated with high calcium levels (Reeve and Shive 1944). *Hollow heart*, blackening of the embryo, and occasionally the cracking of the stem are some of the deficiency symptoms in peanut (Rachie and Roberts 1974). The deficiency can be corrected by the light application of borax both in peanut (Harris and Broalmann 1966; Gillier 1969) and soybean (Mederski and Jones 1961), but the treatment should be made with caution as the high levels of this treatments are toxic to the crops (Gopal 1970, 1971). The boron toxicity can be corrected by adding calcium in the growing medium (Berger 1949).

Ohlrogge (1960) reported a wide range of boron in soybean plants from less than 10 to over 2000 ppm, with concentrations between 20 and 100 ppm indicating apparently normal conditions. Although no sharp lines existed between deficiency and adequency, 16 to 20 ppm represented the transition zone. In peanuts, the main stem generally contained higher amounts than the hypocotyl crown, the stemmy and leafy portions of the first and second lateral branches, or the fruit (Martens et al. 1969). An increase in the boron content with boron application has been observed (Sankaran et al. 1977).

COPPER

Severe copper deficiency in soybean caused the stunting of growth, whereas moderate deficiency merely reduced the yield (Carter and Hartwig 1962). The Cu content in 32 soybean hay samples ranged from 4 to 12 ppm

(McHargue 1925). Martens et al. (1969) observed the highest content of Cu in fruit and hypocotyl-crown in the peanut.

CHLORINE

The work on chlorine with respect to legumes is rare. Jackson et al. (1966) observed higher manganese uptake resulting in the toxicity levels in French bean (*Phaseolus vulgaris*) due to the increasing levels of chlorine.

6.2.2. FUNCTIONAL ELEMENTS

SELENIUM

Selenium increased the phosphorus and decreased the nitrogen contents in forage cowpeas (Singh and Singh 1979).

COBALT

Cobalt is essential for animals. Some responses in leguminous crops have been observed to its application, but when the crop is used to feed animals, extreme care should be taken not to apply cobalt in toxic amounts (Carter and Hartwig 1962).

6.3. Plant Biochemical Aspects

6.3.1. MODES OF ACTION OF ESSENTIAL AND FUNCTIONAL ELEMENTS

There is extensive literature on this subject (Evans and Sorger 1966; Bollard and Butler 1966; Clarkson and Hanson 1980) and recently a special review on boron has also appeared (Gupta 1979). The present knowledge about the modes of action of essential elements is summarized in a tabulated form (Table 6.1).

6.3.2. MINERAL COMPOSITION OF LEGUMES

Legumes being rich in protein, are also the source of minerals. The mineral content of various leguminous seeds is given in Table 6.2, as reported by various workers. The legumes are a richer source of calcium and phosphorus than most cereals. Soybeans and some varieties of chick pea contains more than 200 mg of calcium/100 g whereas, in groundnut the values are much below the average. Calcium values vary widely within species in relation to factors such as variety, climate, cultural methods, and mineral content of the soil. The wide variation reported in the literature might also be due to the different analytical procedures employed. Legumes, though they contain a good amount of phosphorus, it is mostly present as phytic acid which may effect the absorption and utilization of calcium, through the precipitation of

TABLE 6.1. Modes of action of ess ential elements

Elements	Main functions or modes of action	References
1	2	3
Covalently bonded		
Phosphorus	Phosphorus is essential in supplying phosphate, a trivalent resonating tetra oxyanion which serves as linkage unit or binding site. Phosphate plays an important role in energy capture, transfer and recovery in biochemical-biophysical processes.	Mildvan and Krisham (1974); Mitchell (1968)
Sulphur	Components of protein as a constituent of amino acids, cystine, cysteine and methionine. Sulphydryl group participates in redox reactions as with lipoic acid or glutathione. Vital cellular constituent Thiamine or biotin, coenzyme and number of plant secondary products.	Miller and Flemion (1973); Ettlinger and Kjaer (1968)
Complexes and Salt		
Potassium	Highly mobile. Functions as a univalent cation activator for a wide variety of important enzymes. Nutritional need centres on (1) enzyme activation, (2) membrane transport processes, (3) anion neutralization, (4) osmotic potential.	Suelter (1970); Evans and Sorger (1966)
Sodium	Essential for halophytes specially certain salt tolerant plants with C_4 metabolism and possibly it has some role in the related crassulacean acid metabolism. Shows marked genetic variability in amount absorbed.	Shomer-Ilan and Waisel (1973); Brownell and Crossland (1972, 1974)
Magnesium	Constituent of chlorophyll, acts as metal activator of enzymes. Magnesium is strongly electropositive divalent cation, readily mobile and coordinates with strongly nucleophilic ligands through largely ionic bonding.	Eichhorn (1973); Portis and Heldt (1976)
Calcium	Complements potassium in maintaining cell organisation, hydration and permeability. Cofactor for some enzyme (Phospholipases, ATPases). Ca^{2+} dependent modulator for protein in peas. Some reports about Ca^{2+}/ hormone interactions in plants.	Nason and McElroy (1963); Anderson and Cormier (1978); Leopold (1977)

1	2	3
Manganese	Metal activator for many enzyme systems. Required in photosystem II involving change in manganese oxidation state.	Radmer and Cheniae (1977); Cheniae (1970) Govindjee et al. (1978)
Iron	Constituent of hemo proteins (cytochromes) and also of non-heme iron proteins involved in photosynthesis nitrogen fixation and respiratory linked dehydrogenases. It plays a role in aconitase.	Hewitt et al. (1976); Price (1968); Villafrancea (1974)
Copper	Bound copper participates in redox enzymes. Component of plastocyanin, an electron carrier protein in photochemical system of photosynthesis.	Ochiai (1977); Boardman (1975)
Zinc	Essential constituent of carbonic anhydrase, several dehydrogenases, superoxide dismutase, RNA and DNA polymerases and other cofactor roles.	Vallee and Wacker (1976)
Molybdenum	Essential constituent of nitrate reductase and nitrogenase.	Spence (1976); Stieffel (1973)
Boron	Involved in carbohydrate transfer and/or in flavonoid. Key element in the action of phytochrome and gravity.	Pollard et al. (1977); Tanada (1978)
Chloride	Essential in the photo-production of oxygen in chloroplast.	Cheniae (1970)
Silicon	Functional element of importance in some species of plants, probably with a structural role in cell walls.	Lewin and Reimann (1969).

insoluble salts in the stomach and duodenum. This needs to be continuously examined. As common food, legumes are a good source of micronutrients, especially iron and other nutrients. As with macro-elements widely varying values for micro-elements are reported in the literature, due to the factors already mentioned. Typical legume intakes in many tropical countries provide a sufficient amount of daily needs. The absorption of iron from different foods is, however, a complicated question.

6.3.3. GENETIC SUSCEPTIBILITY

The importance of genetic variability in the plant kingdom cannot be overemphasized in checking the adaptability of the plant to a specific environment. In many cases large quantities of the element are in the soil but in forms that are not adequately available to the species and varieties used. In these cases tailoring the plant to fit the soil appears more practical than changing the soil to fit the plant. The work carried out in different crops has amply demonstrated that the emphasis should be given to plant soil interactions and the breeding

TABLE 6.2. Mineral composition of legumes

Name of legume	Macroelements, per cent						References
	N	P	K	Ca	Mg	S	
Soybean	—	0.659	1.670	0.275	—	—	Carter and Hopper (1942)
	—	0.780	1.830	0.240	0.310	0.240	Beeson (1941)
Wild soybean	7.920	0.910	1.770	0.380	—	—	
Navy bean	—	0.453	0.821	0.136	0.163	—	Meiners et al. (1976)
Cowpeas	3.2–3.9	0.518	0.838	0.069	0.206	—	Minchin et al. (1981) Meiners et al. (1976)
Chickpea	—	0.354	0.692	0.103	0.092	—	Meiners et al. (1976)
	—	—	0.791–	0.155–	0.140–	—	Tiwari et al. (1977)
	—	—	1.028	0.233	0.168	—	
Peas	—	0.348	1.075	0.035	0.087	—	Meiners et al. (1976)
Lentil	—	0.522	0.862	0.047	0.047	—	Meiners et al. (1976)
Peanut (Kernel)	4.08–	0.25–	0.68–	0.02–	0.09–	0.19–	Freeman et al. (1954)
	4.53	0.66	0.89	0.08	0.34	0.24	

	Microelements, ppm						
	Fe	Zn	Mn	Cu	B	Cl	
Soybean	80	18	28	12	19	300	Beeson (1941)
Navy bean	53	22	10	8	—	—	Meiners et al. (1976)
Cowpeas	80	29	13	9	—	—	Meiners et al. (1976)
Chickpea	58	29	17	9	—	—	Meiners et al. (1976)
	46–80	—	—	8.5–10.7	—	—	Tiwari et al. (1977)
Peas	22	20	11	6	—	—	Meiners et al. (1976)
Lentil	96	32	14	9	—	—	Meiners et al. (1976)
Peanut (Kernel)	18–	17–	8–	7–	26–	Traces	Freeman et al. (1954)
	1000	800	500	300	500		

or selection of plants to fit problem soils. The use of genetic variability in plant breeding to tailor plants to fit problem soils is a way of accelerating the natural evolutionary process.

Genetic control has been used widely by fruit growers by grafting Fe-efficient root-stocks on desirable fruit quality scions to control Fe problems in calcareous soils (Wann 1941; Wutscher et al. 1970).

Weiss (1943) showed that a single gene controlled the differential Fe-efficiency in PI-54619–5–1 (PI) soybeans (Fe-inefficient) and Hawkeye (HA) soybeans (Fe-efficient). Brown et al. (1958) found through reciprocal grafting that the root stocks are responsible for the differential Fe-efficiency. The non-susceptibility of HA cultivar rather than PI cultivar was associated with the more reductive capacity and absorbing capacity of iron (Brown et al. 1961). Even more Fe was absorbed by chlorotic than non-chlorotic plants. When both cultivars were grown together in the same solution PI inhibited ^{59}Fe uptake by

HA, but HA had no influence on ^{59}Fe in the PI, in which the Fe uptake was very low (Wallace et al. 1962). Brown et al. (1967) showed the different Fe-efficiency in isolines of PI and HA soybeans. Elmstrom and Howard (1969, 1970) studied four genotypes and found that HA and A 62–9 (E–9) were the efficient genotypes, whereas, PI and A 62–10 (I–10) were inefficient genotypes. The ability of several genotypes to accumulate Fe, at low levels of Fe, was rated HA > E–9 > I–10 > PI. At the highest level of Fe the rated efficiencies were E–9 > HA > I–10 > PI. A reduction in the accumulation of Fe by efficient plants occurred when both types of genotypes were grown together in the same solution.

The Fe-efficient plant adapts more to an Fe-stress than an Fe-inefficient plant. Several products or biochemical reactions are affected when an Fe-efficient plant responds or adapts to the Fe-stress.

1. Hydrogen ions are excreted from the roots.
2. Reducing compounds are excreted from the root of some plants.
3. The rate of reduction (Fe^{3+} to Fe^{2+}) increases at the root.
4. Organic acids (particularly citrate) increases in the root sap.

Each of these factors is associated with more efficient uptake and utilization of Fe by the plant.

Various investigators have shown that the root excretions were important in regulating mineral uptake by different plant species.

Differential susceptibility to molybdenum deficiency in alfalfa has been reported (Andrew and Miligran 1954; Young and Takahashi 1953).

Species do differ characteristically in their feeding power for zinc. Ellis (1965) had shown that Sanilac was more susceptible to zinc deficiency than Saginaw. Ambler and Brown (1969) reported the causes for differential susceptibility in Sanilac and Saginaw navy beans and found that the iron phosphorus enhanced the zinc deficiency. Polson and Adams (1970) showed that Saginaw navy bean is both more zinc efficient and more tolerant to the excess zinc than Sanilac navy bean. Shukla and Raj (1980) reported a considerable variability among pigeon pea genotypes in their response to zinc deficiency in soil.

Foy et al. (1969) reported differential aluminium tolerance between Perry and Chief soybean varieties.

6.3.4. QUALITY OF LEGUMES AFFECTED BY MINERAL NUTRITION

OIL

The mineral nutrition has been found to increase and decrease the oil content in legumes depending upon the species and varieties of the crop and the application of the nutrient in question. Application of nitrogen alone or with potassium decreased the oil content in groundnut (Chopra and Kanwar 1966; Yadava and Singh 1970), but the oil content increased with sulphur application (Laurence and Gibbons 1976). In soybean, application of nitrogen alone (Sadat and Firuzeh 1976) or in combination with phosphorus (Agarwal and Narang 1975) increased the oil content in seed. Sulphur application also had a positive effect (Kumar 1978). The foliar spray of boron and manganese could not increase the oil content (Boswell and Worthington 1971), whereas that of molybdenum decreased oil content in soybeans (Parker and Harris 1962; Boswell and Anderson 1969). Possibly it is due to the increased synthesis of proteins with molybdenum application, as the oil and protein contents are inversely related.

PROTEIN

Legumes are the rich sources of proteins among plants, and the quality of the legume is often a function of the protein content, depending upon their end uses. The application of nitrogen, phosphorus alone or in combination with potassium increased the protein content in pea (Trevino and Murray 1975; Selyutin 1976; Shukla et al. 1977–78), and in the deficiency of nitrogen, phosphorus, sulphur and magnesium, protein content decreased (Klein and Jager 1978). The deficiency of calcium and potassium did not affect the protein content. In green gram, the application of both phosphorus and sulphur sometimes decreased the seed protein (Aulakh and Pasricha 1977). The application of nitrogen and phosphorus had a positive effect on the crude protein in black gram (Rajendran et al. 1974).

The application of nitrogen and phosphorus alone (Sadat and Firuzeh 1976; Kapoor and Gupta 1977a, b) or in combination (Agarwal and Narang 1975) increased the protein in soybean. When trace-elements have also been applied, the increase was observed without NPK also (Baia 1977). Molybdenum application also had a positive effect (Parker and Harris 1962; Boswell and Anderson 1969). Chopra and Kanwar (1966) reported more accumulation of protein at the expense of oil in groundnut with nitrogen application, but the situation reversed with the application of sulphur (Laurence and Gibbons 1976). Sulphur deficient plants of pea, chick pea, groundnut and black gram had less protein than the plants grown on a well-supplied sulphur medium (Dube and Misra 1970).

AMINO ACIDS

Sulphur-containing amino acids, viz. cysteine, cystine and methionine have been found to increase with sulphur application in pea (Evans et al. 1947; Rao and Das 1967); in green gram (Arora and Luthra 1970, 1971); in cowpea (Evans et al. 1977). Haghiri (1966) found higher content of amino acids in plants grown under conditions where the macro- and micro- nutrients were present in sufficient amounts and proper balance.

6.3.5. ENZYME ACTIVITIES

Elements are generally associated with the enzymatic systems in the plant body
and they, to some extent, regulate the activities of several enzymes. Evans
(1963) observed an increase in the activity of pyruvate kinase with the supply
of potassium to its deficient pea plants. The addition of ammonium or sodium
substituted completely or partially for that of potassium. Increased activities
of hexokinase, pyruvate kinase, glucose-6-phosphate dehydrogenase, 6-phos-
phogluconase, dehydrogenase and transketolase have been recorded with nitrate incubation in pea roots (Sarkissan and Fowler 1974). The addition of
ammonium ions stimulated the activities of nitrate reductase and NADH-glu-
tamate dehydrogenase but not of glucose-6-phosphate dehydrogenase (Sihag
et al. 1979).

The activity of glutamate dehydrogenase decreased in nitrogen and sulphur deficiency, but increased in calcium and magnesium deficiency in pea
(Klein and Jager 1978), whereas, no effect was observed in the case of phosphorus and potassium deficiency. Verma and Pant (1978) found an increased
activity of ribonuclease due to potassium deficiency in pigeon pea leaves, but
it could not be interpreted as the hunger sign of the crop for this element.

Dungarwal et al. (1974) observed an increase in the activity of peroxidase
and catalase with sulphur application to the peanut. Del-Rio et al. (1978) reported an increase in the peroxidase activity with both iron deficiency and excess and a decrease with the normal iron supply to the pea. An inverse relationship was observed between peroxidase and catalase activities.

Acid phosphatase activity increased in zinc deficient bean leaves (Jyung
et al. 1972) and the formation of ribulose diphosphate carboxylase increased
with zinc application (Jyung and Camp 1976). The activity of soluble starch
synthetase (ADP-glucose: starch L-4-glucosyl-transferase) decreased due to
the zinc deficiency in navy beans (Jyung et al. 1975).

6.4. Conclusions

The need to understand the whole process and its interaction with growth remains as great as it ever was particularly in view of the demands of an increasing population. Legumes being the poor man's protein and essential for
health, their yield needs to be increased by proper fertilization. Recent economic developments along with the realization that the most significant contribution to the world food production must come from crops grown in relatively nutrient poor soils, have emphasized an entirely different kind of efficiency—that which is measured by maximum production from limited nutrient
inputs. Efforts are necessary to increase their efficiency for mineral utilization
by genetic manipulation which seems to be the only alternative. From the
viewpoint of crop production, there remains many unanswered questions on

the modes of action of essential and functional elements and their states of combination and pathways of assimilation.

Secondly, the legumes in general are deficient in sulphur containing amino acids when compared to that of egg protein. These amino acids can be increased to some extent by sulphur fertilization. In many parts of the world where milk protein or animal protein is neither available nor cheap, the simple expedient of supplementing the indigenous cereals with legume proteins could be an important first step in improving the nutritional status of the native population.

References

Adepetu, J.A. and Akapa, L.K. 1977. Root growth and nutrient uptake characteristics of some cowpeas varieties. *Agron. J.* **69**: 940–943.

Agarwal, S.K. and Narang, R.S. 1975. Effect of levels of phosphorus and nitrogen on soybean varieties. *HAU J. Res.* **5**: 303–308.

Ambler, J.E. and Brown, J.C. 1969. Cause of differential susceptibility to zinc deficiency in two varieties of Navy beans (*Phaseolus vulgaris* L.). *Agron. J.* **61**: 41–43.

Ambler, J.E., Brown, J.C. and Gauch, H.G. 1971. Sites of iron reduction in soybean plants. *Agron. J.* **63**: 95–97.

Anderson, A.J. and Spencer, D. 1950. Molybdenum in nitrogen metabolism of legumes and non legumes. *Aust. J. Sci. Res. Ser. B.* **3**: 414–430.

Anderson, J.M. and Cormier, M.J. 1978. Calcium dependent regulator of NAD Kinase in higher plants. *Biochem. Biophys. Res. Commun.* **84**: 595–602.

Andrew, W.D. and Miligran, R.T. 1954. Different molybdenum requirements of Medics and subterranean clover on a red brown soil at Wagga, New-South Wales. *J. Aust. Inst. Agri. Sci.* **20**: 123–124.

Arora, S.K. and Luthra, Y.P. 1970. Metabolism of sulphur containing amino acids in *Phaseolus aureus* Linn. *Z. Pfl. Ernahr. Bodenk.* **126**: 151–158.

Arora, S.K. and Luthra, Y.P. 1971. Relationship between sulphur content of leaf with methionine, cystine and cysteine contents on the seeds of *Phaseolus aureus* L. as affected by S, P, and N application. *Plant Soil* **34**: 91–94.

Aulakh, M.S. and Pasricha, N.S. 1977. Interaction effect of sulphur and phosphorus on growth and nutrient content of moong (*Phaseolus aureus* L.) *Plant Soil* **47**: 341–350.

Austin, R.H. 1930. Effect of soil type and fertilizer treatment on the composition of soybean plant. *J. Am. Soc. Agron.* **22**: 136–156.

Baia, V. 1977. Possibilities of controlling protein content of soybean by fertiliser application. *Lucr. Stiinti. Inst. Agron. Timisoara, Agron.* **14**: 46–48.

Beeson, K.C. 1941. USDA Misc. Publ. 369.

Berger, K.C. 1949. Boron in soils and crops. *Adv. Agron.* **1**: 321–351.

Biddulph, O. and Woodbridge, C.G. 1952. The uptake of phosphorus by bean plants with particular reference to the effects of iron. *Plant Physiol.* **27**: 431–444.

Bilteanu, G., Dumitru, G., Paun, R. and Nica, Q. 1974. Effect of phosphorus, potassium and water on the uptake of zinc in soybean on red-brown forest soil. *Lucr. Stiinti. Inst. Agron. 'N. Balcescu' A.* **15**: 155–161.

Blondel, D. 1969. *Agron. Trop. (Paris)* **24**: 864–887.

Boardman, N.K. 1975. Trace elements in photosynthesis. In: *Trace Elements in Soil-Plant-Animal Systems.* D.J.D. Nicholas and A.R. Egem, eds. pp. 199–212. Academic Press, New York.

Boawn, L.C. and Brown, J.C. 1968. Further evidence for a P-Zn imbalance in plants. *Proc. Soil Sci. Soc. Amer.* **32**: 94–97.

Bollard, E.G. and Butler, G.W. 1966. Mineral nutrition of plants. *Ann. Rev. Plant Physiol.* **17**: 77–112.

Boswell, F.C. and Anderson, O.E. 1969. Effect of time of application and soybean yield on nitrogen, oil and molybdenum contents. *Agron. J.* **61**: 58–60.

Boswell, F.C. and Worthington, R.E. 1971. Boron and Manganese: effects on protein, oil and fatty acid composition of oil in soybean. *J. Agri. Fd. Chem.* **19**: 765–768.

Boyd, H.W. 1971. Manganese toxicity to peanut in autoclaved soil. *Plant Soil* **35**: 133–144.

Branton, D. and Jacobson, L. 1962. Iron transport in pea plants. *Plant Physiol.* **37**: 539–545.

Bray, R.H. 1961. Better crops with plant food. *Physiol.* **45 (3)**: 18–19, 25–27.

Bromfield, A.R. 1973. Uptake of sulfur and other nutrients by groundnuts (*Arachis hypogaea*) in Northern Nigeria. *Exp. Agri.* **9**: 55–58.

Brown, A.L., Yamaguchi, S. and Leaf-Diaz, J., 1965. Evidence for translocation of iron in plants. *Plant Physiol.* **40**: 35–38.

Brown, J.C. and Chaney, R.L. 1971. Effect of iron on the transport of citrate into the xylem of soybeans and tomatoes. *Plant Physiol.* **47**: 836–840.

Brown, J.C., Holmes, R.S. and Tiffin, L.O. 1958. Iron chlorosis in soybeans as related to the genotype of root stalk. *Soil Sci.* **86**: 75–82.

Brown, J.C., Holmes, R.S. and Tiffin, L.O. 1961. Iron chlorosis in soybeans as related to the genotype of root stalk: 3. Chlorosis susceptibility and reductive capacity at root. *Soil Sci.* **91**: 127–132.

Brown, J.C., Weber, C.R. and Caldwell, B.E. 1967. Efficient and inefficient use of iron by two soybean genotypes and their isolines. *Agron. J.* **59**: 459–462.

Brownell, P.F. and Crossland, C.J. 1972. The requirement of sodium as a micronutrient by species having the Cu decarboxylic acid photosynthetic pathway. *Plant Physiol.* **49**: 794–797.

Brownell, P.F. and Crossland, C.J. 1974. Growth responses to sodium by *Bryophyllum tubiflorum* under conditions inducing crassulacean acid metabolism. *Plant Physiol.* **54**: 416–417.

Burleson, C.A., Dacus, A.D. and Gerard, C.J. 1961. The effect of phosphorus fertilization on the zinc nutrition of several irrigated crops. *Proc. Soil Sci. Soc. Amer.* **25**: 365–368.

Carter, J.L. and Hartwig, E.E. 1962. The management of soybeans. *Adv. Agron.* **14**: 359–412.

Carter, J.L. and Hopper, T.H. 1942. Influence of variety, environment, and fertility level on the chemical composition of soybean seed. *USDA Tech. Bull.* 787.

Carter, O.G., Rose, I.A. and Reading, P.F. 1975. Variation in susceptibility to manganese toxicity in 30 soybean genotypes. *Crop Sci.* **15**: 730–732.

Chahal, R.S. and Virmani, S.M. 1974. Uptake and translocation of nutrients in groundnut (*Arachis hypogaea* L.) III. Sulphur. *Oleagineux* **29**: 415–417.

Chattopadhyay, N.C., Das, S. and Sarkar, A.K. 1976. Effect of phosphorus and molybdenum on dry matter production and nitrogen content of *dhaincha* crop (*Sesbania aculeata*) for green manuring. *Indian Agric.* **20**: 65–66.

Cheniae, G.M. 1970. Photosystem II and O_2 evolution. *Ann. Rev. Plant Physiol.* **21**: 467–498.

Chopra, S.L. and Kanwar, J.S. 1966. Effect of sulphur fertilization on chemical composition and nutrient uptake by legumes. *J. Indian Soc. Soil Sci.* **14**: 69–76.

Christensen, N.W. 1972. A new hypothesis to explain phosphorus induced zinc deficiencies. *Diss. Abstr. Int.* **32B**: 43–48.

Clarkson, D.T. and Hanson, J.B. 1980. The mineral nutrition of higher plants. *Ann. Rev. Plant Physiol.* **31**: 239–298.

Cooper, D.R., Hill-Cottingham, D.G. and Lloyd Jones, C.P. 1976. Absorption and redistribution of nitrogen during growth and development of field bean (*Vicia faba*). *Physiologia Pl.* **38**: 313–318.

Dahiya, S.S., and Singh, M. 1976. Effect of salinity, alkalinity and iron application on the availability of iron, manganese, phosphorus and sodium in pea (*Pisum sativum* L.). *Crop. Plant Soil.* **44**: 697–702.

Dalal, R.C. and Quilt, P. 1977. Effects of N.P. liming and Mo on nutrition and grain yield of pigeon pea. *Agron. J.* **69**: 854–857.

Del-Rio, L.A., Gomes, M., Yanez, J., Leal, A. and Gorge, J.L. 1978. Iron deficiency in pea plants. Effect on catalase, peroxidase, chlorophyll and proteins of leaves. *Plant Soil* **49**: 343–353.

Demeterio, J.L., Ellis, R.J. and Paulsen, G.M. 1972. Nodulation and nitrogen fixation by two soybean varieties as affected by phosphorus and zinc nutrition. *Agron. J.* **64**: 566–568.

Dhillon, N.S. and Dev, G. 1978. Effect of elemental sulphur on soybean (*Glycine max* L. Merill). *J. Indian Soc. Soil Sci.* **26**: 55–57.

Dobereiner, J. 1966. Manganese toxicity effects on nodulation and nitrogen fixation of beans (*Phaseolus vulgaris*) in acid soils. *Plant Soil* **24**: 153–166.

Dube, S.D. and Misra, P.H. 1970. Effect of sulphur deficiency on growth, yield and quality in some of the important leguminous crops. *J. Indian Soc. Soil Sci.* **18**: 375–378.

Dungarwal, H.S., Mathur, P.N. and Singh, H.G. 1974. Effect of foliar spray of sulphuric acid in the prevention of chlorosis in peanut (*Arachis hypogaea* L.). *Comm. Soil Sci. Plant Anal.* **5**: 331–339.

Eaton, S.V. 1935. Influence of sulphur deficiency on the metabolism of the soybean. *Botan. Gaz.* **97**: 68–100.

Egli, D.B., Leggett, J.E. and Duncan, W.G. 1978. Influence of N stress on leaf senescence and N redistribution in soybeans. *Agron. J.* **70**: 43–47.

Eichhorn, G.L., ed. *Inorganic Biochemistry*, Vols. 1 and 2. New York, Elsevier, 1973.

Ellis, B.G. 1965. Response and susceptibility in zinc deficiency—a symposium. *Crop Soil* **18** (1): 10–13.

Elmstrom, G.W. and Howard, F.D. 1969. Iron accumulation, root peroxidase activity, and varietal interactions in soybean genotypes that differ in iron nutrition. *Plant Physiol.* **44**: 1108–1114.

Elmstrom, G.W. and Howard, F.D. 1970. Promotion and inhibition of iron accumulation in soybean plants. *Plant Physiol.* **45**: 327–329.

Ettlinger, M.G. and Kjaer, A. 1968. Sulfur compounds in plants. *Adv. Phytochem.* **1**: 59–144.

Evans, H.J. 1956. Role of molybdenum in plant nutrition. *Soil Sci.* **81**: 199–208.

Evans, H.J. 1963. Effect of potassium and other univalent cations on activity of pyruvate kinase in *Pisum sativum*. *Plant Physiol.* **38**: 397–402.

Evans, H.J. and Sorger, G.J. 1966. Role of mineral elements with emphasis on the univalent cations. *Ann. Rev. Plant Physiol.* **17**: 47–76.

Evans, I.M., Boulter, D., Fox, R.L. and Kang, B.T. 1977. Effect of sulphur fertilizers on content of sulpho-amino acids in seeds of cowpea (*Vigna unguiculata*). *J. Sci. Fd. Agric.* **28**: 161–166.

Evans, R.J., John, J.L. St., Craven, P.M., Haddock, J.L., Wells, D.G. and Swenson, S.P. 1947. Some factors influencing the protein, cysteine and methionine content of dry peas. *Cereal Chem.* **24**: 150–156.

Fageria, N.K. 1977. Effect of phosphatic fertilization on growth and mineral composition of pea plants (*Pisum sativum* L.). *Agrochimica* **21**: 75–78.

Foy, C.D., Fleming, A.L. and Armiger, W.H. 1969. Aluminium tolerance of soybean varieties in relation to calcium nutrition. *Agron. J.* **61**: 505–511.

Freeman, A.F., Morris, N.J. and Willich, R.K. 1954. Peanut butter. *U.S. Dept. Agr.* AIC-370.

Gillier, P. 1969. Secondary effect of drought on groundnuts. *Oleagineux* **24**: 79–81.

Gopal, N.H. 1970. The distribution of normal and toxic amounts of boron in groundnut. *Indian J. Plant Physiol.* **13**: 92–98.

Gopal, N.H. 1971. Effect of boron toxicity on iron status on groundnut. *Indian J. Exp. Biol.* **9**: 524–526.

Govindjee, Wyndrzynski, T. and Marks, S.B. 1978. Manganese and Chloride: their roles in photosynthesis. In: *Photosynthetic Oxygen Evolution.* H. Metzner, ed. pp. 321–344. Academic, New York.

Graves, C.R., McCutchen, T. and Freeland, R. 1978. Soybean varieties responses to potassium. *Tennessee Farm and Home Science* **105**: 6–7.

Gupta, U.C. 1979. Boron nutrition of crops. *Adv. Agron.* **31**: 273–307.

Gurley, W.H. and Giddens, J. 1969. Factors affecting uptake, yield response and carry-over of molybdenum in soybean seed. *Agron. J.* **61**: 7–9.

Hammond, L.C., Black, C.A. and Norman, A.G. 1951. Nutrient uptake by soybean on two Iowa soils. *Iowa Agr. Expt. Sta. Res. Bull. No. 384:* 463–512.

Haghiri, F. 1966. Effect of macro-elements on amino acid composition of soybean plants. *Agron. J.* **58**: 609–612.

Harris, H.C. and Broalmann, J.B. 1966. Comparison of calcium and boron deficiencies of the peanut I. Physiological and yield differences. *Agron. J.* **58**: 575–582.

Hartzook, A., Fitchman, M. and Karstadt, D. 1971. The treatment of iron deficiency in peanuts cultivated in basic and calcareous soils. *Oleagineux* **26**: 391–395.

Hartzook, A., Karstadt, D. and Feldman, S. 1972. Varietal differences in iron absorption-efficiency of groundnuts cultivated on calcareous soils. *SABRO Newsl.* **4**: 91–94.

Hauf, L.R. and Schmid, W.E. 1967. Uptake and translocation of zinc by intact plants. *Plant Soil* **27**: 249–260.

Hewitt, E.J., Hucklesby, D.P. and Notton, B.A. Nitrate metabolism. In: *Plant Biochemistry.* J. Bonner and J.E. Varner, eds. pp. 633–681. Academic, New York, 1976.

Hutton, C.E. and Fiskell, J.G.A. 1963. Zinc response by soybeans and wheat on heavily limed soils in Western Florida. *Proc. Soil Crop Sci. Soc. Flo.* **23**: 61–70.

Jackson, T.L., Westermann, D.T. and Moore, D.P. 1966. The effect of chloride and lime on the manganese uptake by bush beans and sweet corn. *Proc. Soil Sci. Soc. Amer.* **30**: 70–73.

Jacquinot, L. 1967. *Agron. Trop.* (Paris) **22**: 611–615.

Jyung, W.H. and Camp, M.E. 1976. The effect of zinc on the formation of ribulose diphosphate carboxylase in *Phaseolus vulgaris*. *Physiologia Pl.* **36**: 350–355.

Jyung, W.H., Camp, M.E., Polson, D.E., Adams, M.W. and Wittwar, S.H. 1972. Differential response of two bean varieties to zinc as revealed by electrophoretic protein patterns. *Crop Sci.* **12**: 26–29.

Jyung, W.H., Ehmann, A., Schlender, K.K. and Scala, J. 1975. Zinc nutrition and starch metabolism in *Phaseolus vulgaris* L. *Plant Physiol.* **55**: 414–420.

Kapoor, A.C. and Gupta, Y.P. 1977a. Effect of P fertilization on P constituents of soybean. *J. Agri. Fd. Chem.* **25**: 670–673.

Kapoor, A.C. and Gupta, Y.P. 1977b. Changes in proteins and amino acids in developing soybean seed and effect of phosphorus nutrition. *J. Sci. Fd. Agric.* **28**: 113–120.

Keefer, R.F. and Singh, R.N. 1969. Residual effect of zinc sources on plant growth and on phosphorus-zinc interaction in soybeans. *Proc. West Virginia Acad. Sci.* **41**: 27–38.

Klein, H. and Jager, H.J. 1978. Some aspects of relation between nitrogen metabolism in *Pisum sativum* and mineral nutrition. *Plant Soil* **50**: 25–35.

Kumar, V. 1978. Availability of sulphur to plants as affected by sulphur, phosphorus, zinc and molybdenum application. Ph.D. thesis, Haryana Agricultural University, Hissar, India.

Laurence, R.C.N. and Gibbons, R.W. 1976. Changes in yield, protein, oil and maturity of groundnut cultivars with the application of sulphur fertilizers and fungicides. *J. Agric. Sci. Camb.* **86**: 245–250.

Leopold, A.C. 1977. Modification of growth regulatory action with inorganic solutes. In: *Plant Growth Regulators, Chemical Activity, Plant Responses and Economic Potential*. C.A. Stutte, ed. pp. 33–41. *Adv. Chem. Ser. 159*, Amer. Chem. Soc., Washington.

Lewin, J. and Reimann, B.E.F. 1969. Silicon and plant growth. *Ann. Rev. Plant Physiol.* **20**: 289–304.

Lingle, J.C., Tiffin, L.O. and Brown, J.C. 1963. Iron uptake and transport in soybeans as influenced by other cations. *Plant Physiol.* **38**: 71–76.

Maftoun, M. and Bassiri, A. 1975. Effect of phosphorus and Ryzelan on the growth and mineral composition of chickpeas. *Agron. J.* **67**: 556–559.

Martens, D.C., Hallock, D.L. and Alexander, M.W. 1969. Nutrient distribution during development of three market type II: B, Cu, Mn and Zn contents. *Agron. J.* **61**: 86–88.

Martens, D.C., Carter, M.T. and Jones, G.D. 1974. Response of soybeans following six annual applications of various levels of boron, copper and zinc. *Agron. J.* **66**: 82–84.

McHargue, J.S. 1925. *J. Agric. Res.* **30**: 193–196.

Mederski, H.J. and Jones, J.B. 1961. Use caution when adding trace nutrients. *Crop Soils* **13** (8): 23.

Meiners, C.R., Derise, N.L., Lau, H.C., Crews, M.G., Ritchey, S.J. and Murphy, E.W. 1976. The content of nine mineral elements in raw and cooked mature dry legumes. *J. Agri. Fd. Chem.* **24**: 1126–1130.

Melton, J.R., Ellis, B.G. and Doll, E.C. 1970. Zinc, phosphorus and lime interactions with yield and zinc uptake by *Phaseolus vulgaris*. *Proc. Soil Sci. Soc. Amer.* **34**: 91–93.

Mengel, K. and Kirkby, E.A. 1978. Principles of plant-nutrition. Bern; *Ist potash Inst.* pp. 593.

Mildvan, A.S. and Krisham, C.M. 1974. The role of divalent cations in the mechanism of enzyme catalysed phosphoryl and nucleotidyl transfer reactions. *Struct. Bonding* **20**: 1–21.

Miller, L.P. and Flemion, F. 1973. The role of minerals in phyto-chemistry. *Phytochemistry* **3**: 12–19. New York: Van Nostrand Reinholt.

Minchin, F.R., Summerfield, R.J. and Neves, M.C.P. 1981. Nitrogen nutrition of cowpeas (*Vigna unguiculata*): effects of timing of inorganic nitrogen applications on nodulation, plant growth and seed yield. *Trop. Agric.* (*Trinidad*) **58**: 1–12.

Mitchell, P. 1968. Chemiosmotic Coupling and Energy Transduction. *Glynn.* 111, Bodmin.

Nason, A. and McElroy, W.D. 1963. Modes of action of the essential mineral elements. In: *Plant Physiology.* F.C. Steward, ed. Vol. 3, pp. 451–536.

Ochiai, E.L. *Bioinorganic Chemistry.* pp. 515. Allyn and Bacon, Boston, 1977.

Oghoghorie, C.G.O. and Pate, J.S. 1972. Exploration of the nitrogen transport system of a nodulated legume using ^{15}N. *Planta* **104**: 35–49.

Ohki, K. 1976. Manganese, deficiency and toxicity levels for 'Bragg' soybeans. *Agron. J.* **68**: 861–864.

Ohki, K. 1977. Critical Zn levels related to early growth and development in determinate soybeans. *Agron. J.* **69**: 969–974.

Ohlrogge, A.J. 1960. Mineral nutrition of soybeans. *Adv. Agron.* **12**: 230–263.

Parker, M.B. and Harris, H.B. 1962. Soybean response to Mo and lime and the relationship between yield and chemical composition. *Agron. J.* **54**: 480–483.

Parker, M.B., Harris, H.B., Morris, H.D. and Perkins, H.F. 1969. Manganese toxicity of soybeans as related to soil and fertility treatments. *Agron. J.* **61**: 515–518.

Paulsen, G.M. and Rotini, O.A. 1968. Phosphorus-zinc interaction in two soybean varieties differing in sensitivity to phosphorus nutrition. *Proc. Soil Sci. Soc. Amer.* **32**: 73–76.

Polar, E. 1970. The distribution of ^{65}Zn in the cotyledons of *Vicia faba* and its translocation during the growth and maturation of the plant. *Plant Soil* **32**: 1–17.

Pollard, A.S., Parr, A.J. and Loughman, B.C. 1977. Boron in relation to membrane function in higher plants. *J. Exp. Bot.* **28**: 831–841.

Polson, D.E. and Adams, M.W. 1970. Differential response of navy beans (*Phaseolus vulgaris* L.) to zinc. I. Differential growth and elemental composition at excessive zinc levels. *Agron. J.* **62**: 557–560.

Portis, A.R. and Heldt, H.W. 1976. Light dependent changes in the Mg++ concentration in the stroma in relation to the Mg++ dependency of CO_2 fixation in intact chloroplasts. *Biochim. Biophys. Acta* **449**: 434–446.

Price, C.A. 1968. Iron compounds and plant nutrition. *Ann. Rev. Plant Physiol.* **19**: 239–248.

Quintana, R.U. 1972. Zinc studies in peanuts (*Arachis hypogaea* L.) *Distt. Abstr. Int. B.* **32** (8) 4357.

Rachie, K.O. and Roberts, L.M. 1974. Grain legumes of the lowland tropics. *Adv. Agron.* **26**: 2–132.

Radmer, R. and Cheniae, G.M. 1977. Mechanisms of oxygen evolution. In: *Primary Processes of Photo-synthesis*. J. Barber, ed. pp. 303–347. Elsevier/North Holland, New York.

Rajendran, K. and Krishnamoorthy, K.K. 1975. Uptake of nutrients by black-gram. *Madras Agric. J.* **62**: 376–379.

Rajendran, K., Sivappah, A.N. and Krishnamoorthy, K.K. 1974. Effect of fertilization on yield and nutrient concentration of black gram (*Phaseolus mungo* L.). *Madras Agric. J.* **61**: 447–450.

Rao, K.B. and Das, N.B. 1967. Effect of sulphur on methionine and cystine content of pea. *Indian J. Agric. Sci.* **37**: 390–395.

Reddy, K.R., Saxena, M.C. and Pal, U.R. 1978. Effect of iron and manganese on ^{65}Zn absorption and translocation in soybean seedlings. *Plant Soil* **49**: 409–415.

Reeve, E. and Shive, J.W. 1944. Potassium-boron and calcium-boron relationships in plant nutrition. *Soil Sci.* **57**: 1–14.

Reid, P.H., Kamprath, E.J. and Evans, H.J. 1960. *Soybean Digest* **20** (7): 18.

Reisenauer, H.M. 1963. The effects of sulphur on the absorption and utilization of molybdenum by peas. *Proc. Soil Sci. Soc. Amer.* **27**: 553–555.

Robertson, W.K. and Yuan, T.L. 1973. Effect of P and S on yields and S contents of soybean (*Glycine max* L.) and white clover (*Trifolium repens* L.). *Proc. Soil Crop Sci. Soc. Fla.* **32**: 152–154.

Russell, D.A. 1957. In: *Soil: The Yearbook of Agriculture*. A. Stefferud, ed. pp. 121–128. USDA, Washington.

Sadat, N. and Firuzeh, P. 1976. The influence of N fertilization and sowing time on the yield, oil and protein content of soybean in a semi-arid site in Iran. *Landwirtschaftliche Forschung* **29**: 170–176.

Saggar, S. and Dev, G. 1974. Uptake of sulphate by different varieties of soybean. *Indian J. Agric. Sci.* **44**: 345–349.

Sanchez, A.C., Escobar, R.N. and Echegaray, A.A. 1977. Effect of nitrogen, phosphorus, molybdenum, cobalt, and iron fertilizers and two strains of inoculum (*Rhizobium phaseoli*) on the nodulation, nitrogen accumulation, and yield of common bean (*Phaseolus vulgaris* L.), *Agrociencia* (*Mexico*) **27**: 79–94.

Sankaran, N., Sennaian, P. and Morachan, Y.B. 1977. Effect of forms and levels of calcium and levels of boron on the uptake of nutrients and quality of groundnut. *Madras Agric. J.* **64**: 384–388.

Sarkissan, G.S. and Fowler, M.W. 1974. Interrelationship between nitrate assimilation and carbohydrate metabolism in plant roots. *Planta* **119**: 335–349.

Selyutin, A.F. 1976. Effect of different rates and proportions of mineral fertilizers on seed yield and quality of peas. Agrotekhnikai biologiya selskhozyaistvennykh Kultur. *Vypusk 3.* Utyanovsk, USSR : 49–51.

Shanker, H. and Kushwaha, R.P.S. 1971. Chemical composition of urd plants (*Phaseolus mungo*) at various stages of growth. *Indian J. Agric. Res.* **5**: 79–82.

Sheldrake, A.R. and Narayanan, A. 1979. Growth, development and nutrient uptake in pigeon peas (*Cajanus cajan*). *J. Agric. Sci. UK* **92**: 513–526.

Shomer-Ilan, A. and Waisel, Y. 1973. The effect of sodium chloride on the balance between C_3 and C_4 carbon fixation pathways. *Physiol. Pl.* **29**: 190–193.

Shukla, D.N., Singh, S. and Subrahmanyam, T. 1977/78. Effect of phosphorus, boron, molybdenum and inoculation on yield and quality of pea. (*Pisum sativum* L.) Var. T-163. *J. Sci. Res. BHU* **28**: 27–30.

Shukla, U.C. and Prasad, K.G. 1979. Sulphur-zinc interaction in groundnut. *J. Indian Soc. Soil Sci.* **27**: 60–64.

Shukla, U.C. and Raj, H. 1980. Zinc response in pigeon pea as influenced by genotypic variability. *Plant Soil* **57**: 327–333.

Sihag, R.K., Guha-Mukherjee, S. and Sopory, S.K. 1979. Effect of ammonium, sucrose and light on the regulation of nitrate reductase level in *Pisum sativum*. *Physiologia Pl.* **45**: 281–287.

Singh, M. and Dahiya, S.S. 1976. Effect of calcium carbonate and iron on the availability and uptake of iron, manganese, phosphorus and calcium in pea (*Pisum sativum* L.). *Plant Soil* **44**: 511–520.

Singh, M. and Singh, N. 1979. The effect of forms of selenium on the accumulation of selenium, sulphur, and forms of nitrogen and phosphorus in forage cowpea (*Vigna sinensis*). *Soil Sci.* **127**: 264–269.

Singh, S.P. 1975. Nutrioperiodism in different varieties of groundnut with respect to macronutrients and the uptake and translocation of zinc. M.Sc. thesis, Haryana Agricultural University, Hissar.

Snider, H.J. 1943. *Soil Sci.* **56**: 187–195.

Spence, J.T. 1976. Reactions of molybdenum coordination compounds as models for biological systems. In: *Metal Ions in Biological Systems.* H.

Sigel, ed. 1973–1978. Vols. 1–7. Dekker, New York.

Stieffel, E.J. 1973. Proposed molecular mechanism for the action of molybdenum in enzymes: coupled proton and electron transfer. *Proc. Natl. Acad. Sci. USA* **70**: 988–992.

Streeter, J.G. 1978. Effect of N starvation of soybean plants at various stages of growth on seed yield and N concentration of plant parts at maturity. *Agron. J.* **70**: 74–76.

Sudia, W.T. and Green, D.G. 1972. The translocation of ^{65}Zn and ^{134}Cs between seed generations in soybean (*Glycine max* L. Mer.). *Plant Soil* **37**: 695–697.

Suelter, C.H. 1970. Enzymes activated by monovalent cations. *Science* **148**: 789–795.

Takkar, P.N., Mann, M.S. and Randhawa, N.S. 1975. Effect of direct and residual available zinc on yield, zinc concentration and its uptake by wheat and groundnut crops. *J. Indian Soc. Soil Sci.* **23**: 91–95.

Tanada, T. 1978. Boron—Key element in the actions of phytochrome and gravity. *Planta* **143**: 109–111.

Tiffin, L.O. 1970. Translocation of iron citrate and phosphorus in xylem exudate of soybean. *Plant Physiol.* **45**: 280–283.

Tiffin, L.O. and Chaney, R.L. 1974. Translocation of iron from soybean cotyledons. *Plant Physiol.* **51**: 393–396.

Tiwari, S.R., Sharma, R.D. and Ram, N. 1977. Mineral contents of some high yielding varieties of Bengal gram (*Cicer arietinum*). *J. Agri. Fd. Chem.* **25**: 420–421.

Tremblay, F.T. and Bauer, K.E. 1948. A method of determining the potassium requirement of peas. *J. Am. Soc. Agron.* **40**: 945–949.

Trevino, I.C. and Murray, G.A. 1975. Nitrogen effects on growth, seed yield, and protein of seven pea cultivars. *Crop Sci.* **15**: 500–502.

Vallee, B.L. and Wacker, W.E.C. 1976. In: *CRC Handbook of Biochemistry and Molecular Biology Proteins*. G.D. Fasman, ed. **2**: 276–292.

Verma, V. and Pant, R.C. 1978. Potassium deficiency in pigeon pea (*Cajanus cajan* L. type 21) plants. 1. Effect on growth, total activity and molecular forms of ribonuclease. *Indian J. Plant Physiol.* **21**: 118–126.

Viets, F.G. Jr., Boawn, L.C. and Crawford, C.L. 1954a. Zinc contents and deficiency symptoms of 26 crops grown on a zinc deficient soil. *Soil Sci.* **78**: 305–316.

Viets, F.G. Jr., Boawn, L.C. and Crawford, C.L. 1954b. Zinc content of bean plants in relation to deficiency symptoms and yield. *Plant Physiol.* **29**: 76–79.

Villafranca, J.J. 1974. The mechanism of aconitase action. *J. Biol. Chem.* **249**: 6149–6155.

Wallace, A., ElGazzar, A.A., Cha, J.W. and Alexander, G.V. 1974. Phosphorus levels versus concentrations of zinc and other elements in bush bean plants. *Soil Sci.* **117**: 347–351.

Wallace, A., Jaffreys and Hale, V.Q. 1962. Differential ability of two soybean varieties to take up ^{59}Fe from soil added to a nutrient solution. *Soil Sci.* **94**: 111–114.

Wann, F.B. 1941. Control of chlorosis in American grapes. *Utah Agr. Exp. Sta. Bull.* **299**: 1–27.

Webb, J.R., Ohlrogge, A.J. and Barber, S.A. 1954. The effect of magnesium upon the growth and the phosphorus content of soybean plants. *Proc. Soil Sci. Soc. Amer.* 458–462.

Weiss, M.G. 1943. Inheritance and physiology of efficiency in iron utilization in soybeans. *Genetics* **28**: 253–268.

Widdowson, J.P. 1965. Molybdenum uptake by French beans on two recent soils. *N.Z.J. Agric. Res.* **9**: 59–67.

Williams, J.H. 1979. The physiology of groundnuts (*Arachis hypogaea* L. cultivar Egrat). II. Nitrogen accumulation and distribution. *Rhod. J. Agric. Res.* **17**: 49–55.

Wooding, F.J., Paulsen, G.M. and Murphy, L.S. 1970. Response of nodulated and non-nodulated soybean seedlings to sulphur nutrition. *Agron. J.* **62**: 277–280.

Wutscher, H.K., Olson, E.O., Shull, A.V. and Peynado, A. 1970. Leaf nutrient levels, chlorosis, and growth of young grapefruit trees on 16 rootstalks grown on calcareous soil. *J. Amer. Soc. Hort. Sci.* **95**: 259–261.

Yadav, O.P. 1979. Effect of phosphorus and zinc on yield, nutrient uptake, nodulation and nitrogen fixation in gram (*Cicer arietinum*). Ph.D. Thesis. Haryana Agricultural University, Hissar, India.

Yadava, R. and Singh, D. 1970. Effect of gypsum on the chemical composition, nutrient uptake and yield of groundnut. *J. Indian Soc. Soil Sci.* **18**: 183–186.

Younge, O.R. and Takahashi, M. 1953. Response of alfalfa to molybdenum in Hawaii. *Agron. J.* **45**: 420–428.

7

Nutritive Value of Food Legumes

Y.P. GUPTA

7.1. Introduction

Food is the primary necessity of life. It serves three main functions—physiological, social and psychological. Physiological functions are related to (1) supply of energy, (2) building and maintaining the cells and tissues, and (3) regulating the body processes. These needs are satisfied by the nutrients present in the food. The selection of a particular food helps in developing social relationships among people to whom it is served, depending upon their liking. Food also has a psychological effect when it satisfies certain emotional needs of the persons who adjust themselves to the unfamiliar food in a new place.

An individual's selection of food is influenced by many factors such as the habits of his family and associates, his religion and region, his attitude towards various foods, etc. Besides, food must provide sufficient quantity of all nutrients for the proper nutrition of the body to enable it to maintain energy, to perform the vital processes of life and provide the material to replace the essential tissue breakdown.

In India, people are mostly vegetarian, depending largely on cereals and pulses as their staple food which provides the main source of dietary proteins and calories. Cereals occupy the first place as the source of calories and protein, followed by food legumes (Aykroyd and Doughty 1964). Among cereals, rice is the chief dietary staple for half of the world's population and constitutes as much as 80 per cent of the calories for most of Asia's peoples. Wheat ranks second to rice in worldwide use. Maize is widely used in Central and South America. Cereal foods are the primary source of energy for much of the population and also sufficiently contribute to the needs of other nutrients.

In India, pulses meet the needs of proteins of a large section of its people particularly that of the poor since the majority of the people cannot afford animal proteins as they are costly, or do not use them because of religious beliefs. Pulses are considered to be "poor man's meat". Even in the West, animal proteins are becoming costly. But the legume proteins are known to be inferior in quality due to the deficiency of sulphur-containing amino acids as well as due to other factors like poor digestibility, availability of amino acids, anti-nutritional factors, etc. The evaluation of protein quality of a mixed diet consisting of cereals and pulses has proved to be beneficial. Therefore, under the existing conditions of our economic development, their exploitation in the diet in combination with cereals in the proper proportion to make it nutritionally balanced in its amino acid pattern comparable to that of FAO reference protein or egg protein, appears to be the only feasible approach to eliminate protein-calorie malnutrition in the developing countries. A programme for improving their production and nutritional status is being vigorously pursued at different agricultural universities and research institutions, both in India as well as abroad.

In the recent past, nutritional improvement of food legumes by breeding has attracted world attention (Milner 1972). It has been reported that the existence of genetic variability provides a great scope for exploiting the tools of breeding for upgrading of protein properties in developing nutritionally improved strains (Swaminathan et al. 1972, 1969 and 1970; Jain 1969). Improving the quantity and quality of proteins in foods was earlier reported by Gupta (1967, 1980a, 1981a) and others (Aykroyd et al. 1966; Gopalan et al. 1972; Kuppuswamy et al. 1958; FAO 1965, 1970).

7.2. Chemical Composition

In general, pulses contain 20–30 per cent protein and about 60 per cent carbohydrates, and are fairly good sources of thiamine, nicotinic acid, calcium and iron (Table 7.1). Their carbohydrates (mainly starch) are well absorbed and utilised providing an equivalent energy value in terms of calories on unit weight-basis to that of cereals. The potential of soybeans has been recently summed up by Gupta and Kapoor (1978). Soybean is rich in protein, oil and mineral salts, consisting of approximately 40 per cent protein, 26 per cent carbohydrates, 20 per cent oil, 4 per cent minerals, 2 per cent phospholipids and sufficient amount of both water and fat-soluble vitamins. Its oil composition mainly comprises palmitic acid, stearic acid, oleic acid, linoleic acid and linolenic acid, while sugars consist of sucrose, raffinose and stachyose. Soybean is a good source of phosphorus and lecithin. Soy flour in the diet has been specifically useful to the diabetic patients because of low carbohydrates.

Considerable information on the chemical composition of different pulses is available. Recent studies on the separation of soybean seed into its seed coat, cotyledon and embryo by Kapoor and Gupta (1977c) revealed that cotyledon constituted the principal component, mainly accounting for the food value of the whole seed. The seed contained about 8.5 per cent seed coat, 89.5 per cent cotyledon and 2 per cent embryo. The composition of its

TABLE 7.1. Proximate chemical composition of different pulses

Pulses	Calorific value, cal/100 g	Crude protein %	Fat %	Carbohydrates %	Minerals (mg/100 g) Ca	Fe	P	Vitamins (mg/100 g) B₁	B₂	Niacin
1. Bengal Gram (*Cicer arietinum*)	361	20.6	2.2	61.2	190	9.8	280	0.30	0.51	2.6
2. Arhar (*Cajanus cajan*)	343	24.1	1.7	62.9	129	5.8	288	0.52	0.14	2.3
3. Lentil (*Lens culinaris*)	346	25.1	1.8	60.8	130	6.1	250	0.50	0.21	1.8
4. Peas (*Pisum sativum*)	315	26.1	1.5	56.6	150	5.2	300	0.45	0.50	1.3
5. Black gram (*Vigna mungo*)	350	26.2	1.2	56.6	185	8.7	345	0.42	0.37	2.0
6. Green gram (*Vigna radiata*)	334	24.0	1.3	56.6	140	8.4	280	0.47	0.39	2.0
7. Cowpea (*Vigna unguiculata*)	342	23.4	1.8	60.3	76	5.7	430	0.92	0.18	1.9
8. Kidney bean (*Lablab purpureus*)	341	22.1	1.7	61.4	137	6.7	410	0.54	0.18	2.1
9. Horse gram (*Macrolyloma uniflorum*)	322	22.0	0.5	57.3	280	7.6	390	0.42	0.20	1.5
10. Khesari dal (*Lathyrus sativus*)	351	28.2	0.6	58.2	110	5.6	500	—	0.41	—
11. Soybean (*Glycine max L.*)	462	40.0	19.5	26.0	290	12.5	675	1.30	0.76	2.4

Source: Aykroyd and Doughty (1964); Aykroyd et al. (1966); Kapoor and Gupta (1977c).

anatomical parts revealed that the cotyledon contained protein, minerals, and oil ranging from 91 to 93 per cent, 84 to 88 per cent and 90 to 91 per cent of the total amount of respective nutrient of the whole seed. However, seed coat accounted for 34 to 37 per cent of calcium of the whole seed. Studies on the nutrient composition of seed parts of different pulses by Singh et al. (1968) and that of Bengal gram by Lal et al. (1963a, 1968) and Verma et al. (1964, 1966) revealed similar findings.

Work carried out on different pulses and soybean at this Institute by Gupta et al. (1977, 1978) and also Singh et al. (1968), Lal et al. (1963b), Dhingra and Das (1959), and Sharma and Goswami (1971) revealed wide variation in their nutrient composition. Among the pulses, Bengal gram has been widely studied. The biochemistry of Bengal gram was reviewed by Krishnamurti (1975). Certain workers (Rao 1969, 1976; Pant and Kapur 1963) studied the nature of carbohydrates in pulses.

The available information shows that the chemical composition of pulses is largely influenced by different environmental factors such as variety, soil type, climate, location, fertilisers, crop rotation and other agricultural practices. Work from this laboratory (Grover and Gupta 1972; Kapoor et al. 1972; Gupta et al. 1978) showed wide variation in the quantity of protein of different pulses (Table 7.2). Analysis of chemical constituents in green gram,

TABLE 7.2. Variation in the quantity of protein in different strains of pulses

Pulses	Crude protein %
Bengal gram	17.1–28.1
Arhar	20.6–26.3
Lentil	24.8–32.1
Peas	21.0–32.3
Cowpea	21.2–30.6
Black gram	23.3–28.9
Green gram	23.4–32.0
Sem	19.3–27.8
Moth	25.0–31.3
Kulthi	23.0–32.3
Khesari dal	22.7–29.6
French bean	20.2–28.0
Rice bean	18.4–27.0
Soybeans	31.2–45.2

Source: Gupta et al. (1978); Kapoor et al. (1972); and unpublished data.

black gram, *kulthi*, *moth*, rice bean and soybean revealed that non-protein nitrogen was maximum in *moth* and *kulthi*, constituting 12.0 to 19.6 per cent of the total nitrogen. Non-protein nitrogen in pulses was reported to be significant (Kulkarni and Sohonie 1956). Studies on the distribution of different phosphorus compounds in soybean seed revealed that soybean was rich in total phosphorus containing phytin phosphorus as the major form of phosphorus followed by nucleic acid and phospholipid phosphorus, but rice bean contained the lowest amount of phytin phosphorus among the pulses analysed. It was also observed that soybeans which were rich in protein were poor in their oil content and vice-versa. The EC 4496 (N-49) variety of soybean having the lowest amount of protein as 31.2 per cent, contained the highest amount of oil as 26.4 per cent.

The influence of genetic variation on the chemical composition of pulses was observed by many other workers (Lal et al. 1963b; Dhingra and Das 1959; Chandra and Arora 1968; Esh et al. 1959; Das 1960; Gupta et al. 1976a). Bengal gram, black gram, green gram, lentil, arhar and pea were reported to vary in their protein content from 20.1 to 32.4 per cent and in their phytin phosphorus from 22.6 to 50.2 per cent of the total phosphorus. Kabuli types of Bengal gram were found to contain higher amount of protein, ether extract and iron than the common types. Soybeans were reported to vary in their content of protein and ether extracts from 35.7 to 45.1 per cent and from 18 to 28 per cent respectively. Akyroyd et al. (1966) reported the oil content of soybean to be 19.5 per cent whereas Pant and Kapur (1963) found it to be 16.9 to 18.2 per cent. They also observed that crude protein, mineral matter, calcium, iron and phosphorus contents of soybeans varied from 36.2 to 44.9 per cent, 3.2 to 4.2 per cent, 320 to 350 mg, 9.2 to 14.9 mg and 580 to 630 mg per 100 g material respectively. Gupta et al. (1976a) and others (Kapoor et al. 1975; Deodhar et al. 1973) compared the chemical composition of vegetable and grain type soybeans. Their mineral composition, calcium, phosphorus, iron, potassium, zinc, copper and manganese were reported by Gupta et al. (1976a). Sood et al. (1977) observed that soybean variety PK 71–5 under late-sown trials contained the highest protein as 53.4 per cent, and Harsoy deciduous in spring trial contained the maximum fat as 25.1 per cent. The protein content of Bengal gram was reported to vary widely by Chandra and Arora (1968) ranging from 18.4 to 29.8 per cent, while its mineral matter was found to range from 4.1 to 4.7 per cent. Bengal gram was found to be rich in phosphorus, potassium and magnesium but deficient in calcium (Narayana Rao et al. 1959). Studies on cowpea by Arora and Das (1976) showed that its protein, mineral matter, total soluble sugars and starch contents varied from 17.9 to 27.5 per cent, 3.1 to 4.6 per cent, 13.8 to 19.8 per cent and 50.7 to 67.0 per cent respectively. Chemical composition of Bengal gram, arhar, peas, black gram, green gram, lentil, khesari dal and horse gram for their protein, fat, carbohydrates, mineral matter, calcium, phosphorus, iron, magnesium, sodium, potassium, copper and sulphur was studied by Balasubramanian et al. (1962). Sharma et al. (1977) analysed 22 cultivars of arhar for their proteins, minerals, sugars, and starch, and found significant varietal differences in these constituents. Analysis of 25 varieties of green gram and black gram revealed wide variation in their mineral composition (Kadwe et al. 1974).

Vitamins (thiamine, riboflavin, nicotinic acid, vitamin C, carotene, etc.) in pulses have been studied by various workers. The thiamine content in different pulses was found to vary from 3.1 to 5.0 μg/g by Gupta and Das (1956, 1959), from 2.8 to 5.0 μg/g by Bashir Ahmed et al. (1948), and from 0.8 to 4.9 μg/g by Chitra et al. (1955). Bashir Ahmed et al. (1948) observed a loss of 35.3 to 53.3 per cent of thiamine during curry preparation, while Vallidevi et al. (1972) reported a loss of 20 to 35 per cent and a further loss of 10–15 per cent during storage. Narayana Rao et al. (1959) found that Bengal gram contained 0.4 mg of thiamine, 0.3 mg of riboflavin and 2.5 mg of nicotinic acid per 100 g material. Besides, it was found to be a good source of pyridoxine, pantothenic acid and choline. Vitamin C in Bengal gram was found to vary from 2.2 to 20.0 mg per 100 g material by different workers (Chandra and Arora 1968; Bhagvat and Narsinga Rao 1942).

Studies by various workers showed that germinated pulses contained significantly increased amounts of carotene, tocopherol, Vitamin C, pantothenic acid, biotin nicotinic acid, thiamine, riboflavin, choline, and Vitamin B_{12} (Chattopadhyaya and coworkers 1950, 1951 a, b, 1952; Nandi and Banerjee 1950, a and b; Rohtagi et al. 1955; Banerjee and coworkers 1954, 1955). However, germinated pulses contained less folic acid. Sprouted soybean widely varied in its ascorbic acid content from 2.4 to 42.5 mg/100 g (Gupta et al. 1976a). A decrease in the non-protein nitrogen of green gram on germination after five hours, and an increase in its protein nitrogen after 15 hours were reported by Srivastava and Kooner (1972). Germinated pulses contained less amount of phytin phosphorus and higher amount of water soluble inorganic phosphorus (Balavady and Banerjee 1953). On the other hand, phytase activity which was absent in ungerminated pulses, appeared on germination and phosphatase activity increased in germinated pulses. Arora and Gandhi (1970) observed that total phosphorus and other phosphorus fractions of soybean cotyledon decreased with germination time. A decline of about 25 per cent in the levels of nucleic acids (RNA and DNA) in Bengal gram occurred during the first three days of germination, accompanied by an increase in the activities of ribonuclease and deoxyribonuclease (Hadi 1966).

Studies from this laboratory (Kapoor et al. 1971; Gupta et al. 1978) revealed that location influenced the quantity of protein in black gram and soybean. Simla grown strains of black gram contained higher amount of protein than those grown at Delhi and Amravati, while soybean grown at Amravati and Jabalpur were higher in proteins than those grown at Delhi. Studies by Krober (1969) revealed that location influenced the protein content of arhar, black gram, green gram, and cowpea. Location, genotype and other environmental factors were reported to govern the chemical composition of soybean (Anonymous 1971; Lal et al. 1973; Kapoor et al. 1975).

Studies carried out by Kapoor and Gupta (1977d) in this laboratory on the effect of fertilisation revealed a favourable response to the application of phosphorus on the chemical composition of soybean. Phosphorus application enhanced the quantity of protein, iron, total phosphorus, phytin phosphorus, nucleic acid phosphorus, phospholipid phosphorus and inorganic phosphorus of soybean. In another study by Gupta and Gupta (1972), application of sulphur and molybdenum had a beneficial effect in increasing the amount of pea protein. Nitrogen and phosphatic fertilisers were reported to have considerably improved the protein, phosphorus and carbohydrate contents of Bengal gram (Singh 1971). According to Ranjan et al. (1962), protein synthesis was adversely affected under phosphorus deficiency because of the accumulation of arginine, asparagine and glutamine in the tissues of leguminous plants.

Studies carried out in this laboratory with sun-dried and dehydrated varieties of pea by Grover et al. (1971, 1972) revealed that dehydration did not affect its amount of crude protein and oil. In another communication from this laboratory by Kapoor and Gupta (1977a), roasting of soybean seed at 120°C for 12 hours reported slight reduction in its amount of crude protein.

Nutrients in pulses have been found to be affected by the maturity of the seed. Kapoor and Gupta (1977e) observed that the amount of nonprotein nitrogen in soybean was substantial at the early stage of seed formation, but it generally decreased and true protein was the major form at the mature stage of seed development. However, the amount of crude protein did not alter much, but the amount of oil gradually increased as the seed developed maturity. Similar changing pattern in the quantity of non-protein nitrogen and true protein was obtained during the developing stages of lentil seed formation. Chatterjee et al. (1977 and 1978) also did not observe much alteration in the amount of protein of Bengal gram and pea during their seed formation from 15 days to mature stage. Arora and Gandhi (1970) found that acid soluble phosphorus, lipid phosphorus, nucleic acid phosphorus and protein phosphorus fractions of soybean increased with the developing seed. Studies with cowpea by Arora et al. (1973) showed that its total phosphorus and acid soluble phosphorus fractions decreased with the developing seed. Considerable changes in the phosphorus fractions of Bengal gram and its seed parts during the development were reported by Lal et al. (1963a, 1968). Their studies revealed a decrease in the various phosphorus fractions as the seed develops.

7.2.1. PROTEINS AND AMINO ACIDS

Pulses and soybeans have been shown to be rich in protein. However, their contribution in a diet does not depend on its quantity alone but on its quality as well. The quality of a protein is known to be affected by (1) essential amino acid composition, (2) amino acid imbalance, (3) biological availability of essential amino acids, (4) digestibility, and (5) interference in protein utilisation by anti-nutritional factors.

TABLE 7.3: Essential amino acid composition of pulses (g/100 g protein)

Amino acid	Bengal gram	Arhar	Lentil	Peas	Beans	Black gram	Green gram	Cowpea	Horse gram	Khesari dal	Soybean	Reference protein	FAO protein	Egg protein
Lysine	6.3	6.8	5.1	8.9	6.8	6.5	7.3	6.7	8.3	7.4	6.3	4.2	7.2
Threonine	3.4	3.8	3.0	4.2	3.3	3.9	3.4	4.1	3.8	2.3	4.1	2.8	5.2
Valine	5.5	4.8	5.1	6.5	5.4	5.6	6.9	5.2	5.4	4.7	4.7	4.2	7.4
Leucine	8.2	6.8	5.5	9.5	8.9	7.2	7.7	7.4	7.9	6.6	7.1	4.8	7.8
Isoleucine	6.0	5.7	5.8	7.4	6.0	5.8	6.3	4.9	6.7	6.7	4.3	4.2	6.8
Methionine	1.2	1.1	0.6	1.3	1.0	1.1	1.5	1.3	0.8	0.6	1.2	2.2	3.4
Tryptophan	0.8	0.8	0.6	0.7	1.0	0.5	0.4	1.0	0.6	0.4	1.2	1.4	1.5
Phenylalanine	4.9	9.0	4.0	4.6	5.5	5.5	5.3	5.7	8.5	4.2	4.9	2.8	5.8
Arginine	6.9	5.4	7.0	13.4	9.2	5.7	6.3	6.9	5.4	7.8	6.7	—	6.7
Histidine	2.3	3.4	2.1	2.7	2.8	2.7	2.7	3.1	3.0	2.5	3.3	—	2.4

Source: Milner (1972); Patwardhan and Ramachandran (1960).

The amino acid composition of pulses has been widely studied. It has been observed that pulse proteins are mainly deficient in sulphur containing amino acids and tryptophan, but are rich in lysine in which cereals are relatively deficient. Table 7.3 shows the pattern on the relative proportion of essential amino acids in different pulses in comparison to those of egg or FAO reference protein (FAO 1957). Pulses exhibited wide strain variation in their essential amino acid make-up (Table 7.4).

TABLE 7.4. Variation range in levels of essential amino acids in different pulses

Amino acids	Range g/g N
Lysine	0.34–0.71
Threonine	0.16–0.31
Valine	0.24–0.49
Leucine	0.20–0.68
Isoleucine	0.32–0.62
Methionine	0.03–0.11
Tryptophan	0.02–0.08
Phenylalanine	0.15–0.49
Arginine	0.36–0.57
Histidine	0.08–0.21

Source: Milner (1972).

A large number of strains of different pulses and soybeans have been screened (Gupta and Das 1955; Singh et al. 1960; Gupta, 1969a, b, 1971; Kapoor et al. 1972; Grover and Gupta, 1972; Kapoor and Gupta 1977; Gupta et al. 1978) for their methionine and tryptophan contents as the limiting amino acids. These studies revealed wide genetic variability among them suggesting that the genetic constitution of the seed greatly influences these amino acids. The variation in their methionine and tryptophan contents ranged from 0.5 to 1.9 and from 0.4 to 1.5 g per 16 g nitrogen respectively. These pulses were found to be deficient both in methionine and tryptophan; peas, lentil and khesari dal being the poorest in comparison to Bengal gram, arhar, green gram, black gram and cowpea. *Annigeri-1* variety of Bengal gram developed by the University of Agricultural Sciences, Bangalore, was found to contain higher amounts of methionine and tryptophan as 1.38 and 0.92 per cent in the protein in comparison to that of *Chaffa* variety which was found to contain 1.26 and 0.71 per cent as the corresponding values. Methionine, cystine and tryptophan composition of pure strains of Bengal gram, green gram, black gram, arhar, lentil, pea, khesari dal was reported by Gupta and Das (1955). Studies by Gupta (1969 a and b, 1971) showed that these amino acids were greatly influenced by genetic factor. Studies on methionine, lysine, threonine and tryptophan composition of arhar and black gram by Singh et al. (1960) revealed that black gram had higher amounts of these amino acids as compared to arhar.

Studies on the overall amino acid pattern of *Bragg, Panjab I* and *Lee* varieties of soybean by Kapoor and Gupta (1977) revealed that soy protein contained a balanced amino acid pattern with the exception of methionine which was found to be deficient as the major limiting amino acid, but it was found to be rich in lysine, leucine, arginine, glutamic and aspartic acids. These results revealed that methionine caused imbalance in the amino acid make-up of soy protein. However, these three varieties of soybean did not exhibit much differences in their amino acid pattern.

Studies on the distribution of methionine and tryptophan in the anatomical parts of soybean seed by Kapoor and Gupta (1977) revealed that cotyledon, being the major component of the seed, accounted for about 93 per cent of methionine and tryptophan of the whole seed, while the seed coat was the poorest. The embryo was rich in methionine and tryptophan, but it contributed only about 2.5 per cent of their total quantity in the seed.

Considerable work on the amino acid composition of Bengal gram, black gram, green gram, arhar, lentil, pea, cowpea and khesari dal has been done by different workers. The amino acid composition of different pulses has been summarised by Patwardhan and Ramachandran (1960). Various studies revealed that pulses widely differed in their amino acid composition and methionine, in general, was the major deficiency limiting amino acid in legume proteins. Bengal gram was reported to be deficient in methionine and tryptophan (Chatterjee and Abrol 1975). Hanumantha Rao and Subramanian (1970) concluded that the major deficiency in pulses was tryptophan followed by sulphur containing amino acids. They found that the amount of total essential amino acids in pulses (g/16 g nitrogen) ranged from 34 in Bengal gram to 41 in black gram, and the pulses were good sources of lysine, valine, leucine and aromatic amino acids. They also found that black gram and green gram were relatively better balanced in their amino acid composition than Bengal gram, arhar and lentil. Ramachandran and Phansalkar (1956) observed that pulse proteins were deficient in methionine and tryptophan but these were good sources of lysine, pea containing 11.4 per cent. Tara and Rama Rao (1972) found that methionine followed by isoleucine were the limiting amino acids in the protein of arhar dal (raw and processed). Manage and Sohonie (1972) compared the essential amino acid composition of horse gram to that of casein and whole egg protein. They observed the methionine and cystine content of horse gram to be 48 to 50 per cent of that of egg protein. Studies by Nagpal and Bhatia (1971) showed that the amount of tryptophan in arhar, lentil, black gram, green gram, and Bengal gram ranged from 0.56 to 1.07 g/16g nitrogen. Evaluation of 14 genotypes of arhar by Singh et al. (1973) revealed wide variation in their methionine and tryptophan contents showing significant varietal differences. Their values ranged from 1.20 to 1.88 g and from

0.44 to 0.62 g per 100 g protein respectively. Vijayaraghavan and Srinivasan (1953) found that arhar proteins were low in tryptophan. Phenylalanine was reported to be deficient in broad beans (*Vicia faba*), threonine in broad beans and Bengal gram, and valine in peas (Kuppuswamy et al. 1958). Banerjee (1960) studied the essential amino acid composition of Bengal gram, arhar, black gram, green gram, peas, lentil and khesari dal, and found that khesari dal was quite rich in essential amino acids. Inamdar and Sohonie (1960) found that double bean was markedly deficient in valine, leucine and tryptophan along with methionine. The influence of strain variation in the amino acids of pulses was reported by Das (1960)—see Table 7.5.

TABLE 7.5. **Variation range in levels of limiting amino acids in different pulses (g/100 g protein)**

Pulses	Amino acids	
	Methionine	Tryptophan
Bengal gram	0.5–1.8	0.5–1.4
Arhar	1.3–1.6	0.7–1.2
Lentil	0.5–0.7	0.6–0.8
Peas	0.5–0.8	0.4–0.6
Black gram	1.0–1.6	0.6–1.3
Green gram	1.0–1.5	0.7–1.2
Cowpea	0.8–1.9	1.0–1.1
Commom bean	0.5–1.9	0.5–1.5
Sem	1.0–1.4	0.7–1.3
Khesari dal	0.5–0.9	0.4–0.7
French bean	1.2–1.6	0.7–1.2
Soybean	1.0–1.8	0.8–1.4

Source: Gupta et al. (1978); Kapoor et al. (1972); Grover and Gupta (1972); Gupta and Das (1955).

Correlations among amino acids were found to be significant (Table 7.6). However, valine and methionine, methionine and isoleucine, methionine and threonine, and isoleucine and lysine did not show significant correlation with each other. Correlation studies could be used as an aid in a selection programme for more than one amino acid at a time.

Studies have shown that location influenced methionine and tryptophan contents of soybean (Kapoor et al. 1971), but the phosphorus application did not affect them (Kapoor and Gupta 1977 d). Gupta and Gupta (1972) also found that the application of molybdenum increased the amount of methionine of pea protein while sulphur application had no effect. It was, however, reported earlier (Kuppuswamy et al. 1958) that sulphur containing fertilisers increased the cystine content of pea proteins. Studies by Arora and Luthra (1970) revealed that increasing levels of sulphur up to 90 ppm increased the

TABLE 7.6. Correlation between essential amino acids in bean seeds

	Methionine	Isoleucine	Leucine	Phenyl alanine	Lysine	Threonine
Valine	.24	.87**	.66**	.66**	.48*	.67**
Methionine		.17	.59**	.63**	.78**	.30
Isoleucine			.68**	.61**	.37	.44*
Leucine				.90**	.64**	.56**
Phenylalanine					.73**	.53*
Lysine						.60**

* and **=significant at 5% and 1% levels, respectively.
Source: Milner (1972).

sulphur containing amino acids in Bengal gram, but higher levels of sulphur decreased them.

The stage of seed formation was found to greatly influence the amino acid composition. The concentrations of free amino acids in soybean seed, lysine, hydroxyproline, and tryptophan decreased and that of glutamic acid and alanine increased as the seed approached maturity. However, the total concentration of methionine and tryptophan in the protein was not much affected by maturity. Studies with lentil seed showed a decreasing tendency in the amino acids (methionine and tryptophan) with the maturity of the seed. Chatterjee et al. (1977, 1978) with Bengal gram and pea seed revealed that the free amino acids, in general, declined as the seed developed maturity. The total concentration of threonine, methionine, valine and tryptophan in the Bengal gram protein decreased with maturity, but that of arginine increased and that of lysine, histidine, isoleucine and phenylalanine remained constant, while the concentration of lysine and histidine were maximum at maturity in the pea protein. The changing pattern and variation in the concentration of free amino acids and total amino acids during the seed development reflected the metabolic activity and inter-conversion of amino acids at the stage of synthesis of storage proteins of the seed.

Studies on free amino acids in pulses have been made by different workers (Pant and Kapoor 1964; Pant and Tulsiani 1968). *Rajmah*, black gram and *moth* were found to differ in their pattern of free amino acids from that of beans, kidney bean and lima bean seeds. The total free amino acid content of the seed varied from 0.15 to 0.36 g. The free amino acid concentration of cotyledon of Bengal gram showed a four-fold increase on the seventh day of germination as compared to dormant seeds (Azhar et al. 1972). The amount of free amino acids of green gram gradually decreased after 30 hours of germination (Srivastava and Kooner 1972). Aspartic and glutamic acids were predominant in the cotyledon.

Pulse proteins chiefly comprise globulins with lesser amount of albumins. The main constituent of Bengal gram proteins was reported to be globulin accounting for 88 per cent of the total nitrogen (Narayana Rao et al. 1959).

Studies on the solubility classes of soy proteins by Kapoor and Gupta (1977) showed that glycinin was the major fraction accounting for 63 per cent of the total protein, while albumins, prolamines and glutelins constituted 12, 3 and 8 per cent of the total protein respectively. Fractionation of proteins from lentil seeds revealed that globulins accounted for 59 per cent of total protein while albumins, prolamines and glutelins were only 10, 3 and 14 per cent of the total protein respectively (Gupta 1981a). The nature of protein components in Bengal gram separated on the basis of solubility criteria was studied (Esh et al. 1960; Narayana Rao and Rajgopal Rao 1974). Papain treatment was reported to improve the solubility of arhar proteins (Tara et al. 1972).

Studies on the distribution of phosphorus compounds in the protein fraction of soybean by Kapoor and Gupta (1977c) revealed that glycinin being the major component of soy protein, contained the maximum amount of total phosphorus, phytin phosphorus, nucleic acid phosphorus and inorganic phosphorus while albumin fraction was quite low. Phosphorus application enhanced the concentration of phosphorus compounds in the protein fractions. The association of nucleic acid phosphorus, mainly with glycinin as the major component of soy protein, reflected their close participation in the metabolic activity.

Different protein fractions were found to differ in their amino acid composition. Albumins of pea differed from globulins in having a higher amount of tryptophan and lysine (Kuppuswamy et al. 1958). Studies from this laboratory by Kapoor and Gupta (1977c) showed that glycinin fraction of soybean contained 73 and 71 per cent of the amount of methionine and tryptophan of the seed respectively but it was quite deficient in methionine. Albumin fraction contributed 16 and 20 per cent of the quantity of methionine and tryptophan of the seed respectively, and it was also deficient in methionine but was rich in tryptophan. The prolamine fraction was rich both in methionine and tryptophan while the glutelin fraction was the poorest in these amino acids. The amino acid pattern of isolated proteins (Table 7.7), albumin and globulin fractions from pea, globulin fraction from cowpea and glycinin fraction from soybean revealed that all these fractions were deficient in methionine in comparison to that of FAO reference protein. Studies in the amino acid pattern of protein fractions of Bengal gram by Mathur et al. (1968) showed that albumin and glutelin fractions were higher in cystein, lysine, methionine and tryptophan, while globulin fraction was rich in arginine and glutamic acid.

There was a shift in the nature of protein fractions during the developing stage of seed formation (Kapoor and Gupta 1977e). The data on the changing pattern of protein fractions of soybean and lentil seeds are shown in Table 7.8. These studies showed that albumin and glutelin fractions of soybean were predominant at the early stage but glycinin became predominant at the ma-

300 Chemistry and Biochemistry of Legumes

TABLE 7.7. Amino acid composition of isolated protein (g/16 g N)

| Amino acid | Pea | | Cowpea | Soybean |
	Albumin	Globulin	Globulin	Glycinin
Lysine	10.6	7.9	6.6	6.2
Histidine	2.9	2.3	3.0	3.2
Arginine	5.7	8.8	7.2	6.2
Threonine	5.2	3.2	3.6	3.4
Valine	5.5	5.5	6.4	5.8
Methionine	1.1	0.5	1.6	1.3
Isoleucine	6.1	5.9	5.4	5.1
Leucine	5.9	8.7	9.2	8.0
Phenylalanine	5.7	3.1	7.9	5.8
Aspartic acid	11.2	9.9	12.5	14.3
Serine	3.8	3.6	5.9	5.1
Glutamic acid	15.9	23.9	19.0	17.8
Proline	4.8	4.3	4.3	5.4
Glycine	5.0	3.6	3.2	5.7
Alanine	5.2	3.7	4.2	3.5
Tyrosine	5.2	3.1	3.9	3.5

Source: Milner (1972); Kapoor and Gupta (1979).

TABLE 7.8. Distribution of protein fractions (solubility classes) during the developing seed (expressed as percentage of total protein)

Seed and stage	Albumin	Glycinin/Globulin	Prolamine	Glutelin
Soybean				
Early milky	42.7	20.0	4.7	16.9
Mid milky	19.4	16.2	4.4	41.9
Mature	9.0	65.0	4.6	7.8
Lentil				
Early milky	18.5	25.5	9.5	32.0
Mid milky	10.5	44.5	5.8	24.5
Mature	10.4	58.7	3.3	14.2

Source: Kapoor and Gupta (1977); Gupta (1981a).

ture stage. These observations suggested that glycinin and glutelin fractions of soy protein were inversely synthesised and glutelin was converted to glycinin at the time of synthesis of storage proteins. Similar observations on the nature of protein fractions of lentil seed during its development were obtained.

The changing pattern of protein fractions of soybean during its development was, however, not affected by phosphorus application. The higher level of albumin fraction at the initial stages of seed formation was reflected by the higher amount of non-protein nitrogen at that stage.

The pattern of amino acids and protein fractions at the early stage of seed formation revealed that the immature seed of soybean was better in its protein quality than the mature seed. Seeds of Bengal gram, pea and lima bean were also reported to be superior in their protein quality at the unripe stage (Kuppuswamy et al. 1958; Chatterjee et al. 1977, 1978).

Studies on the electrophoretic fractions of proteins from soybean, Bengal gram and green gram (Kapoor and Gupta 1971, 1977b) revealed distinct differences in their protein pattern. However, pea and black gram did not exhibit differences in their electrophoretic protein components. Albumin fractions isolated from soybeans also differed in its electrophoretic pattern but there was no difference in the glycinin fraction (Kapoor and Gupta 1977b). Narayana Rao and Rajagopal Rao (1974) observed that globulin protein fractions isolated from Bengal gram revealed distinct differences in their pattern by acrylamide gel electrophoresis but they were similar in nature on gel chromatography and sedimentation analysis. The electrophoretic characterisation of Bengal gram globulin fractions was studied by Tawade and Cama (1962).

The protein chemical score of different pulses based on amino acid data have been worked out by Hanumantha Rao and Subramanian (1970) and that of soybean by Kapoor and Gupta (1977) according to the principle of Mitchell and Block (1946) and FAO (1970) taking egg protein as the reference protein. The values are summarised in Table 7.9. The ratios of each essential

TABLE 7.9. Protein chemical score of common pulses based on egg protein

Pulses	Score
Bengal gram	32
Arhar	26
Lentil	19
Black gram	29
Green gram	32
Soybean	45

Source: Hanumantha Rao and Subramanian (1970); Kapoor and Gupta (1977b).

amino acid in food protein to its total essential amino acids have been expressed as the percentage of ratios between the corresponding amino acid of egg and total essential amino acids of egg. The lowest percentage has been taken as the chemical score, and the amino acid showing the lowest score has been taken as the limiting amino acid. These pulses were found to differ in their chemical score and thus exhibited differences in their protein quality. Bengal gram, green gram and soybean were high in their chemical score, lentil being the poorest and soybean, the highest.

The requirement index, according to Rama Rao et al. (1964), expressed as the percentage of the most limiting amino acid to the corresponding amino acid requirement pattern of the rat, and the essential amino acid index, according to Oser (1959), expressed as the percentage of the geometric mean of the ratios of essential amino acids in food protein relative to their respective amounts in the whole egg protein, have also been used as an index of chemical score for evaluating the nutritive value of a protein.

7.2.2. BIOLOGICAL EVALUATION OF PROTEINS

Biological evaluation has been recognised as a test criterion for assessing the quality of dietary proteins. Considerable work on the biological evaluation of pulse proteins has been done in India. Their digestibility coefficient (DC), biological value (BV) and protein efficiency ratio (PER) were reported to vary from 70 to 90 per cent, 45 to 74 per cent and 0.7 to 1.1 per cent respectively (Phansalkar 1960). Different pulses were found to vary widely in their biological parameters (Table 7.10); lentil, peas and khesari dal being the poorest, while

TABLE 7.10. **Biological parameters of pulses (variation range)**

Pulses	Biological value (%) (BV)	Digestibility coefficient (%) (DC)	Protein efficiency ratio (PER)
Bengal gram	52–78	80–89	0.7–1.8
Arhar	61–74	67–90	0.7–1.7
Lentil	41–58	78–90	0.6–1.1
Black gram	60–64	78–85	1.0–1.9
Green gram	39–66	74–91	0.8–1.8
Peas	48–69	60–91	1.0–1.5
Cowpea	45–72	58–79	—
Kidney bean	62–68	64–76	1.2–1.6
Khesari dal	50	90	0.3
Soybean	64–80	76–87	1.3–2.0

Source: Milner (1972); Kuppuswamy et al. (1958); Phansalkar (1960).

soybean and Bengal gram were relatively higher. Studies by Kapoor and Gupta (1975) from this laboratory revealed that steamed soybean was higher in its biological parameters when fed at 10 per cent level of protein. However, different pulses were found to be quite low in their PER values ranging from 0.5 to 1.7 (Dhingra and Das 1959; Singh et al. 1960). Bengal gram and black gram being relatively higher in comparison to arhar, lentil, peas and green gram. Strain variation existed among these pulses.

The biological evaluation of Bengal gram proteins was studied by many workers (Esh and Som 1955; Joseph et al. 1958; Rama Rao et al. 1953). Esh and Som (1955) found the PER value of Bengal gram and lentil to be 1.29 and 0.67 at 12 per cent protein level. Joseph et al. (1958) found the PER value of Bengal gram as 1.27 at 10 per cent protein level, while Rama Rao et al. (1953) found it to be 1.48.

Pant and Kapoor (1963) studied the biological parameters of soybean (raw), Bengal gram, green gram and arhar. They found their values of PER, BV and DC as 0.50, 1.30, 0.50 and 0.82; 57.5, 61.1, 47.7 and 64.8; and 91.4, 86.6, 84.9 and 86.2 respectively. Bengal gram was reported to be superior in its nutritive value than others including raw soybean. Cowpea fed at 18 per cent protein level was reported to promote higher growth response in rats than arhar (Sivaraman and Menachery 1967). Banerjee (1960) reported the PER values of different pulses varying from 1.16 to 1.87; lentil being the poorest and green gram, the highest. Differences in the values obtained by various authors were attributed to the experimental condition being not identical (Patwardhan and Ramachandran 1960).

Raw pulse proteins have been found to possess low nutritive value because of various factors including heat-labile trypsin inhibitor. Heat processed soybean was reported to contain proteins of high nutritive value comparable to milk proteins (Kuppuswamy et al. 1958). Hirwe and Mager (1953) reported that autoclaved Bengal gram promoted better growth than raw gram. Esh and Som (1955) found improved digestibility, biological and growth-promoting values under autoclaved conditions. Ray (1970) reported an increase in the BV of horse gram from 66 (raw) to 88 (autoclaved). Krishnamurty et al. (1958) found that raw soybean did not adequately support the growth of rats; PER being 0.5, but steaming of soaked soybean containing 60 per cent moisture increased the value of PER to 2.4. Heat processing had a significant beneficial effect in improving the protein quality of Bengal gram, green gram, and horse gram (Chandrasekhra and Jayalakshmi 1978). Venkat Rao et al. (1964) found that heat treatment markedly increased the PER of pulses. Similar effect of heat on Bengal gram proteins was observed by Rao (1974).

Parching was reported to improve the nutritive value of proteins in Bengal gram, green gram, black gram, horse gram and dried peas (Sivaraman and Menachery 1967). Kapoor and Gupta (1977a) found that roasting at 120°C for 12 hours slightly lowered the PER value of soybean protein. However, Chandrasekhra and Jayalakshmi (1978) observed that roasted seeds of Bengal gram, green gram and horse gram had higher PER and BV values.

Supplementation of pulses with amino acids was found to improve the nutritive quality of their proteins (Tables 7.11 and 7.12). Addition of methionine considerably increased the PER, BV and DC of arhar, black gram and soybean protein (Singh et al. 1960; Kapoor and Gupta 1975). The addition of methionine was equally beneficial in improving the nutritive quality of Bengal gram, lentil and green gram proteins (Esh and Som 1953; Esh and Som 1955; Banerjee et al. 1956). Supplementation with other amino acids like tryptophan or threonine or lysine was ineffective but in the presence of methionine, tryptophan or threonine significantly enhanced the nutritive value. Maximum

TABLE 7.11. Biological parameters of pulses with amino acid supplementation

Amino acid supplement	BV (%)	DC (%)	PER
Arhar			
None	68.6	82.9	0.92
Methionine (0.3%)	81.6	77.5	1.67
Tryptophan (0.1%)	77.3	82.2	0.97
Threonine (0.1%)	74.7	86.5	0.93
Lysine (0.2%)	79.0	84.5	1.17
Black gram			
None	68.4	85.1	1.25
Methionine (0.3%)	83.2	83.1	2.18
Tryptophan (0.1%)	69.3	93.9	1.17
Threonine (0.1%)	73.0	90.1	1.35
Lysine (0.2%)	78.6	90.0	1.42
Soybean			
None	78.4	87.1	2.47
Methionine (0.3%)	84.2	90.9	3.05
Tryptophan (0.1%)	80.2	88.3	2.57
Methionine (0.3%)+			
Tryptophan (0.1%)	84.9	91.6	3.32

Source: Singh et al. (1960); Kapoor and Gupta (1975).

TABLE 7.12. Protein efficiency ratio of Bengal gram and lentil
with amino acid supplementation

Amino acid supplement	PER
Bengal gram	
None	1.29
Methionine	1.67
Tryptophan	1.21
Threonine	1.39
Lysine	1.30
Methionine + Tryptophan	1.87
Methionine + Threonine	2.29
Methionine + Lysine	1.22
Methionine + Threonine + Tryptophan	2.34
Methionine + Threonine + Tryptophan + Lysine	2.35
Lentil	
None	0.67
Methionine	1.72
Tryptophan	0.67
Lysine	0.67
Threonine	0.64
Methionine + Tryptophan	2.04
Methionine + Threonine	1.74
Methionine + Lysine	1.71
Methionine + Tryptophan + Threonine	2.59
Methionine + Tryptophan + Threonine + Lysine	2.60

Source: Esh and Som (1955).

improvement was obtained with a combination of methionine, tryptophan and threonine, raising the PER of Bengal gram from 1.3 to 2.3 and that of lentil from 0.7 to 2.6 (Esh and Som 1955). Studies by Devdas et al. (1967) revealed that the protein quality of arhar supplemented with methionine and tryptophan when fed at 15 per cent protein level, did not equal that of standard protein probably due to the limited availability of either lysine or any other limiting amino acid in the diet. Joseph et al. (1960) reported an increase in the PER value of soybean from 2.1 to 2.9 when supplemented with 0.6 per cent methionine.

Supplementation with Vitamin B_{12} was also reported to improve the nutritive quality of pulse proteins. Singh et al. (1960) found an increase in the PER and BV of arhar protein at 12 per cent level when supplemented with 20 or 30 μg of Vitamin B_{12} but no effect was observed at 17 per cent protein level. However, studies by Esh (1955) revealed that Vitamin B_{12} had a supplementary effect in increasing the PER value of lentil proteins at 18 per cent protein level and not at 12 per cent level. Lower concentration of Vitamin B_{12} was reported to have no effect on the BV of lentil protein (Banerjee et al. 1956).

These findings revealed that pulses having the same level of deficiency in certain amino acids responded differently on supplementation. Their different behaviour was evaluated by working out their relative deficiency in those amino acids taking the minimum daily amino acid requirements for the growing rats from Rose (1937)—see Table 7.13. Black gram and arhar were found

TABLE 7.13. Deficiency of amino acids in pulses compared to their daily requirements

Amino acids	Essential amino acid requirement for growth of rats* as % of diet	Arhar		Black gram		Soybean	
		Present in the diet (%)	Deviation from requirement (%)	Present in the diet (%)	Deviation from requirement (%)	Present in the diet (%)	Deviation from requirement (%)
Methionine	0.6	0.15	−75	0.08	−86	0.10	−83
Tryptophan	0.2	0.039	−80	0.024	−88	0.125	−38
Threonine	0.5	0.39	−22	0.40	−20	—	—
Lysine	1.0	0.66	−34	0.65	−35	—	—

*Source: Rose (1937) for minimum daily amino acid requirements for the growing rats.

to be maximally deficient in both methionine and tryptophan, with a slight difference in their degree of deficiency, while threonine and lysine were not deficient to that extent (Singh et al. 1960). The deficiency of methionine in soybean was of the same order while it was not so with tryptophan (Kapoor and Gupta 1975).

The variation in the nutritive quality of pulse proteins has also been attributed to the availability of amino acids (Table 7.14). The availability of methionine was higher in soybean than arhar (Singh et al. 1960; Kapoor and Gupta 1975). At the same time, the availability of methionine and threonine in arhar was lower than that of tryptophan and lysine. The availability of threonine was the lowest. Supplementation was found to increase the availability of the corresponding amino acid. The variation in the availability of methionine in different pulses was reported by Esh and Som (1953) who observed the availability of tryptophan to be high.

TABLE 7.14. Availability of amino acids in arhar and soybean

Supplement	Methionine %	Tryptophan %	Threonine %	Lysine %
Arhar	75.0	82.0	66.5	81.8
Arhar + Methionine	93.9	—	—	—
Arhar + Tryptophan	—	97.5	—	—
Arhar + Threonine	—	—	81.8	—
Arhar + Lysine	—	—	—	91.8
Soybean	92.0	80.0	—	—
Soybean + Methionine	98.8	—	—	—
Soybean + Tryptophan	—	93.6	—	—

Source: Singh et al. (1960); Kapoor and Gupta (1975).

Pulses when supplemented with cereals were reported to enhance the overall nutritive value of the proteins in the mixed diet since pulses provide lysine in which cereals are deficient. The supplementary effect of beans in combination with maize, providing 50 per cent of the protein, was found to increase the PER value by 122 per cent, while combination of beans with wheat providing 10 per cent of the protein increased the PER by 90 per cent (Table 7.15). Studies by Kapoor and Gupta (1979) from this laboratory on the supplementary effect of soy protein isolate providing 50 per cent protein in the diet revealed that the PER of wheat increased by 29 per cent and that of sorghum by 21 per cent. Phansalkar et al. (1957, 1960) and Patwardhan and Ramachandran (1960) showed marked improvement in the nutritive value of dietary protein in a combined diet consisting of pulses and cereals in the proportion of 3 : 7 having a higher PER value than when cereal or pulse formed the sole source of protein. Supplementary effect of Bengal gram proteins on bajra proteins was reported by Rama Rao et al. (1953). However, Bengal gram proteins did not show any supplementary effect with groundnut (Joseph et al. 1958). Heat processed soybean was superior to Bengal gram in its supplementary value to the poor rice diet (Desikachar et al. 1956). Daniel et al. (1969) observed that a blend of 80 parts of wheat with 20 parts of Bengal gram providing 14 per cent protein on fortification with vitamins and minerals served as a good supplement for growing children. However, the incorporation of soybean at 5–6 per cent in the poor Indian rice, ragi, maize, and pearl millet diet had the same supplementation effect in improving the diet as with 15–16 per cent supplementation with Bengal gram or arhar (1968).

TABLE 7.15. Protein values of cereal-pulse combination

Distribution of protein in the diet (%)		PER	Increase (%)
From cereal	From pulse		
100 Rice	0 Beans	2.25	
80 Rice	20 Beans	2.62	16.4
100 Maize	0 Beans	0.90	
50 Maize	50 Beans	2.00	122.2
100 Maize	0 Cowpea	1.22	
50 Maize	50 Cowpea	1.84	50.8
100 Wheat	0 Beans	1.05	
90 Wheat	10 Beans	1.73	64.7
100 Maize	0 Soybean	1.50	
40 Maize	60 Soybean	2.85	90.0
100 Sorghum	0 Soy protein isolate	1.30	
50 Sorghum	50 Soy protein isolate	1.68	29.2
100 Wheat	0 Soy protein isolate	1.59	
50 Wheat	50 Soy protein isolate	1.92	20.7

Source: Milner (1972); Kapoor and Gupta (1979).

Subrahmanyan et al. (1957) showed that Bengal gram could be used with groundnut flour in the preparation of Indian multipurpose food fortified with vitamins and minerals to serve as a low cost protein food. Joseph et al. (1957) studied the nutritive value of Indian multipurpose food and found it to be comparable to the American multipurpose food. The incorporation of Indian multipurpose food at 12.5 per cent level was reported to have a significant supplementary value to the poor vegetarian diet based on cereals and millets (Kuppuswamy et al. 1957). An improved nutritional status of children supplemented with the Indian multipurpose food was observed (Joseph et al. 1959).

A comparative study with Bengal gram, Indian multipurpose food (blend of 75 per cent of groundnut and 25 per cent of Bengal gram fortified with vitamins and minerals) and skim milk powder by Joseph et al. (1957) revealed that Indian multipurpose food was superior to Bengal gram but inferior to skim milk powder in meeting the protein requirements of the protein depleted rats. Vegetable diets based on Bengal gram or Bengal gram plus rice were successfully used for the treatment of more than 70 cases suffering from Kwashiorkor, although the response to the treatment was more favourable with skim milk powder (Venkatachalam et al. 1956). Protein foods based on blends of groundnut, Bengal gram, soybean and sesame flours, fortified with minerals and vitamins were found to be nutritionally superior for improving the nutritional status of school going children (Daniel et al. 1969; Guttikar et al. 1965).

Germination affects the nutritive value of pulse proteins. Devadatta et al. (1951) reported that germination of Bengal gram significantly improved the digestibility coefficient of the proteins but lowered their BV. Studies with sprouted green gram and horse gram by Chandrasekhar and Chitre (1978) revealed that sprouting had a beneficial effect in promoting growth of rats and improving PER values. Availability of lysine from black gram seed was optimum on germination after 42 hours (Venugopal and Rama Rao 1978).

Pulse digestibility has been found to be affected by the methods of processing and cooking, by the quantities consumed and by the state of the digestive tract. In the raw state, pulses contain substances which are indigestible or antagonistic to digestion such as saponins, glycosides, alkaloids, conjugates of protein with phytin or hemi-cellulose and others which inhibit the action of the digestive enzyme trypsin. Proper cooking or preliminary soaking has been observed to eliminate the action of all such substances. Pancreatic digestion of proteins *in vitro* in Bengal gram, green gram, black gram, arhar, lentil and pea was much slower than that of meat meal and casein. Among the pulses, Bengal gram, lentil and pea were more easily digested than arhar and black gram. *In vitro* availability of essential amino acids and PER of cooked Bengal gram were reported by Rao (1973). *In vitro* digestion of soybeans with trypsin showed wide variation among them (Gupta et al. 1976). Autoclaving markedly improved the *in vitro* digestion of soybean seed. Studies by Manage and Sohonie (1972) showed that the rate of *in vitro* proteolytic digestion of horse gram was not comparable to that of casein. Trypsin inhibitor was reported to be interfering with the proteolytic digestion. Similar results were obtained with double bean (*Phaseolus lunatus*) and field bean (*Dolichos lablab*) (Inamdar and Sohonie 1961; Phadke and Sohonie 1962).

7.2.3. ANTI-NUTRITIONAL FACTORS

Pulses are known to contain toxic substances which include trypsin inhibitors, phytohaemagglutinins (substances which agglutinate red cells and destroy them), lathyric factors causing lathyrism, compounds causing favism, cyanogenetic factors, goitrogenic factors, saponins and alkaloids. It was reported by Liener (1962) that these toxic substances are generally eliminated by soaking and subsequent discarding of the liquid and or by heat treatment at relatively elevated temperatures. Various toxic factors associated with legume proteins have been reviewed by Liener (1974). It was suggested that the nutritionists, food scientists, and plant breeders should be prepared to meet the challenge of the potential carriers of toxic constituents when the world population would be faced with a limited selection of protein foods, most of which will be of plant origin. These substances have been briefly described by Gupta (1980b and 1981b). A detailed account of the toxic constituents of legumes is given in Chapter 5 of this book.

7.2.4. UNIDENTIFIED TOXIC FACTORS

Legumes contain estrogenic factors (Liener 1974). Derivatives of isoflavone combined with a sugar residue were isolated from soybean. These compounds were shown to possess estrogenic activity when tested on female rats and mice.

Incorporation of isolated soy protein in the diet was reported to cause a decrease in the availability of certain trace elements such as zinc, manganese, copper and iron (Odell and Savage 1960). Peas were found to contain a factor which interfered with the availability of zinc for chicks (Kienholz et al. 1962). Autoclaving the peas eliminated the requirements for supplemental zinc.

Toxic factors in raw horse gram (*Macrotyloma uniflorum*) were studied by Ray (1968, 1969). He observed that experimental rats fed on raw seeds exhibited poor growth response and toxic symptoms leading ultimately to death. Fractionation of the toxic factor by ethanol extraction was reported to be lipid-like material possibly bound to a protein moiety. Varshney (1969) reported the presence of saponins in pulses.

7.3. Flatulence

Food legumes are reported to be the notorious producer of flatulence (Calloway 1966; Murphy 1964; Steggerda and Dimmick 1968; Hellendoorn 1969), which causes considerable inconvenience when consumed in large quantities, especially by young children. Ingestion of cooked dry beans causes human flatulence (Calloway 1966; Calloway and Murphy 1968). This characteristic discourages broader use of these low cost high protein foods. The dominant components of flatus gases are carbon dioxide and hydrogen but some persons also pass large amounts of methane (Calloway 1966).

The specific factor or factors responsible for flatulence have not yet been established. However, certain oligosaccharides—sucrose, stachyose, and raffinose—have been identified to be associated with the gas-producing factor. It was reported that stachyose and raffinose in the beans were not rapidly digested. Administration of as little as 2 g of stachyose was shown to cause measurable increase in breath hydrogen (Calloway and Murphy 1968). The pulmonary excretion of hydrogen from 5 g raffinose was almost equal to that from 100 g feed of beans but flatus egestion was not increased appreciably when the sugar was consumed.

bean. On the other hand, Sessa et al. (1969) reported that odour or flavour associated with volatile carbonyl compounds contributed little to the overall soybean flavour. Mattick and Hand (1969) isolated and identified ethyl vinyl ketone, a volatile compound which attribtuted the green beany odour and flavour to the raw bean. They postulated that this compound was derived from linolenic acid through lipoxidase enzyme action. But Khaleque et al. (1970) reported a relatively small contribution by lipoxidase activity to the beany flavour. According to Arai et al. (1966, 1967), phenolic acids, volatile fatty acids and volatile neutral compounds contributed to the beany flavour.

Studies were undertaken to make soybean free from objectionable flavour, and to make it palatable and acceptable as food for human consumption. It was found that soybean seed could be made free from beany flavour and bitter taste by roasting irrespective of its prior soaking in $NaHCO_3$ or not, while soaking and cooking with $NaHCO_3$ were effective by 80 per cent (Kapoor and Gupta 1976; Gupta 1978). Mustakes et al. (1967) observed that soaking in 1 per cent solution of $NaHCO_3$ reduced the soybean flavour. Similar results were obtained by Chandrasekhara et al. (1966).

7.6. Utilisation Problems

According to FAO Production Year Book (1978), next to China, India is the largest producer of food legumes in the world, its estimates being 11.8 million metric tons out of the world total of 62 million metric tons. Their consumption in different parts of the world varies depending on dietary patterns, availability and local conditions. Dietary surveys in 63 countries of the world revealed that food legumes intake per capita per day was 2–13 g in 33 countries, 14–24 g in 10 countries, 25–35 g in 15 countries, 35–36 g in one country and 47–57 g in four countries. In some developing countries, such as Togoland, South of Sahara and India, their dietary intake per capita per day was 13–140 g, 10–150 g and 14–114 g respectively.

Dietary survey on per capita consumption of pulses in different States within India by Gopalan et al. (1971) revealed that in Jammu and Kashmir, Tamil Nadu, it is as low as 8 and 16 g per day respectively and in Uttar Pradesh and Madhya Pradesh, it was as high as 55 g per day (Table 7.17).

In India, pulses are mostly consumed as 'dal' processed by dehusking and milling. The milling quality of food legumes in this country is reported to vary widely due to varietal and agroclimatic characteristics (Parpia 1972). The presence of gums between the husk and cotyledons determines the extent to which the husk adheres to the endosperm. Large-seeded varieties of pulses can be milled more easily than small and medium seeded types and require less drastic premilling treatments. Traditional processing methods were reported to be defective and gave low yields because of powdering and breakage. Khare et al. (1966a) reported a loss of 8.6 per cent in the yield of arhar dal

TABLE 7.17. Consumption pattern of pulses in India

State	Consumption (g/capita/day)
Madhya Pradesh	55
Uttar Pradesh	55
Gujrat	43
Rajasthan	43
Bihar	42
Punjab	35
Maharashtra	33
West Bengal	30
Andhra Pradesh	29
Mysore	26
Kerala	24
Tamil Nadu	16
Jammu & Kashmir	8
All India	34

Source: Gopalan et al. (1971).

during milling and a loss of 4 per cent in case of pea.

Some attempts have been made to improve the traditional methods of processing by the introduction of mechanical devices at CFTRI, Mysore (Desikachar and Subrahmanyan 1961). The adoption of improved technology increased the milling yields of dal by 10 to 15 per cent (Table 7.18), and could add about 1.5 million tons of food legumes to the food supplies.

TABLE 7.18. Per cent average yields of dal from pulses by different processes

Pulse	Home scale process	Conventional commercial methods	Improved CFTRI process	Maximum theoretical yield
Bengal gram	75	75	82	89
Arhar	68	75	83	89
Black gram	69	71	82	88
Green gram	64	65	82	89

Source: Parpia (1972).

Insect and rodent infestation and mold growth during post-harvest handling, storage and distribution of pulses, cause substantial losses, both quantitative and qualitative. Khare et al. (1966b) observed that birds and rodents accounted for considerable losses during drying and storage of pulses. The products of pea were found to be heavily infested. Pulse beetle (*Bruchus chinensis*) reportedly caused heavy destruction of pulse seeds to make them unhygienic because of the presence of excreta (Parpia 1971). Venkat Rao (1960) observed that seeds infested with these pests were unfit for human consumption as these were contaminated with metabolites such as uric acid, and had and mineral matter content in some indigenous and exotic varieties of gram (*Cicer arietinum* L.). *Curr. Sci.* **37**: 237–238.

Chandrasekhara, M.R., Shurpalekar, S.R., Subbaran, B.H., Kurien, S. and Shurpalekar, K.S. 1966. Development of infant food based on soybean. *J. Fd. Sci. Tech.* **3**: 94–97.

Chandrasekhar, U. and Chitre, S. 1978. Evaluation of the protein quality of sprouted horse gram and gram on albino rats. *Indian J. Nutr. Dietet.* **15**: 223–227.

Chandrasekhar, U. and Jayalakshmi, K. 1978. Evaluation of protein quality of sprouted, roasted and autoclaved legumes on albino rats. *Indian J. Nutr. Dietet.* **15**: 414–421.

Chatterjee, S.R. and Abrol, Y.P. 1975. Amino acid composition of new varieties of cereals and pulses and nutritional potential of cereal pulse combinations. *J. Fd. Sci. Tech.* **12**: 221–227.

Chatterjee, S.R., Pokhriyal, T.C. and Abrol, Y.P. 1977. Protein content and amino acid composition of developing seeds of Bengal gram (*Cicer arietinum* L.). *Plant Biochem. J.* **4**: 62–71.

Chatterjee, S.R., Pokhriyal, T.C. and Abrol, Y.P. 1978. Changes in the content of protein and amino acids in developing pea seeds. *Plant Biochem.* **5**: 69–76.

Chernick, A. and Chernick, B.A. 1963. Studies of factors affecting cooking quality of yellow peas. *Canad. J. Pl. Sci.* **43**: 174–183.

Chitre, R.G., Desai, D.B. and Raut, V.S. 1955. Nutritive value of pure bred strains of cereals and pulses. Part I. Thiamine, riboflavin and nicotinic acid contents of 107 pure bred strains of cereals and pulses. *Indian J. Med. Res.* **43**: 575–583.

Crean, D.E.C. and Haisman, D.R. 1963. The interaction between phytic acid and divalent cations during the cooking of dried peas. *J. Sci. Fd. Agric.* **14**: 824–833.

Daniel, V.A., Desai, B.L.M., Subramanya Raj Urs, T.S., Venkat Rao, S., Swaminathan, M. and Parpia, H.A.B. 1968. The supplementary value of Bengal gram, red gram, soybean as compared with skim milk powder to poor Indian diets based on ragi, kaffir corn and pearl millet. *Indian J. Nutr. Dietet.* **5**: 283–291.

Daniel, V.A., Desai, B.L.M., Venkat Rao, S., Swaminathan, M. and Parpia, H.A.B. 1969. Mutual and amino acid supplementation of proteins. IV. The nutritive value of the proteins of blends of wheat and Bengal gram fortified with limiting amino acids. *Indian J. Nutr. Dietet.* **6**: 15.

Das, N.B. 1960. Influence of variety and cultural conditions on the protein content in foodstuffs. In: *Symposium on Proteins*. Chemical Research Committee and Society of Biological Chemists (India), Mysore, pp. 431–434.

Datta, S. and Datta, S.L. 1978. Available lysine in cooked pulses. *Indian J. Nutr. Dietet.* **15**: 128–130.

Deodhar, A.D., Lal, M.S., Sharma, Y.K. and Mehta, S.K. 1973. Chemical composition of vegetable type varieties of soybean. *Indian J. Nutr. Dietet.* **10**: 134–138.

Desikachar, H.S.R., Sankaran, A.N. and Subrahmanyan, V. 1956. The comparative value of soybean and Bengal gram as supplements to the poor South Indian rice diet. *Indian J. Med. Res.* **44**: 741–748.

Desikachar, H.S.R. and Subrahmanyan, V. 1961. The effect of flaking on the culinary quality of pulses. *J. Sci. Industr. Res.* **20D**: 413–415.

Devdas, R.P., Girija Bai, R. and Snehlata, N. 1967. Effect of methionine and tryptophan supplementation to two improved strains of red gram on protein utilisation by albino rats. *Indian J. Nutr. Dietet.* **4**: 300.

Devadatta, S.C., Acharya, B.N. and Nadkarni, S.B. 1951. Effect of germination on the nutritional qualities of some of the vegetable proteins. *Proc. Indian Acad. Sci.* **33B**: 150–158.

Dhingra, P.K. and Das, N.B. 1959. Nutritive value of pure strains of Indian pulses. *Ann. Biochem.* **19**: 245–248.

Esh, G.C. and Som, J.M. 1953. Nutritional survey of available food materials. Part IV. Availability of methionine, cystine and tryptophan in pulses. *Indian J. Physiol. Allied Sci.* **7**: 158–162.

Esh, G.C. and Som, J.M. 1954. Nutritional survey of available food materials. VI. Lysine, threonine and phenylalanine contents of pulses (Bengal gram). *J. Instn. Chem. India* **26**: 147–152.

Esh, G.C. 1955. Studies on the nutritive value of plant proteins. Part II. Influence of Vitamin B_{12} on the nutritive value of pulse proteins. *Indian J. Physiol. Allied Sci.* **9**: 129–133.

Esh, G.C. and Som, J.M. 1955. Studies on the nutritive value of plant proteins. Part I. Pulse proteins—their improvement by amino acid supplementation. *Proc. Natl. Inst. Sci. (India)* **21B**: 68–73.

Esh, G.C., De, T.S. and Basu, K.P. 1959. Influence of genetic strain and environment on the protein content of pulses. *Science* **129**: 148–149.

Esh, G.C., De, T.S. and Basu, K.P. 1960. Nutritive value of proteins of Bengal gram of high and low protein content. *Brit. J. Nutr.* **14**: 425–431.

FAO 1957. Protein Requirements. Food Agriculture Organisation, U.N. *FAO Nutr. Studies No. 16*, FAO, Rome.

FAO/WHO, Joint Expert Committee. 1965. Protein Requirements. *FAO WHO Tech. Rep. Series No. 301*, Rome.

FAO 1970. Amino Acid Contents of Foods and Biological Data on Proteins. *FAO Nutr. Studies No. 24*, FAO, Rome.

Fujimaki, M., Arai, S., Kirigaya, N. and Sakurai, Y. 1965. Studies on flavour components in soybean. Part I. Aliphatic carbonyl compounds. *Agr. Biol. Chem.* **29**: 855–863.

Gopalan, C., Balasubramanian, S.C., Rama Sastry, B.V. and Visweshwara

Rao, K. 1971. *Diet Atlas.* National Institute of Nutrition, Hyderabad, India, p. 37.

Gopalan, C., Ramashastri, B.V. and Balasubramanian, S.C. 1972. *Nutritive Value of Indian Foods.* National Institute of Nutrition, Hyderabad, India.

Grover, H.L., Kapoor, H.C. and Gupta, Y.P. 1971. Effect of dehydration on the quality of pea. *Proc. 58th Indian Sci. Cong.* Part III, p. 765.

Grover, H.L. and Gupta, Y.P. 1972. Estimation of tryptophan in cereals and pulses. *Indian J. Agri. Res.* **6**: 267–272.

Grover, H.L., Kapoor, H.C. and Gupta, Y.P. 1972. Note on quality characters of sundried and dehydrated varieties of pea. *Indian J. Agron.* **17**: 247–249.

Gupta, Y.P. 1967. Recent trends in improving the quantity and quality of proteins in foods. *Proceedings of the Symposium on Science and India's Food Problem*, sponsored by ICAR and INSA, New Delhi, pp. 488–493.

Gupta, Y.P. 1969a. Improving the quantity and quality of proteins through genetic manipulation. *Symp. on New Trends in Agriculture.* ICAR, Kanpur, pp. 18–19.

Gupta, Y.P. 1969b. Protein quality of pulses. *Proc. Third Annual Workshop Conference on Pulse Crops* by ICAR, New Delhi, p. 157.

Gupta, Y.P. 1971. Influence of genetic factor on the quantity and quality of proteins in cereals and pulses. *First Asian Congress of Nutrition, Research Communications.* Hyderabad, p. 10.

Gupta, Y.P. 1978. Nutritional evaluation of soybeans. *47th Annual General Meeting of Soc. Biol. Chem. (India)*, New Delhi.

Gupta, Y.P. 1980a. Improving nutritional quality of pulses. *Indian Fmg.* **30**: 9–11.

Gupta, Y.P. 1980b. Khesari dal consumption—a health hazard. *Indian Fmg.* **30**: 7–9.

Gupta, Y.P. 1981a. Factors influencing nutritive value of pulses. *Pulse Crops Newsletter* **1** (2): 76–77.

Gupta, Y.P. 1981b. Toxic substances in raw pulses. *Indian Fmg.* **31**: 9–11.

Gupta, Y.P. and Das, N.B. 1955. Amino acid content of pure strains of Indian pulses. I. Methionine, cystine and tryptophan. *Ann. Biochem.* **15**: 75–78.

Gupta, Y.P. and Das, N.B. 1956. Thiamine content of pure strains of pulses. *Proc. 43rd Indian Sci. Cong.*, Part III, p. 590.

Gupta, Y.P. and Das, N.B. 1959. Thiamine content of cereals and pulses. *Indian J. Agric. Sci.* **29**: 27–33.

Gupta, J.K. and Gupta, Y.P. 1972. Note on effect of sulphur and molybdenum on quality of pea. *Indian J. Agron.* **17**: 245–247.

Gupta, Y.P., Grover, H.L. and Kapoor, A.C. 1978. Preliminary studies on the quality characters of soybeans. *Curr. Agric.* **2**: 39–43.

Gupta, D.P. and Gupta, A.K. 1977. Biochemical evaluation and cooking qualities of some soybean varieties grown at different locations of Madhya Pradesh (India). *Curr. Agric.* **1**: 50–55.

Gupta, Y.P. and Kapoor, A.C. 1978. Potential of soybean for human consumption. *Indian Fmg.* **27**: 10–12.

Gupta, A.K., Kapoor, M. and Deodhar, A.D. 1976a. Chemical composition and cooking characteristics of vegetable and grain type soybeans. *J. Fd. Sci. Tech.* **13**: 133–137.

Gupta, A.K., Wahie, N. and Deodhar, A.D. 1976b. Protein quality and digestibility *in vitro* of vegetable and grain type soybeans. *Indian J. Nutr. Dietet.* **13**: 244–251.

Guttikar, M.N., Panemangalore, M., Narayana Rao, M., Rajagopalan, R. and Swaminathan, M. 1965a. Studies on processed protein foods based on blends of peanut, Bengal gram, soybean and sesame flours and fortified with minerals and vitamins. I. Preparation, chemical composition and shelf-life. *Indian J. Nutr. Dietet.* **2**: 21–23.

Guttikar, M.N., Panemangalore, M., Narayana Rao, M., Rajalakshmi, D., Rajagopalan, R. and Swaminathan, M. 1965b. Studies on processed protein foods based on blends of peanut, Bengal gram, soybean and sesame flours and fortified with minerals and vitamins. II. Amino acid composition and nutritive value of proteins. *Indian J. Nutr. Dietet.* **2**: 24–27.

Hadi, S.M. 1966. Quantitative changes in nucleic acids and nucleases during germination of *Cicer arietinum*. *Indian J. Biochem.* **3**: 203–204.

Halstead, R.L. and Gfeller, F. 1964. The cooking quality of field peas. *Canad. J. Pl. Sci.* **44**: 221–228.

Hanumantha Rao and Subramanian, N. 1970. Essential amino acid composition of commonly used Indian pulses by paper chromatography. *J. Fd. Sci. Tech.* **7**: 31–34.

Hellendoorn, E.N. 1969. Intestinal effects of following ingestion of beans. *Fd. Tech.* **23** (6): 87–92.

Hirwe, R. and Mager, N.G. 1953. Effect of autoclaving on the nutritive value of pulses. *Indian J. Med. Res.* **41**: 191–200.

Inamdar, A.N. and Sohonie, K. 1960. Studies on nutritive value of double bean (*Vicia faba* Moench.). In: *Symposium on Proteins.* pp. 375–380. Chemical Research Committee and Society of Biological Chemists (India), Mysore,

Inamdar, A.N. and Sohonie, K. 1961. Studies on nutritive value of double bean. Part I. Digestibility of double bean flour *in vitro*. *Ann. Biochem.* **21**: 191–198.

Jain, H.K. 1969. Genetic improvement of food grains for better nutrition. *World Science News* **6**: 30–32.

Joseph, K., Narayana Rao, M., Sankaran, A.N., Swaminathan, M. and Subrahmanyan, V. 1957a. The relative value of the proteins of Indian multipurpose food, Bengal gram (*Cicer arietinum*) and skimmed milk powder in meeting the protein requirements of protein depleted animals. *Ann. Biochem.* **17**: 103–106.

Joseph, K., Narayana Rao, M., Swaminathan, M., Sankaran, A.N. and Subrahmanyan, V. 1957b. Nutritive value of Indian multipurpose food. *Fd. Sci.* **6**: 80–83.

Joseph, K., Narayana Rao, M., Swaminathan, M. and Subrahmanyan, V. 1958. Supplementary values of the proteins of Bengal gram (*Cicer arietinum*) and sesame to groundnut proteins. *Fd. Sci.* **7**: 186–187.

Joseph, K., Narayana Rao, M., Swaminathan, M., Sankaran, A.N. and Subrahmanyan, V. 1959. Effect of long-term feeding of diets containing proteins from low fat groundnut flour, Bengal gram (*Cicer arietinum*) and their blends on the growth and composition of blood and liver of albino rats. *Ann. Biochem.* **19**: 131–138.

Joseph, K., Naryana Rao, M., Swaminathan, M., Indiramma, K. and Subrahmanyan, V. 1960. The nutritive value of protein blends similar to FAO reference protein pattern in amino acid composition. In: *Symposium on Proteins*. pp. 369–374. Chemical Research Committee and Society of Biological Chemists (India), Mysore.

Kadwe, R.S., Thakare, K.K. and Badhe, N.N. 1974. A note on the protein content and mineral composition of twenty-five varieties of pulses. *Indian J. Nutr. Dietet.* **11**: 83–85.

Kapoor, H.C., Grover, H.L. and Gupta, Y.P. 1971. Effect of variety and location on the quality of soybean. *Proc. 58th Indian Sci. Cong. Part III*, p. 803.

Kapoor, H.C. and Gupta, Y.P. 1971. Electrophoretic differences in seed proteins among varieties of soybean and pulses. *Annual General Meeting of the Society of Biological Chemists (India)*, Bangalore.

Kapoor, A.C. and Gupta, Y.P. 1975. Biological evaluation of soybean protein and effect of amino acid supplementation. *J. Fd. Sci.* **40**: 1162–1164.

Kapoor, A.C. and Gupta, Y.P. 1976. Note on cookability and beany flavour in soybean. *Indian J. Agric. Sci.* **46**: 546–548.

Kapoor, A.C. and Gupta, Y.P. 1977a. Note on the effect of roasting on the nutritional quality of soybean protein. *Indian J. Agric. Sci.* **47**: 365–367.

Kapoor, A.C. and Gupta, Y.P. 1977b. Chemical evaluation and electrophoretic pattern of soy proteins. *J. Fd. Sci.* **42**: 1558–1561.

Kapoor, A.C. and Gupta, Y.P. 1977c. Distribution of nutrients in the anatomical parts of soybean seed and different phosphorus compounds in the seed and its protein fractions. *Indian J. Nutr. & Dietet.* **41**: 100–107.

Kapoor, A.C. and Gupta, Y.P. 1977d. Effect of phosphorus fertilization on phosphorus constituents in soybeans. *J. Agric Fd. Chem.* **25**: 670–673.

Kapoor, A.C. and Gupta, Y.P. 1977e. Changes in proteins and amino acids in developing soybean seed and effect of phosphorus nutrition. *J. Sci. Fd. Agric.* **28**: 113–120.

Kapoor, A.C. and Gupta, Y.P. 1978. Trypsin inhibitor activity in soybean seed as influenced by stage of its development and different treatments and the distribution in its anatomical parts. *Indian J. Nutr. Dietet.* **15**: 429–433.

Kapoor, A.C. and Gupta, Y.P. 1979. Amino acid pattern of soy protein isolate and its supplementary effect on biological parameters of cereal proteins. *Curr. Agric.* **3**: 173–179.

Kapoor, U., Kushwah, H.S. and Datta, I.C. 1975. Studies in gross chemical composition and amino acid content of soybean varieties. *Indian J. Nutr. Dietet.* **12**: 47–52.

Kapoor, V.P., Raina, R.M., Samiuddin, Tripathi, R.S., Khan, P.S.H. and Farooqi, M.I.H. 1971. Chemical analysis of seeds from 60 leguminous species. *Sci. & Cult.* **37**: 349–352.

Kapoor, H.C.; Srivastava, V.K. and Gupta, Y.P. 1972. Estimation of methionine in black gram (*Phaseolus mungo* Roxb.), green gram (*P. aureus* Roxb.) and soybean (*Glycine max* (L) Merr.). *Indian J. Agric. Sci.* **42**: 296–299.

Khaleque, A., Bannatyne, W.R. and Wallace, G.M. 1970. Studies on the processing and properties of soy milk. I. Effect of processing conditions on the flavour and composition of soy milk. *J. Sci. Fd. Agric.* **21**: 579–583.

Khare, R.N., Krishnamurthy, K. and Pingale, S.V. 1966a. Milling losses of food grains. I. Studies on losses of red gram (*Cajanus cajan*) during milling. *Bull. Grain. Technol.* **4**: 125–133.

Khare, R.N., Krishnamurthy, K. and Pingale, S.V. 1966b. Milling losses of food grains. II. Studies on losses of peas (*Pisum sativum*) during milling. *Bull. Grain. Technol.* **4**: 169–176.

Kienholz, E.M., Jensen, L.S. and McGinnins, J. 1962. Evidence for a growth inhibitor in several legume seeds. *Poult. Sci.* **41**: 367.

Kon, S. 1968. Pectic substances of dry beans and their possible correlation with cooking time. *J. Fd. Sci.* **33**: 437–438.

Krishnamurthy, K., Taskar, P.K., Ramakrishnan, T.N., Rajagopalan, R., Swaminathan, M. and Subrahmanyan, V. 1958. Effect of heat processing on the trypsin inhibitor and nutritive value of the proteins of soybean. *Ann. Biochem.* **18**: 153–156.

Krishnamurti, C.R. 1975. Biochemical studies on Bengal gram. *Biochem. Rev.* **45**: 43–58.

Krober, O.A. 1968. Nutritional quality in pulses. *J.P.G. School* (IARI) **6**: 157–160.

Krober, O.A. 1969. Quality testing of materials from the coordinated variety field trials. *Third Annual Workshop Conference on Pulse Crops.* pp. 152–155. IARI, New Delhi.

Kulkarni, K. and Sohonie, K. 1956. Nonprotein nitrogen in vegetables.

Indian J. Med. Res. **44**: 511–518.

Kuppuswamy, S., Joseph, K., Narayana Rao, M., Rama Rao, G., Sankaran, A.N., Swaminathan, M. and Subrahmanyan, V. 1957. Supplementary value of Indian multipurpose food to poor vegetarian diets based on different cereals and millets. *Fd. Sci.* **6**: 84–86.

Kuppuswamy, S., Srinivasan, M. and Subrahmanyan, V. 1958. Proteins in foods. *Indian Council of Medical Research Special Report Series No. 33.* New Delhi.

Lal, B.M., Prakash, V. and Verma, S.C. 1963a. The distribution of nutrients in the seed parts of Bengal gram. *Experientia* **19**: 154–155.

Lal, B.M., Rohewal, S.S., Verma, S.C. and Prakash, V. 1963b. Chemical composition of some pure strains of Bengal gram. *Ann. Biochem.* **23**: 543–548.

Lal, B.M. and Verma, S.C. 1968. Physiology of Bengal gram seed. III. Changes in the phosphorus compounds of the seed parts during ripening of the seed. *J. Sci. Fd. Agri.* **19**: 113–116.

Lal, M.S., Mehta, S.K., Deodhar, A.D. and Sharma, Y.K. 1973. Protein and oil content, their correlations and phenotypic stability in soybean as influenced by different environments in Madhya Pradesh. *Indian J. Agric. Sci.* **43**: 14–17.

Liener, I.E. 1962. Toxic factors in edible legumes and their elimination. *Am. J. Clin. Nutr.* **11**: 281.

Liener, I.E. 1974. Toxic factors associated with legume proteins. *Indian J. Nutr. Dietet.* **10**: 303–322.

Manage, L. and Sohonie, K. 1972. Proximate composition, amino acid make-up and *in vitro* proteolytic digestibility of horse gram. *J. Fd. Sci. Tech.* **9**: 35–36.

Mathur, K.S., Sharma, R.D. and Ram, N. 1968. Chromatographic amino acid pattern of various protein fractions of Bengal gram. *Indian J. Med. Res.* **56**: 863–866.

Mattson, S., Akerberg, E., Eriksson, E., Koutler, A.E. and Vahtras, K. 1950. Factor determining the composition and cookability of peas. *Acta. Agr. Scand.* **1**: 40–61.

Mattick, L.R. and Hand, D.B. 1969. Identification of a volatile component in soybeans that contributes to the raw bean flavour. *J. Agric. Fd. Chem.* **17**: 15–17.

Milner, M. 1972. Nutritional improvement of food legumes by breeding. *Proceedings of a Symposium sponsored by Protein Advisory Group of United Nations System*, New York.

Mitchell, H.H. and Block, R.J. 1946. Some relationship between amino acid contents of proteins and their nutritive values for the rat. *J. Biol. Chem.* **163**: 599–620.

Murphy, E.L. 1964. Flatus. Conference on nutrition in space and related waste problems. *Natl. Aeronaut. and Space Adm. Rept. NASA. SP.* **70**: 255–259.

Murphy, E.L. 1972. The possible elimination of legume flatulence by genetic selection. In: *Nutritional Improvement of Food Legumes by Breeding.* pp. 273–276. M. Milner, ed. *Proc. Symp. Protein Advisory Group.* New York.

Mustakes, G.C., Albrecht, W.J., Bookwalter, G.N. and Griffin, E.L. 1967. The effect of preprocessing conditions on the flavour of soybean. USDA Publication ARS—71–34, Washington D.C.

Nagpal, M.L. and Bhatia, I.S. 1971. Tryptophan content of some Indian foods and feeds. *Indian J. Nutr. Dietet.* **8**: 183–185.

Naik, M.S. and Narayana, N. 1959. Relation of maturity to the composition of seeds of gram (*Cicer arietinum*). *Indian J. Appl. Chem.* **22**: 239–242.

Nandi, N. and Banerjee, S. 1950. Effect of germination on the Vitamin C content of pulses grown in Bengal. *Indian Pharmacist* **5**: 63–68.

Nandi, N. and Banerjee, S. 1950. Effect of germination on the riboflavin content of pulses grown in Bengal. *Indian Pharmacist* **5**: 202–205.

Narayana Rao, M., Rajagopalan, R., Swaminathan, M. and Subrahmanyan, V. 1959. The chemical composition and nutritive value of Bengal gram (*Cicer arietinum*). *Fd. Sci.* **8**: 391–395.

Narayana Rao, D. and Rajagopal Rao, D. 1974. The nature of major protein components of the newer varieties of *Cicer arietinum. J. Fd. Sci. Tech.* **11**: 139–140.

Odell, B.L. and Savage, J.E. 1960. Effect of phytic acid on zinc availability. *Proc. Soc. Exptl. Biol. Med.* **103**: 304.

Oser, B.L. 1959. An integrated essential amino acid index for predicting the biological value of protein. In: *Protein and Amino Acid Nutrition.* A.A. Albanese, ed. pp. 281–295. Academic Press, New York.

Pant, R. and Kapur, A.S. 1963a. The soluble carbohydrates of some Indian legumes. *Naturwissenschaften* **50**: 95.

Pant, R. and Kapur, A.S. 1963b. A comparative study of the chemical composition and nutritive value of some Indian pulses and soybean. *Ann. Biochem.* **23**: 457–460.

Pant, R. and Kapur, A.S. 1964. Free amino acids in some edible and non-edible Indian legumes. *Z. Physiol. Chem.* **338**: 39–41.

Pant, R. and Tulsiani, D.R.P. 1968. Free amino acid analysis of *Phaseolus* seeds. *J. Fd. Sci. Tech.* **5**: 138–139.

Parpia, H.A.B. 1971. Increased production and utilization of legumes for supplementing human diets. *Report of International Conference*, MIT, Boston, Mass, p. 103.

Parpia, H.A.B. 1972. Utilisation problems in food legumes. In: *Nutritional Improvement of Food Legumes.* M. Milner, ed. pp. 281–295. Proc. Symp. Protein Advisory Group, New York.

Patwardhan, V.N. and Ramachandran, M. 1960. Vegetable proteins in nutrition. *Sci. & Cult.* **25**: 401–407.

Phadke, K. and Sohonie, K. 1962. Nutritive value of field bean (*Dolichos lablab*). I. *In vitro* digestibility of raw and autoclaved beans. *J. Sci. Industr. Res.* **21C**: 155–158.

Phansalkar, S.V., Ramachandran, M. and Patwardhan, V.N. 1957. Nutritive value of vegetable proteins. I. Protein efficiency ratios of cereal and pulses and the supplementary effect of the addition of a leafy vegetable. *Indian J. Med. Res.* **45**: 611–621.

Phansalkar, S.V. 1960. Nutritive evaluation of vegetable proteins. In: *Symposium on Proteins.* pp. 345–354. Chemical Research Committee and Society of Biological Chemists (India), Mysore,

Ramachandran, M. and Phansalkar, S.V. 1956. Essential amino acid composition of certain vegetable foodstuff. *Indian J. Med. Res.* **44**: 501–509.

Rama Rao, G., Murthy, H.B.N. and Swaminathan, M. 1953. Supplementary relations of Bengal gram and groundnut proteins to bajra (*Pennisetum typhoideum*) proteins. *Bull. Cent. Food. Technol. Res. Inst.* **3**: 44.

Rama Rao, P.B., Norton, H.W. and Johnson, B.C. 1964. The amino acid composition and nutritive value of proteins. V. Amino acid requirements as a pattern for protein evaluation. *J. Nutr.* **82**: 88–92.

Ranjan, S., Pandey, R.M., Srivastava, R.K. and Laloraya, M.M. 1962. Effect of phosphorus deficiency on the metabolic changes in free amino acids in certain leguminous plants. *Nature* (London) **103**: 997–908.

Rao, P.S. 1969. Studies on the digestibility of carbohydrates in pulses. *Indian J. Med. Res.* **57**: 2151–2157.

Rao, S.P. 1976. Nature of carbohydrates in pulses. *J. Agric. Fd. Chem.* **24**: 95.

Rao, G.R. 1974. Effect of heat on the proteins of groundnut and Bengal gram. *Indian J. Nutr. Dietet.* **11**: 268–275.

Rao, M. 1973. *In vitro* availability of essential amino acids and the protein efficiency ratio of cooked *Cicer arietinum. Diss. Abst. Int. B.* **33**: 4369.

Ray, P.K. 1968. A comparison of growth of rats fed with the raw seeds of horse gram (*Dolichos biflorus*). *Sci. & Cult.* **34**: 350–352.

Ray, P.K. 1969. Toxic factor(s) in raw horse gram (*Dolichos biflorus*) *J. Fd. Sci. Tech.* **6**: 207–208.

Ray, P.K. 1970. Nutritive value of horse gram (*Dolichos biflorus*). III. Determination of biological value, digestibility and net protein utilisation. *Indian J. Nutr. Dietet.* **7**: 71–73.

Rohtagi, K., Banerjee, M. and Banerjee, S. 1955. Effect of germination on B_{12} values of pulses. *J. Nutr.* **56**: 403–408.

Rosenbaum, T.M., Henneberry, G.O. and Baker, B.E. 1966. Constitution of leguminous seeds. VI. The cookability of field peas (*Pisum sativum* L.). *J. Sci. Fd. Agric.* **17**: 237–240.

Rose, W.C. 1937. The nutritive significance of the amino acids and certain related compounds. *Science* 17: 298–300.

Sessa, D.J., Honig, D.H. and Rackis, J.J. 1969. Lipid oxidation in full fat and defatted soybean flakes as related to soybean flavour. *Cereal Chem.* 46: 675–686.

Sharma, K.P. and Goswami, A.K. 1971. Chemical evaluation of some high yielding varieties of Bengal gram (*Cicer arietinum*). *Indian J. Agri. Res.* 5: 109–111.

Sharma, Y.K., Tiwari, A.S., Rao, K.C. and Mishra, A. 1977. Studies on chemical constituents and their influence on cookability in Pigeon pea. *J. Fd. Sci. Tech.* 14: 38–40.

Singh, R.G. 1971. Response of gram (*Cicer arietinum* L.) to the application of nitrogen and phosphate. *Indian J. Agri. Sci.* 41: 101–106.

Singh, D.K., Gupta, Y.P. and Das, N.B. 1960. Effect of amino acids and Vitamin B_{12} on the nutritive value of pulse protein. *Ann. Biochem.* 20: 1–6.

Singh, L., Sharma, D., Deodhar, A.D. and Sharma, Y.K. 1973. Variation in protein, methionine, tryptophan and cooking period in Pigeon pea (*Cajanus cajan* (L.) Mill sp.). *Indian J. Agri. Sci.* 43: 795–798.

Singh, S., Singh, H.D. and Sikka, K.C. 1968. Distribution of nutrients in the anatomical parts of common Indian pulses. *Cereal Chem.* 45: 13–18.

Sivaraman, E. and Menachery, M. 1967. Studies on the nutritive value of cowpea (*Vigna catiang*) and tur dal (*Cajanus cajan*). *Indian Vet. J.* 44: 107.

Shurpalekar, S.R., Chandrasekhar, M.R., Swaminathan, M. and Subrahmanyan, V. 1961. Chemical composition and nutritive value of soybean and soybean products. *Fd. Sci.* 10: 52–64.

Sood, D.R., Wagle, D.S., Nainawatee, H.S. and Gupta, V.P. 1977. Varietal differences in the chemical composition of soybean (*Glycine max* L.) *J. Fd. Sci. Tech.* 14: 177–178.

Srivastava, A.K. and Kooner, N.K. 1972. Physiological and biochemical studies in seed germination of mung (*Phaseolus aureus* Roxb). *Indian J. Exptl. Biol.* 10: 304–306.

Steggerda, F.R. and Dimmick, J.F. 1968. Effects of bean diets on concentration of carbon dioxide in flatus. *Am. J. Clin. Nutr.* 19: 120–124.

Subrahmanyan, V., Rama Rao, G., Kuppuswamy, S., Narayana Rao, M. and Swaminathan, M. 1957. Standardisation of conditions for the production of Indian multipurpose food. *Fd. Sci.* 6: 76–80.

Swaminathan, M.S., Austin, A., Kaul, A.K. and Naik, M.S. 1969. Genetic and agronomic enrichment of the quantity and quality of proteins in cereals and pulses. In: *New Approaches to Breeding for Improved Plant Protein.* pp. 71–86. International Atomic Energy Agency, Vienna.

Swaminathan, M.S., Naik, M.S., Kaul, A.K. and Austin, A. 1970. Choice of strategy for the genetic upgrading of protein properties in cereals, millets and pulses. In: *Improving Plant Protein by Nuclear Techniques.*

8

Genetic Improvement of Grain Legumes

J. SMARTT

In a review of this nature it is obviously impracticable to cover all aspects of the genetic improvement of all leguminous crops. What can be attempted broadly is to consider the major areas of concern to grain legume breeders and to suggest appropriate strategies for implementation. As a first step, it is necessary to examine carefully the objectives which breeders have defined and to make any modifications which recent advances in research suggest. Secondly, it is necessary to review the genetic resources which are available in individual crops and which will determine whether the objectives set are likely to be attainable.

In the last decade, two highly significant events forced legume breeders to reconsider their selection objectives. The first was the world energy crisis which has produced enormous increases in the costs of agricultural chemicals, especially nitrogenous fertilizers. This has placed a considerable premium on the ability of the legumes to support the *Rhizobium* nitrogen fixing symbiosis and it behoves plant breeders to evaluate very critically their own selections in terms of their nitrogen fixing ability. The second was a drastic reappraisal of human protein requirements in nutrition (Payne 1977). This obliges us to determine the optimal rather than the maximum protein content and quality in materials produced for direct human consumption. Different considerations obviously apply to animal feeding stuffs where a high protein content is generally desirable but where the protein quality required is determined by its ultimate destination. The amino acid profile of feeding stuffs fed to nonruminants is obviously more critical than that which is fed to ruminants.

In common with other plant breeders, the legume breeders' prime objective is to develop genotypes capable of producing optimal crop yields of a satisfactory quality in suitable areas of production. This comparatively simple statement of aims has very wide implications for the breeder in the determination and definition of his selection objectives.

First, it is essential that the selected genotype be physiologically efficient. It must have a vegetative morphology which permits it to intercept and fix light energy efficiently. It must also be able to use this energy effectively in the production of dry matter. The distribution pattern of assimilates must be such as to maximize useful dry matter production, i.e. it must have a *high harvest index*. The uptake and utilization of water and soil nutrients must be efficient and effective as also the symbiosis with appropriate strains of *Rhizobium*. Secondly, high intrinsic resistance to prevalent pest and disease attack must also be incorporated so that as far as possible the cost of protectants is saved and the useful dry matter produced is used in the greatest possible measure for its intended purpose. Thirdly, the quality of the produce should be satisfactory for the use which is to be made of it. It is obviously wasteful to select for a standard of quality substantially in excess of what is required. It is important that the level of quality achieved should be maintained in storage and in shipping for a reasonable length of time after harvest.

Bearing in mind these considerations we can therefore consider the position in terms of the following: (i) physiological efficiency, (ii) resistance to pests and diseases, (iii) quality, and (iv) germplasm resources. Appropriate breeding strategies can be considered under each heading.

8.1. Physiological Efficiency

It has often been stated that the plant physiologist is able to tell the plant breeder after the event why his varieties are successful but has been able to give little useful advice on selection criteria. This situation is now changing and the plant physiologist can often give effective aid when alternative choices are open to a plant breeder in carrying out selection. The physiological processes which are of most concern to the breeder are *light interception* and *photosynthesis*, the uptake of raw materials for assimilation (carbon dioxide, mineral nutrients and water), subsequent biosynthesis, and the partitioning of assimilates between various sinks in the plant itself.

Light interception and photosynthesis are the most important and here it is apparent that the breeder can more easily manipulate the efficiency of light interception than photosynthesis. In legumes such as the groundnut and the common bean (*Phaseolus vulgaris*) a range of morphological variants occur which clearly differ from each other in their efficiency of light energy interception. Determinate dwarf varieties of *Phaseolus vulgaris* and the fastigiate sequentially branched forms of the groundnut produce a leaf area smaller than the remaining forms found in the two species. Individual leaflet size is another variable. Genetic manipulation of the individual leaf area and leaf number is thus possible. When the optimal leaf size and number has been determined for a particular locality, selection standards can be set for that locality. These standards will be influenced by local climatic and meteorological conditions. The canopy geometry appropriate for the production of common beans under irrigation in North America is unlikely to be equally appropriate in the United Kingdom for example. However, the consideration of selection criteria imposed in cultivar production may permit a saving of effort when introductions are made from similar climatic zones; such introductions may be exploited after relatively little preliminary evaluation and assessment.

The efficiency of mineral uptake, water and carbon dioxide is commonly measured indirectly by the overall dry weight gains (Duncan et al. 1976), and similarly partitioning of assimilates. It is obvious that a genotype in which life span coincides with the duration of the crop season will be at an advantage, other things being equal, compared with those with shorter or longer growing seasons. The former will waste part of the growing season while the latter will perhaps be unable to mature much of the crop. The pattern of assimilate partitioning is also exceedingly important. In order to produce a high harvest index a minimum of overlap between the vegetative and reproductive phases of development is desirable. Ideally the vegetative framework of the plant should be established quickly which will effectively intercept the incident light energy so that all the leaf area operates above the compensation point. After this is achieved, no superfluous vegetative aerial growth should occur. Similar conditions apply to the root system; excessive vigour of growth in the root system is as undesirable in its way as in the canopy. An unduly aggressive or luxuriant growth in either aerial parts or the root system is highly undesirable. The growth of the canopy is of course more readily monitored than that of the root system. A study of the physiological efficiency of root systems in different genotypes might be of value in particular cases.

By these tokens, determinate growth patterns or approximations to them are likely to be the most efficient in the production of maximum yields in a fixed time. When the canopy has been established the bulk of assimilates can be diverted to reproductive channels and produce maximal seed yields. In circumstances where sustained production at a less intense level is required, the indeterminate growth habit, where both vegetative and reproductive growth and development continue indefinitely, may be advantageous. This is particularly so in the humid tropics where it may be more convenient to produce and use fresh material rather than store it.

In practice therefore the plant breeder can produce the most physiologically efficient types by selecting the most desirable canopy geometry for his area and by carrying out further selection both for high yield and high harvest index. These can be performed without the active involvement of a plant physiologist. However, if a plant physiologist can be involved in this work so much the better since this can help determine the existence of intrinsic physiological variability within a population and indicate scope for further selection.

8.1.1. NODULATION

The almost unique feature of legumes is the nitrogen fixing symbiosis they establish with *Rhizobium*. This feature should make the legumes the most highly prized crop plants on earth since with effective nodulation a legume crop is not only self-sufficient for its own nitrogen needs but may well also raise the soil nitrogen status to the advantage of a succeeding non-leguminous crop. Furthermore, this nitrogen is released slowly in the soil and will not produce any nitrogen polluted run-off. With these inbuilt advantages it is astonishing that plant breeders working on grain and oilseed legumes have largely ignored this aspect. In terms of the energy-cost effectiveness in agriculture *Rhizobium* nitrogen fixation is energy sparing and it could be argued that money invested in breeding efficient nitrogen fixing genotypes would save both capital investment costs of a nitrogen fertilizer plant and transport costs on fertilizers.

It is perhaps unfortunate that in the development of major grain legume crops so little attention has been paid to the selection of efficient host genotypes and effective *Rhizobium* strains on these hosts. This has imposed perhaps an unnecessarily low ceiling on the yields attainable in the cultivation of grain legumes in certain advanced agricultural economies. The utilization of nitrogen by legumes is less efficient when absorbed from the soil than when fixed by *Rhizobium*. It is therefore not economically satisfactory to supplant nodular fixed nitrogen by fertilizer nitrogen (de Mooy et al. 1973). Recent reports of high groundnut yields in Central Africa, ca 6 tonnes kernels/hectare (Hildebrand and Smartt 1980), have indicated that the comparison made between the levels of cereal and legume yields are in need of revision. Like is not being compared with like in that high legume yields are not produced by the addition of artificial nitrogenous fertilizers and also that the energy content of their oil and protein is higher than that of carbohydrate.

It can be argued that the selection for efficiently nodulating genotypes of grain legumes is of great importance. There are indications that a highly effective nodulation occurs in some grain legumes, e.g. *Psophocarpus tetragonolobus* (Masefield 1961) and it is probable that if, inadvertently, selection under domestication has reduced nodulation and its effectiveness, this could be restored by crossing to more primitive forms and selection.

8.2. Resistance to Pests and Diseases

Resistance to the attacks of pests and diseases is important since it enables a crop to withstand satisfactorily the onslaught of pests and pathogens, maintaining stable levels of crop production and saving costs of protectants and pesticides. It is of little use to develop a highly productive cultivar without the best genetic protection against pests and diseases that can be provided. Unfortunately such protection may be less than perfect. Sometimes new problems arise quite suddenly, e.g. Cylindrocladium black rot in groundnuts, first recognized in 1965 (Garren and Jackson 1973).

On the whole, resistance to disease attack has attracted more attention from plant breeders than pest resistance. Much of this work has been concerned with the rusts of cereals and much has been learned regarding how best to handle genetic resistance to disease in breeding programmes. Van der Plank (1968) distinguishes two types of disease resistance, namely, vertical and horizontal resistance. Vertical resistance is typically high grade but specific to a single or restricted range of biotypes, while horizontal resistance may be less effective but is non-specific. It is fortunate that the effects of genetically distinct resistance factors are often additive. Whereas the use of vertical resistance genes singly is often effective for only a short time, two or more such factors in combination usually give a much longer term protection. In terms of durable resistance to disease, the cumulative effects of different horizontal resistance factors are often more effective in the long term though less spectacular than those of vertical resistance factors. The latter are usually quite spectacular initially in controlling a pathogen but unfortunately breakdown of resistance may be equally spectacular. The work of Ogle and Johnson (1974) in Australia suggests that the situation regarding bean rust (*Uromyces appendiculatus*) on *Phaseolus vulgaris* beans is similar to that of rusts on cereals. Major gene resistance (vertical resistance) does not appear to be very durable. The durability of genetically controlled host resistance of the vertical type obviously depends on the mutability of loci controlling virulence in the pathogen and the effectiveness of the fungal genetic system in producing recombinants. Mutability is perhaps the more important of the two.

The spectacular, though short-lived, successes in major gene disease resistance breeding has unfortunately led to the neglect of minor gene resistance which probably would be of more lasting effectiveness. In the case of the groundnut, *Arachis hypogaea*, some resistance to the leafspot pathogens (*Mycosphaerella* spp.) is found in the alternately branched cultivars (subsp. *hypogaea*), the nature of this resistance has been little studied (Hemingway 1957) and no attempts have been made to intensify its expression by hybridization and selection. Instead, efforts have been directed to the incorporation of higher grade resistance found in wild diploid species in section *Arachis* of the genus. Smartt et al. (1978) have commented on the possible limitations of this approach. It seems sensible to incorporate and utilize all possible genetic resources in controlling diseases and not to neglect any.

Considerable work has already been undertaken on the identification and evaluation of disease resistance in important legumes (Garren and Jackson 1973; Athow 1973; Kennedy and Tachibana 1973; Dunleavy 1973; Bean Pro-gram (CIAT) Report 1978). While the genetic control of the resistances has not been elucidated in most instances, all promising selections can be tested for reaction to locally important pathogens.

The situation is broadly similar for nematode and insect pests. Useful resistance has been reported for soybeans (Good 1973; Turnipseed 1973), for groundnuts (Garren and Jackson 1973; Bass and Arant 1973) and for the common bean (CIAT Annual Report 1978). The situation is broadly similar in other edible legumes.

8.2.1. Breeding Strategies for Pest and Disease Resistance

The genetic control of resistance to the pests and diseases of most grain legumes has not been studied in detail. Sufficient study has been carried out on some diseases such as bean rust (*Uromyces appendiculatus*) to suggest that the concepts of vertical and horizontal resistance might have some useful application in formulating breeding strategy. The dangers of reliance on vertical resistance to bean rust in *Phaseolus vulgaris* are already apparent. Horizontal resistance could be most usefully employed when the progeny from parents showing rather low levels of resistance show transgressive segregation. If this process can be repeated, accumulation of horizontal resistance factors could be achieved which might equal resistance conferred by the major genes but be more stable. It is possible to combine the vertical and horizontal resistance (or major and minor gene resistance) in a single genotype. This combination is not achieved without some difficulty. The presence of both types of resistance has to be confirmed by the genetic analysis of progeny from the selected line crossed with a parent carrying neither vertical nor horizontal resistance factors.

At the present time, breeding for disease and pest resistance in legumes is an empirical process. It may not always be possible to incorporate high grade resistance into new cultivars; where this cannot be done, tolerance of the pest or disease can be exploited (Coyne and Schuster 1976). With more extensive collection and evaluation of germplasm occurring in the grain legumes, the range of exploitable genetic resources can be expected to increase. This includes not only advanced cultivars and primitive land races but also wild species.

8.3. Quality

Some of the most difficult problems encountered by the legume breeder are encountered in the definition of breeding objectives concerned with quality. The aspect of quality which is of particular interest concerns the chemical composition of the seed and the properties of its various constituents. The biochemist can often give considerable guidance in determining the ideal selection objectives but may not always be able to advise on how best to achieve them. Much, of course, depends on the use made of the crop in question. As an example, the objective of selection of the soybean might equally be to increase either the protein or the oil content (but not both). According to Brim (1973) little apparent progress has been achieved in increasing oil content while genetic control of quality is complex. The situation regarding protein content is somewhat different, substantial increases up to $\pm$ 50 per cent protein content have been achieved while maintaining yield at acceptable levels. In general there is a trend for yield and protein content to be inversely related. The genetic control of protein content and quality does appear to be even more complex than is the case with the oil. It is apparent that selection for reduced linolenic acid would improve the quality of soybean oil but as Brim (1973) has reported although variation in linolenic acid content does occur, this is of small magnitude. The situation regarding protein quality and nutritional properties in the soybean and other grain legumes is complicated and will be considered in the section immediately following.

8.3.1. NUTRITIONAL ASPECTS

Legume seeds are almost without a rival as concentrated protein sources. Protein contents range from 17.1 per cent at 9 per cent moisture for seeds of chickpea to 29.6–50.3 per cent at 5–9.4 per cent moisture for soybeans (Smartt 1976)—typical contents for soybean are probably in the range 35–50 per cent and similarly for lupins. Even the lowest protein contents of legume seeds are higher than and frequently double those of cereals. The utilization of this protein is not always simple and straightforward. While virtually no problems attend the use of the garden pea (*Pisum sativum*) considerable problems are encountered in the direct use for human food of the soybean (*Glycine max*) seed. The incidence and nature of these problems has been ably reviewed by Liener (1978). He considers the following classes of compounds, protease inhibitors, phytohaemagglutinins, glycosides, metal binding constituents, anti-vitamins and the incitants of lathyrism and favism. These compounds can be considered as being either anti-metabolites or toxins or both since the pathological effects may take an appreciable time to develop and result from antagonistic effects on some metabolic process or other.

8.3.2. PROTEASE INHIBITORS

The most widely studied of these inhibitors has been the *trypsin inhibitor* of the soybean. This apparently combines with the enzyme to produce an inactive complex. Since the secretion of the pancreatic enzymes (trypsin and chymotrypsin) is controlled by feedback inhibition, inactivation by protease inhibitors stimulates additional secretion by the pancreas. This, if sustained, produces pancreatic hypertrophy and since the pancreatic secretion is rich in cystine, the loss of sulphur-amino acids (deficient in the soybean protein) also ensues. However, protease inhibitors are themselves proteinaceous and are inactivated by heat or by microbiological fermentation and their undesirable effects can be neutralized in these ways. Brim (1973) noted that the variants of the trypsin inhibitors had been reported, but whether these have different effects on digestibility had not been determined. Until the role of protease inhibitors in the plant is better understood it is probably not possible to make a realistic assessment of whether their concentration in the seed should be reduced by selective breeding or not.

8.3.3. PHYTOHAEMAGGLUTININS

Phytohaemagglutinins or lectins are proteinaceous and have the ability to agglutinate red blood cells. They lose this ability on heating. Physiologically these materials decrease the rate of absorption of digested food in the gut. Their occurrence and effects have been studied most extensively in *Phaseolus vulgaris* where they appear to be an important toxic-antimetabolic factor in addition to the trypsin inhibitors which are also present (Honovar, Shih and Liener 1952). By way of contrast phytohaemagglutinins appear to be absent in some other pulses, e.g. *Cicer arietinum*, *Cajanus cajan* and *Vigna radiata*, these are also much lower in antitryptic activity. Systematic studies have been made of the incidence of phytohaemagglutinins in *Phaseolus vulgaris* by Jaffé and co-workers (Jaffé, Brücher and Palozza 1972) and a polymorphism was found. This polymorphism has been studied serologically by Klozová et al. (1976) who have found that the cultivar Krupnaya sakharnaya was lacking haemagglutinating activity (Klozová and Turková 1978a and b). In place of their proteins I and II, proteins of different serological specificity but similar electrophoretic mobility occur instead. It would appear therefore that there is considerable scope for selection against phytohaemagglutinins since polymorphism for these factors is already established and the genetic control also appears to be simple (Kloz and Klozová 1968). Breeding for the absence of phytohaemagglutinins is of particular importance if the material is to be used uncooked in livestock rations. It is of less importance when beans are first cooked since this usually inactivates the phytohaemagglutinins. Liener (1978) recommends the use of standard trypsinated cow blood cells to test for agglutinating activity.

8.3.4. CYANOGENIC GLYCOSIDES

Cyanogenic glycosides are found in a wide variety of plants in different plant organs, in the foliage of sorghum and white clover, in the tubers of cassava and in the seed of some legumes, most notably the lima bean (*Phaseolus lunatus*). According to Montgomery (1969) samples of lima bean implicated in poisoning humans fatally produced 210.0–312.0 mg HCN/100 g seed, normal levels are in the range 14.4–16.7 for non-toxic strains. Even this level is considerably higher than that found in other common pulses (± 2 mg/100 g) in *Vigna unguiculata, Pisum sativum, Phaseolus vulgaris, Cicer arietinum* and *Cajanus cajan*.

There would seem to be little cause for concern regarding the cyanogenic glycoside content of any pulse other than *Ph. lunatus* where the situation does require constant monitoring.

8.3.5. METAL-BINDING CONSTITUENTS

The requirements for certain metals such as zinc, manganese and calcium in the diet are increased when materials containing appreciable amounts of phytic acid are fed. The formation of metal conjugates is reduced by heat inactivation, the use of phytase or the removal of phytins. Metal-binding constituents have been found in soybean (Liener 1969) and garden pea (Kienholz et al. 1962). The problem is of minor significance unless large quantities of seed or meal are fed uncooked.

8.3.6. ANTI-VITAMINS

Soybean seed contains materials which interfere with the utilization of vitamins A, D, E and B_{12} while antagonists of vitamin E have been reported from *Phaseolus vulgaris*. All except the anti-vitamin A factor appear to be thermolabile (Liener 1978).

8.3.7. INCITANTS OF LATHYRISM AND FAVISM

Neurolathyrism, a condition in which the central nervous system is affected producing weakness and paralysis of leg muscles and even death in extreme cases, has been associated with the consumption of *Lathyrus sativus*, the grass pea. The incitant has been tentatively identified as β-N-oxalyl-a, β diaminopropionic acid which possibly acts antagonistically in amino acid metabolism (Adiga et al. 1962). A comprehensive survey of landraces and wild accessions of *Lathyrus sativus* is indicated, in addition to study of other *Lathyrus* species which might be crossed with it. Liener (1978) however points out that other neurotoxins are found in the genus and in addition incitants of osteolathyrism in which skeletal abnormalities are produced.

Favism is a haemolytic disease produced in susceptible individuals by ingestion of *Vicia faba* beans. The incitants appear to be the pyrimidines, divicine and isouramil (Mager et al. 1965). These act by diminishing levels of reduced glutathione and glucose-6-phosphate dehydrogenase activity which in turn reduce the stability of blood cell membranes and bring about haemo-lysis followed by anaemia in severe cases. A similar effect is exerted by dopaquinone (Beutler 1970). However Liener (1978) casts some doubt on the role of divicine as an incitant of favism since this material occurs in other legumes which do not produce the disease. There is no evidence available at present to suggest that selection could be effective in reducing the concentrations of the favism factor or factors in *Vicia faba*, the only plant known to produce this condition.

8.3.8. ALKALOIDS

The concentration of alkaloids in the seed of established pulse and legume oilseed crops does not present any problem today. However, in recent years, considerable interest has been generated in developing *Lupinus* species as a grain crop. One of the major difficulties is the occurrence of alkaloids in the seed. Genotypes do occur in which alkaloids are virtually absent, the "sweet" lupins, which offer a real prospect for development of a high protein grain crop where the soybean cannot be grown (Gladstones 1970; Hill 1977).

8.3.9. FLATULENCE FACTORS

As Calloway (1973) stated, legumes are notorious inducers of flatulence. Where they constitute a small proportion of the diet little inconvenience results, but where they are a major component of the diet this can be considerable. The gas produced in flatulence comprises hydrogen and carbon dioxide largely but methane may also be produced depending on the microbial flora of the gut. This situation is produced when man ingests food containing materials he cannot digest but which some members of his gut flora can. Indigestible oligosaccharides, stachyose, raffinose and verbascose are thought to be major flatulence factors.

There does appear to be variation within species in concentration of the flatulence factors (Murphy 1973). These are particularly high in *Phaseolus vulgaris* beans on the whole; appreciably lower in *Phaseolus lunatus*, some cultivars of which have very low flatulence inducing properties. Green gram, soybean and *Pisum sativum* appear to be generally intermediate while the groundnut and some *Pisum sativum* cultivars are low in flatulence factors. It would seem that there is scope for selection for a reduction in the flatulence factor concentration especially in those pulses likely to be consumed in quantity.

8.3.10. BREEDING STRATEGY FOR IMPROVEMENT OF NUTRITIONAL QUALITY

The improvement of the nutritional quality of legume seeds has two aspects—

the largely negative aspect of reducing the concentration of or eliminating toxic and other undesirable material and the positive aspect of improving content and quality of the protein present. Progress in the improvement of legumes by the elimination or reduction of phytohaemagglutinins and alkaloids does seem to be promising. Other facets of "negative" improvement such as the reduction of the protease inhibitor contents require more extensive investigation before the prospects can be assessed satisfactorily. Careful evaluation of the more positive aspects of improvement is necessary in order to direct the effort into the most appropriate channels.

From the standpoint of nutritional value, the protein component of the legume seed must be considered in terms of the proportion it constitutes of the total dry matter and secondly in terms of its amino acid composition. Certain broad generalizations can be made. The protein content of most legume seeds lies in the range 20–40 per cent which is considerably higher than that found in cereals. In amino acid composition, lysine is usually present at a level higher than found in reference protein (casein) while the sulphur amino acids occur at sub-optimal levels. The level of tryptophan often tends to be sub-optimal. Our objective in selection must be determined by the intended use to which the material is to be put. Plant proteins produced by grain legumes are destined usually to be fed either directly to humans or livestock. It is also unlikely that they will be the sole protein source although they might well be the most significant. The implications which these facts have for the breeder are important.

Payne (1977) has argued very strongly that efforts made to improve the protein content and composition of crops are a waste of time. Since protein requirements for human nutrition have been revised downward in recent years, it would seem that most human diets in fact contain adequate protein both in amount and quality. The limiting essential amino acids, determined by comparison with the amino acid profile of reference protein, are probably present in most diets in adequate amounts. Payne also points out that since there is usually an inverse correlation between yield and protein content, plant breeders would be best advised to select for high yield at a moderate protein content. As far as meeting human nutritional needs this argument carries much weight. It is somewhat reassuring that protein deficiency is not as serious as had previously been thought. Nevertheless with the increasing cost and scarcity of animal protein it seems highly probable that increasing amounts of plant protein will be used in the production of meat substitutes (textured vegetable protein, TVP). In all probability the favoured raw material for production of TVP and meat extenders will be high protein seeds such as soy or winged bean (*Psophocarpus tetragonolobus*). It appears that existing protein contents in soy and the winged bean are adequate but if other species were to be used e.g. *Vicia faba* or *Phaseolus* spp. then higher protein genotypes might be desired.

Protein content and composition are of prime importance in animal feeding stuffs. The growth rates in farm livestock are much more rapid than in humans and dietary protein requirements during the growth phase are consequently much higher. Therefore for optimal growth, both amount and composition of protein fed should be ideal. The question to be resolved is how best and most economically can the ideal protein content and quality in livestock rations be achieved. Can this best be done by compounding proteins from different sources, from plants and animals or from different plant species, or has this to be effected by producing a major protein concentrate which is to be diluted by a primarily carbohydrate energy source? If different sources of plant proteins are available it is probable that a satisfactory blend could be produced in terms of both protein content and quality. If a single major protein source is to be used then the better the balance of amino acids that can be achieved the higher will be its nutritional and economic value.

In terms of selection it is possible to obtain genotypes with high protein contents but as has already been noted this is commonly achieved at the expense of total useful yield of dry matter (Brim 1973; Payne 1977). The range in some pulses may be very considerable. Kelly (1973) reports protein contents ranging from 17–35 per cent in *Phaseolus vulgaris*, from 18–28 per cent in both pigeon pea (*Cajanus cajan*) and chickpea (*Cicer arietinum*), 23–36 per cent in lentil (*Lens culinaris*) and 22–37 per cent in *Vicia faba* beans (Hawtin et al. 1977). Considerable variation has also been recorded in levels of limiting essential amino acids in legume seeds (Bressani and Elias 1977). Recorded levels in genotypes producing high levels of sulphur amino acids may be twice to three times those of low lines. In the case of tryptophan relatively little variation has been found in *Glycine max* and *Vigna unguiculata* but considerably more in *Phaseolus vulgaris*, *Arachis hypogaea* and *Cicer arietinum*. There would thus appear to be potentially exploitable variation assuming that heritability of amino acid composition is high and that it was held to be necessary to improve the amino acid profile of the legume in question rather than look to other plant proteins in the diet to correct the deficiency.

Other nutritionally related factors which must be considered are the components of acceptability of a crop genotype to the consumer. In the first place the appearance must be acceptable, in *Phaseolus* beans testa colour is extremely important, white seeded navy beans are the only acceptable beans for the production of canned baked beans while beans with red testas are preferred in Mexican cooking, numerous other examples could be quoted. The next important consideration to consider is what can be termed "cookability". It is important that the consistency of the seed be such that it becomes tender after a reasonable cooking period. In many areas of the developing world fuel supplies are short and economy of use is important. Ideally tenderness should be maintained at an acceptable level in storage for a period of one year where a single crop is taken per annum. Some legumes, e.g. *Voandzeia subterranea*

and *Phaseolus acutifolius* have seeds which quickly lose tenderness on maturation and can be used effectively often for only a short period before the seed
dries out on maturation. Effectively, the use as a conventional pulse is denied
to potential consumers but the possibility remains of producing a good food
by germinating such seeds and using the sprouted seedlings. This mode of use
could well be extended beyond the Asiatic grams and pulses, where it is custo-
mary, to others where it is not attempted at present.

No potentially useful legume species nor any mode of utilization of its
seed should be ignored in the attempt to improve world food production.

8.4. Germplasm Resources

The prime consideration of plant breeders engaged in the improvement of
any cultigen is the identification of the source or sources of appropriate characters to incorporate into their improved cultivars. For the most part collec-
tions of cultivars and landraces are an adequate source of the necessary
germplasm but in some instances they may be lacking. This situation which is
encountered not infrequently in the legumes extends the search for germplasm to the wild relatives of cultigens. These include conspecific wild forms
as well as closely related wild species. This subject has been reviewed by
Smartt (1980) who has designated primary, secondary, tertiary and quaternary gene pools in grain legumes as a whole. The magnitude of these different
gene pools varies considerably from species to species. In all cases the primary gene pool is the major germplasm resource, that is to say most of the
usable germplasm in grain legumes resides in the cultivar and landrace collec-
tions of the cultigens. As well as being the most accessible these germplasm
resources are the most easily and conveniently exploited. The problem of re-
introducing undesirable genetic traits from the wild population to the cultigen
is completely avoided.

In the exploitation of other germplasm resources the problem of affecting
clean transfers of specific characters can be difficult. In most instances a
backcross programme may be effective. It is perhaps appropriate to consider
the nature of characters most likely to be sought in gene pools other than the
primary. It is unlikely that any important agronomic character such as
growth habit will be sought in other than a primary gene pool, it is however
very probable that the other gene pools could provide useful source of genetic
resistance to pests and diseases. Improvement of quality might also be sought
in these higher order gene pools, exploitable seed protein polymorphisms
might well occur which could enable biochemical changes to be effected in the
cultigen. The problems which do arise concern the ease or otherwise of effect-
ing gene transfer. These would normally be minimal when a secondary gene
pool (i.e. from conspecific wild forms) was being exploited but would be
considerably more difficult in the cases of tertiary gene pools and impractica-

showed that cross-compatibility was confined to species placed by Krapovic-kas (1974) and Gregory et al. (1973) in the same section of the genus. A more complete study embracing cross-compatibility between all sections of the genus has recently been completed by Gregory and Gregory (1979). This has shown that interspecific cross-compatibility is almost completely restricted to species within the same section. Some intersectional crosses have been made and such crosses could possibly form bridges between sections facilitating gene transfer. No intersectional crosses involving *A. hypogaea* were produced, other reports to the contrary notwithstanding. The major useful role that interspecific gene transfer could serve in *Arachis hypogaea* is the improve-ment of resistance to the *Mycosphaerella* leafspots ("Cercospora"). Resistant species have already been identified and incorporated in breeding programmes (*A. cardenasii*, *A. chacoense* and *A. sp.* HLK 410) with promising results. It is possible that resistance to virus diseases, e.g. rosette, might also be transfer-red to the cultigen from a wild species, this might be difficult as it could invol-ve intersectional gene transfer. It is possible that protoplast fusion might have some future role to play particularly where failure to hybridize arises from hybrid embryo abortion rather than genetic incompatibility of genomes in somatic tissues.

If the examination of all gene pools fails to identify appropriate germ-plasm then a programme of mutation breeding might be attempted. In the past however the effects of such programmes while of considerable interest have been transient (Gregory and Gregory 1967).

8.5. Conclusions

In terms of interest in the genetic improvement of grain legumes the seventies can be considered as the decade of the legumes. This can be related in part at least to the establishment and commissioning of research facilities at the new international research institutes particularly, ICRISAT, CIAT, IITA and ICARDA, all of which have strongly supported legume research programmes. These have amplified, very considerably, established research programmes in the developed and developing worlds. It is important now that objectives in research and development be very carefully assessed and evaluated objecti-vely. We should not be afraid to reject outdated axioms. The prime example of the latter is that efforts can be directed usefully to the improvement of protein content and some aspects of quality of legume seeds. This is probably only valid when the staple diet is excessively deficient of protein (e.g. cassava) and no protein supplementation other than from legumes is possible. In the nutritional field the most important objectives should be the improvement of digestibility of legume foods and the elimination as far as possible of anti-metabolic and toxic factors.

Physiological efficiency is of obvious importance in the improvement of legumes, considered broadly in the economic context. An optimal level of output for an optimal level of input is desirable. The inputs which include labour, fertilizers, agrochemicals and capital equipment must usually be assumed to be limited and the breeder must take account of this. Climatic considerations must not be ignored, different plant types are appropriate in the humid and the semi-arid or arid tropics under labour intensive production systems. The accepted wisdom of selecting for the highest level of yield consistent with the practicable level of inputs cannot be challenged nor can that of preserving the useful dry matter produced from attack by pests and diseases. One proviso that must be borne in mind in the latter context is that the resistance to pests must not be achieved at the expense of acceptability to the human consumer. Toxic materials such as certain non-protein amino acids might well confer resistance to or immunity from pest attack, but if these materials are found in the parts eaten by man, they will have to be removed or neutralized.

Possibly one of the most encouraging developments of the past decade has been the increased awareness of the serious problem of erosion of our crop plants' genetic resources and the counter measures currently being taken by the FAO through its International Board for Plant Genetic Resources. The production of the valuable works of Frankel and Bennett (1970) and Frankel and Hawkes (1975) mark milestones in our appreciation of the importance of this problem.

One problem which does remain and which is of its nature perhaps intractable is the formulation of realistic long-term selection objectives. As Smartt (1978) suggested one might apply Vavilov's Law of Homologous Series as an approximate measure of probable end-points of selection. For example protein contents of ± 50 per cent have been recorded in soybeans (Brim 1973) it seems probable that this represents a selection end-point. The soybean crop is still troubled by the problem of pod shattering, it seems probable that further selection could be maintained as the end-point of this particular line of selection is exemplified by the stringless (and indehiscent) pod type found in some *Phaseolus vulgaris* cultivars. The papilionoid legumes show remarkably parallel trends in their evolutionary response to selection pressure under domestication and there is evidence to suggest that the potential for genetic improvement in these crop plants is nowhere near exhausted.

References

Adiga, P.R., Padmanaban, G., Rao, S.L.N. and Sarma, P.S. 1962. The isolation of a toxic principle from *Lathyrus sativus* seeds. *Journal of Scientific and Industrial Research*, New Delhi, **21C**: 284–286.
Athow, K.L. 1973. Fungal diseases. In: *Soybeans*: *Improvement, Production*

Klozová, E., Kloz, J. and Winfield, P.J. 1976. Atypical composition of seed proteins in cultivars of *Phaseolus vulgaris* L. *Biologia Plantarum* **18**: 200–205.

Krapovickas, A. 1969. The origin, variability and spread of the groundnut. In: *The Domestication and Exploitation of Plants and Animals*. P.J. Ucko and G.M. Dimble, eds. pp. 427–441. Duckworth, London.

Krapovickas, A. 1974. Evolution of the genus *Arachis*. In: *Agricultural Genetics: Selected Topics*. R. Moav, J. ed. Wiley, London.

Krapovickas, A. 1975. Estudio y conservación del germoplasma en el genero *Arachis*. Miscelanea, Facultad de Agronomia y Zootecnia, Universidad Nacional de Tucuman, no 54: 9–12.

Krapovickas, A. and Rigoni, V.A. 1949. Cromosomas de un especie silvestre de *Arachis*. Idia, Buenos Aires **2**: 23–24.

Krapovickas, A. and Rigoni, V.A. 1956. Noroeste argentino y Bolivia probable centro de origen del maní. *Darwiniana* **11**: 197–228.

Krapovickas, A. and Rigoni, V.A. 1957. Nuevas especies de *Arachis* vinculadas al problema del origen del maní. *Darwiniana* **11**: 431–455.

Kumar, L.S.S., D'Cruz, R. and Oke, J.C. 1957. A synthetic allohexaploid in *Arachis*. *Current Science* **26**: 121–122.

Liener, I.E. 1969. Miscellaneous toxic factors. In: *Toxic Constituents of Plant Foodstuffs*. I.H. Liener, ed. Chapter 13. Academic Press, New York.

Liener, I.E. 1978. Protease inhibitors and other toxic factors in seeds. In: *Plant Proteins*. G. Norton, ed. Chapter 7. Butterworths, London.

Mager, J., Glaser, G., Razin, A., Izak, G., Bien, S. and Noam, M. 1965. Metabolic effects of pyrimidines derived from fava bean glycosides on human erythrocytes deficient in glucose–6–phosphate dehydrogenase. *Biochemical and Biophysical Research Communications* **20**: 235–240.

Masefield, G.B. 1961. Root nodulation and agricultural potential of the leguminous genus *Psophocarpus*. *Tropical Agriculture* **38**: 225–229.

Montgomery, R.D. 1969. Cyanogens. In: *Toxic Constituents of Plant Foodstuffs*. I.E. Liener, ed. Chapter 5. Academic Press, New York.

Murphy, E.L. 1973. The possible elimination of legume flatulence by genetic selection. In: *Nutritional Improvement of Food Legumes by Breeding*. M. Miller, ed. pp. 273–276. Protein Advisory Group of the United Nations System, New York.

Ogle, H.J. and Johnson, J.C. 1974. Physiologic specialization and control of bean rust (*Uromyces appendiculatus*) in Queensland. *Journal of Agricultural and Animal Sciences* **31**: 71–82.

Orf, J.H. and Hymowitz, T. 1979. Inheritance of the absence of the Kunitz trypsin inhibitor in seed proteins of soybeans. *Crop. Sci.* **19**: 107–109.

Payne, P.R. 1977. Human protein requirements. In: *Plant Proteins*. G. Norton, ed. pp. 247–263. Butterworths, London.

Raman, V.S. 1957. Studies in the genus *Arachis*. I. Observations on the morphological characters of certain species of *Arachis*. *Indian Oilseeds Journal* **1**: 235–246.

Raman, V.S. 1976. *Cytogenetics and Breeding in Arachis*. Today and Tomorrow's Printers and Publishers, New Delhi.

Smartt, J. 1965. Cross-compatibility relationships between the cultivated peanut *Arachis hypogaea* L. and other species of the genus *Arachis*. Ph.D. thesis, N.C. State University, No. 65—8968 University Microfilms Inc. Ann Arbor, Michigan.

Smartt, J. 1976. *Tropical Pulses*. Longman, London.

Smartt, J. 1978. The evolution of pulse crops. *Economic Botany* **32**: 185–198.

Smartt, J. 1980. Evolution and evolutionary problems in food legumes. *Economic Botany* **34**: 219–235.

Smartt, J. and Gregory, W.C. 1967. Interspecific cross-compatibility between the cultivated peanut *Arachis hypogaea* L. and other members of the genus *Arachis*. *Oleagineux* **22**: 455–459.

Smartt, J., Gregory, W.C. and Gregory, M.P. 1978. The genomes of *Arachis hypogaea* 2. The implications in interspecific breeding. *Euphytica* **27**: 677–680.

Turnipseed, S.G. 1973. Insects. In: *Soybeans: Improvement, Production and Uses*. B.E. Caldwell, ed. Chapter 17. American Society of Agronomy, Madison, Wisconsin.

Van der Plank, J.E. 1968. *Disease Resistance in Plants*. Academic Press, London.